Nuclear Energy: Concepts, Techniques and Applications

Nuclear Energy: Concepts, Techniques and Applications

Edited by Matt Fulcher

SYRAWOOD
PUBLISHING HOUSE

New York

Published by Syrawood Publishing House,
750 Third Avenue, 9th Floor,
New York, NY 10017, USA
www.syrawoodpublishinghouse.com

Nuclear Energy: Concepts, Techniques and Applications
Edited by Matt Fulcher

International Standard Book Number: 978-1-68286-532-3 (Hardback)

Cataloging-in-Publication Data

Nuclear energy : concepts, techniques and applications / edited by Matt Fulcher.
     p. cm.
Includes bibliographical references and index.
ISBN 978-1-68286-532-3
1. Nuclear energy. 2. Nuclear engineering. I. Fulcher, Matt.
TK9145 .N83 2018
621.48--dc23

TABLE OF CONTENTS

PREFACE

Nuclear energy is developed with the help of nuclear reactions, especially through the process of nuclear decay, as geothermal energy. Nuclear energy is the leading source of low carbon power generation methods of electricity generation. Other advantages of nuclear energy include low operating costs and relatively longer operating lifetime. The various studies that are constantly contributing towards advancing technologies and evolution of this field are examined in detail. It attempts to assist those with a goal of delving into the field of nuclear energy.

Various studies have approached the subject by analyzing it with a single perspective, but the present book provides diverse methodologies and techniques to address this field. This book contains theories and applications needed for understanding the subject from different perspectives. The aim is to keep the readers informed about the progresses in the field; therefore, the contributions were carefully examined to compile novel researches by specialists from across the globe.

Indeed, the job of the editor is the most crucial and challenging in compiling all chapters into a single book. In the end, I would extend my sincere thanks to the chapter authors for their profound work. I am also thankful for the support provided by my family and colleagues during the compilation of this book.

Editor

Why nuclear energy is essential to reduce anthropogenic greenhouse gas emission rates

Agustin Alonso[1*], Barry W. Brook[2], Daniel A. Meneley[3], Jozef Misak[4], Tom Blees[5], and Jan B. van Erp[6]

[1] University Politecnica de Madrid, Madrid, Spain
[2] University of Tasmania, Hobart TAS 7005, Australia
[3] CEI and AECL, Ontario, Canada
[4] UJV-Rez, Prague, Czech Republic
[5] Science Council for Global Initiatives, Chicago, Il, USA
[6] Illinois Commission on Atomic Energy, Chicago, Il, USA

Abstract. Reduction of anthropogenic greenhouse gas emissions is advocated by the Intergovernmental Panel on Climate Change. To achieve this target, countries have opted for renewable energy sources, primarily wind and solar. These renewables will be unable to supply the needed large quantities of energy to run industrial societies sustainably, economically and reliably because they are inherently intermittent, depending on flexible backup power or on energy storage for delivery of base-load quantities of electrical energy. The backup power is derived in most cases from combustion of natural gas. Intermittent energy sources, if used in this way, do not meet the requirements of sustainability, nor are they economically viable because they require redundant, under-utilized investment in capacity both for generation and for transmission. Because methane is a potent greenhouse gas, the equivalent carbon dioxide value of methane may cause gas-fired stations to emit more greenhouse gas than coal-fired plants of the same power for currently reported leakage rates of the natural gas. Likewise, intermittent wind/solar photovoltaic systems backed up by gas-fired power plants also release substantial amounts of carbon-dioxide-equivalent greenhouse gas to make such a combination environmentally unacceptable. In the long term, nuclear fission technology is the only known energy source that is capable of delivering the needed large quantities of energy safely, economically, reliably and in a sustainable way, both environmentally and as regards the available resource-base.

1 Introduction

The need to reduce anthropogenic greenhouse gas (AGHG) emissions is of great urgency if catastrophic consequences caused by climate change are to be prevented. However, while the United Nations Framework Convention on Climate Change (UNFCCC), through its various meetings of the Conference of the Parties (COP), has emphasized the role of *renewable* energy sources, it barely mentions nuclear energy and the important contribution that it is already making in reducing AGHG emissions and could increasingly be making in the future. This is difficult to understand because nuclear fission is the only major energy source that could sustainably, reliably and economically provide the large quantities of clean energy that will be needed to make substantial progress in reducing AGHG emissions.

When addressing issues related to the long-term energy policy, two important questions need to be asked, namely:

– Is it possible to replace all or most fossil-derived energy with *renewables* and, if so, would this be sustainable and economically viable?
– Is nuclear energy sustainable and what should its role in the energy mix be?

The term *sustainable* is generally understood, Brundtland Commission [1], to mean *"meeting the needs of the present without compromising the ability of future generations to meet their own needs"*. In the context of energy options, 'sustainable' implies the ability to provide energy for indefinitely long time periods (i.e., on a very large civilization spanning time scale) without depriving future generations and in a way that is environmentally friendly, economically viable, safe and able to be delivered reliably. It should thus be concluded that, in this context, the term 'sustainable' is more restrictive than the term 'renewable', as large scale renewable

*e-mail: agustin.alonso@nexus5.com

systems backed by fossil fuels cannot be considered clean sources of electricity. On the other hand, nuclear energy from fission of uranium and plutonium is sustainable, meeting all of the above-mentioned criteria as discussed later.

The energy consumption in industrial nations may be roughly divided in three equal parts, namely:

– generation of electrical energy;
– heat in industrial processes and space heating;
– and transportation.

Nuclear fission is a low AGHG emission energy source that is already widely deployed for generation of electrical energy. Therefore, one effective way to reduce fossil fuel consumption and AGHG emissions would be by increasing the number of nuclear power plants for electrical energy generation.

It would be well within realistic limits to aim for replacement of the major part of the world's fossil fuel-based electrical energy generating capacity. Industrial nations should take the lead in this change because they are more capable of doing so, having already developed the necessary technological and mature economic base. In parallel to this major change in the generation of electrical energy, the use of fossil fuels for transportation should be reduced by greater reliance on nuclear-derived electrical energy as well as on liquid fuels produced synthetically by means of nuclear power plants. Also the use of nuclear-derived process heat for industrial application and services should be encouraged [2]. Gradual conversion of the electrical generating capacity from fossil fuel-based to nuclear fission would be the way offering least economic disturbance.

2 Intermittent 'renewables' when applied to the electric grid

Wind and solar energy have served humanity well during centuries and in many applications, including grinding wheat, pumping water, sawing wood, drying foods and producing sea salt. Wind also served as an important energy source for transportation, making possible the exploration of the entire world by means of ships propelled by the wind. The common characteristic of these applications is that they are not time-constrained: if there is no wind today, the tasks can wait to be finished tomorrow or the ships will arrive somewhat later. This is not possible if intermittent renewable energy sources are used for base-load delivery of electrical energy to the grid, as strict demands have to be fulfilled instantaneously and completely.

2.1 Grid-connected 'renewables' with gas-fired backup are not sustainable

Intermittent 'renewables' are, in *certain* applications, not 'sustainable' because not all necessary criteria are being met. Intermittent 'renewable' energy sources, *when used for large-scale delivery of energy to the electric grid*, require the availability of energy storage facilities or flexible backup power plants capable of rapid output adjustments. This is

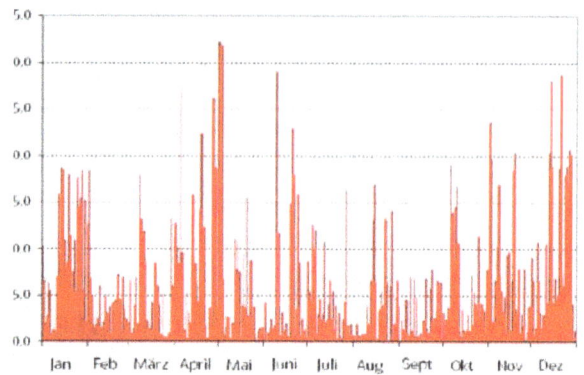

Fig. 1. Intermittence of wind energy in E.ON-grid in Germany (from Ref. [3]).

because wind turbines and solar/photovoltaic plants will vary their output between 0% and 100% of nameplate capacity, as it can be observed in the typical example given in Figure 1.

As energy from the grid is generated and consumed simultaneously, there can be no mismatch if grid stability and frequency are to be maintained within strict tolerances. The backup power is usually provided by gas-fired stations because technology for storing large amounts of electricity is not yet available. Although reversible pumped hydro-power stations can be used to store potential energy, there are siting, technical and economic limitations that prohibit their widespread use. Gas-fired plants emit carbon dioxide and are associated with leakage of methane (the primary component of natural gas) into the atmosphere, which is a strong AGHG emitter. Only if the backup energy is delivered by hydro-electrical energy plants or similar means to store and control the generated energy, then grid-connected intermittent 'renewables' can be qualified as sustainable.

2.2 Grid-connected 'renewables' are not economically viable

Averaged over a year, wind/solar photovoltaic systems deliver from 25% to 45% of their nameplate production capacity. Therefore, the backup power plants or energy storage facilities will have to deliver the remaining 75% to 55% of the energy. Seasonal variability is another major, yet rarely acknowledged, impediment to all-renewables scenarios, as it is seen in Table 1.

Advocates often dismiss the issue of seasonal variability, pointing out that the wind blows more in the winter when solar output is minimal, and asserting that wind and solar balance out on a daily basis because wind blows more at night. However, these generalizations do not hold up to scrutiny. While some areas of the world do have more wind in the winter, others do not.

The backup power for wind/solar photovoltaic plants depends in most cases on combustion of less expensive natural gas. Storage may be of various types: potential energy storage capacity may be created by pumping up water

Table 1. Seasonal variability of wind-generated electrical energy in Texas, USA. Highest and lowest monthly generation values (GWh).

Year	Highest value (month)	Lowest value (month)	Ratio (high/low)
2009	1,993 (April)	1,341 (July)	1.44
2010	2,721 (April)	1,589 (Sept.)	1.75
2011	3,311 (June)	1,694 (Sept.)	1.95
2012	3,131 (March)	1,821 (Aug.)	1.74
2013	3,966 (May)	2,023 (Sept.)	1.96

Source: Private communication, P. Peterson, Prof. Nuclear Engineering, Univ. of California at Berkeley, USA

or compressing air, small scale storage could be achieved in condensers and batteries. However, most energy storage facilities are not cost-effective for base-load application and often have undesirable environmental impacts. Also, storage is associated with energy losses. Consequently, grid-connected wind/solar photovoltaic installation will usually rely on gas-fired backup power plants.

Many wind and solar photovoltaic installations are far removed from the load centers, requiring additional long-distance transmission lines, sized for their peak output, which are then under-utilized by from 55% to 75%. Furthermore, the backup power plant will have to operate in stand-by mode, ready to adapt to the varying output (from 0% to 100%) of the intermittent energy source. This results in a penalty on the overall thermal efficiency of the backup plant, which can be as high as 20%. Grid-connected wind and solar photovoltaic installations will thus be dependent on subsidies because redundant and under-utilized investments are necessary (i.e., for the intermittent energy source, for the backup source and for the additionally required transmission capability). In view of the above-given reasons, it has to be concluded that the combination of an intermittent energy source and its back-up power plant will not be able to achieve economic viability, as illustrated in Table 2. However, in isolated

Table 2. Average power plant operating expenses for USA electric utilities (mS/kWh).

	2008	2009	2010	2011	2012
Nuclear					
Operation	9.9	10.0	10.5	10.9	11.6
Maintenance	6.2	6.3	6.8	6.8	6.8
Fuel	5.3	5.4	6.7	7.0	7.1
Total	21.5	21.7	24	24.7	25.5
Intermittent plus gas turbine					
Operation	3.8	3.0	2.8	2.8	2.5
Maintenance	2.7	2.6	2.7	2.9	2.7
Fuel	64.2	52.0	43.2	38.8	30.5
Total	70.7	57.6	48.7	44.5	35.7

Source: USA Energy Information Administration

locations and some processes without access to a large electric grid, intermittent energy sources either directly or combined with storage capacity may be economically viable.

Much confusion exists concerning the generating cost per kWh for wind and solar plants. In this respect, it is of interest to distinguish clearly between the 'bare' cost of a kWh generated by wind or solar photovoltaic installations that is consumed or stored locally and the cost of a kWh delivered to the electrical grid. In the latter case, it is necessary to account for the investments in the backup power and transmission capacity. The difference between these two prices is very substantial; the cost per kWh delivered to the grid in most cases being several hundred percent higher than the 'bare' cost. As an example, Table 2 shows that for the combination of intermittent energy source with gas-fired backup power, the cost for fuel per kWh varies between 5 and 12 times the cost for operation and maintenance.

2.3 Grid-connected 'renewables' have deleterious consequences

Grid-connected intermittent energy sources will cause grid disturbances that will deleteriously affect the grid's reliability, particularly if the installed capacity of the intermittent sources becomes a high percentage of the grid's total capacity. Delivery unreliability of the electrical grid can have serious economic and social consequences as has been observed when long-lasting blackouts occurred in large urban areas. To date, in most grids, 'renewables' have only reached a relatively low market penetration and so have been able to rely mostly on existing marginal capacity, or on large import–export capacity of interconnected other grids.

Problems will emerge when the percentage of grid-connected intermittent energy sources exceeds the existing marginal capacity (without availability of adequate dedicated back-up power capacity) and it becomes necessary for the base-load plants to function as back-up plants. This mode of forced 'accommodative' operation penalizes nuclear power plants more than it does fossil-fired plants because the capital-cost component of the generating cost for the former is relatively high and the fuel cost component is low, whereas for the latter the reverse is true, as shown in Table 3.

This practice of distorting the energy market by subsidies and supporting regulations has serious and undesirable consequences, resulting in closure of base-load

Table 3. Generation cost breakdown (%).

Component	Nuclear	Coal	Gas
Capital	59	42	17
Fuel	15	41	76
Operation & Maintenance	26	17	7

Source: OECD/International Energy Agency

generating capacity (including nuclear power plants), loss of grid reliability and higher net greenhouse gas emissions. This issue is of particular relevance for countries having an interconnected grid with an adjacent country that is relying (or is planning to rely) to a large extent on intermittent 'renewable' energy sources. In this respect, the question should be raised whether a country with a large installed wind/solar electrical generating capacity should be required to pay a connection fee to compensate adjacent countries for the use of their interconnected electric grids for providing backup power capacity.

It is often claimed by advocates of 'renewables' that the problems associated with the intermittency of wind and solar energy can be overcome by performing more research and carrying out more engineering development. Unfortunately, no level of research and development will be able to overcome the fact that the sun does not always shine and that the wind does not always blow. Not even the much-praised 'smart grid' can change this inconvenient fact.

2.4 The relevance of methane as a greenhouse gas

Methane, CH_4, the main component of natural gas, is a potent greenhouse gas as compared to carbon dioxide, CO_2; making it one of the six gases considered in the Kyoto Protocol, the second in importance. To measure the relative climate importance of the two gases, the International Panel on Climate Change (IPCC) has introduced the concept of *global warming potential* (GWP) [4] which is defined (glossary) as:

"**Global warming potential (GWP),** index based on radiative properties of greenhouse gases measuring the radiative forcing following a pulse emission of a unit of gas of a given greenhouse gas in the present day atmosphere integrated over a chosen time horizon, relative to that of carbon dioxide. The GWP represents the combined effect of the different times these gases remain in the atmosphere and their relative effectiveness in causing radiative forcing."

The *radiative forcing* of a greenhouse gas is itself defined [4] (glossary) as:

"**Radiative forcing,** change in the net, downward minus upward, radiative flux (expressed in $W.m^{-2}$) at the tropopause or top of the atmosphere due to a change in an external driver of climate change, such as, for example in the change in the concentration of a gas or the output of the sun."

The GWP of any gas is calculated through the expression

$$GWP_m(t) = \frac{\int_{t_r}^{t_h} a_m C_m(t)dt}{\int_{t_r}^{t_h} a_c C_c(t)dt},\qquad(1)$$

where sub-index m represents methane and c carbon dioxide; a is the radiative forcing of the gas and $C(t)$ the time function, which represents the evolution of the gas in the atmosphere after the release of a pulse emission of a unit of gas. The integration goes from the time of release, t_r, to

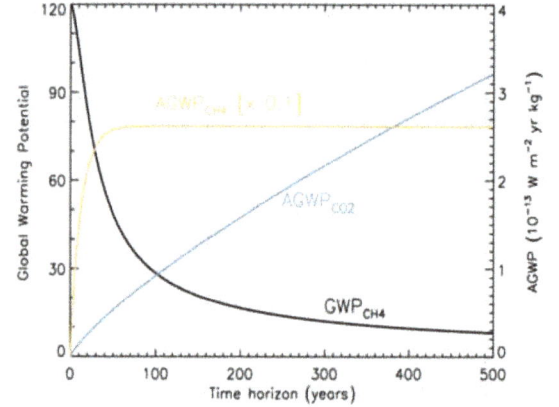

Fig. 2. Value of methane global warming potential, GWP_{CH_4}, as a function of time horizon (taken from Ref. [5]).

the selected time horizon, t_h. Function $C(t)$ takes into account the rather complicated chemical reactions and other removal processes that take place among the different constituents in the atmosphere causing the disappearance of the released gases.

Each integral term in the definition is also called the *absolute global warming potential* (AGWP) of the concerned and the reference gas and is measured in $W/m^2/y/kg$. To estimate the magnitudes defined above, the IPCC has provided the graph reproduced in Figure 2.

It is accepted that a pulse release of methane in the atmosphere will be removed exponentially with time by getting involved in chemical reactions with hydroxyl radicals (OH) present in the atmosphere. The coefficient in the exponential function is the inverse value of the so-called *turn over* or *global atmospheric lifetime* of methane, represented by symbol T. This symbol is given the value of $11.2+1.3$ years. The $AGWP_{CH_4}$ is then obtained by the equation:

$$AGWP_{CH_4} = \int_0^t a_m e^{-\frac{t}{T}}dt = a_m T\left(1 - e^{-\frac{t}{T}}\right).\qquad(2)$$

In less than a century, the $AGWP_{CH_4}$ reaches an asymptotic value, $a_m T$, which is the product of the *radiative forcing* of methane multiplied by the assumed lifetime of methane in the atmosphere measured in $W/m^2/y/kg$. Note that the graph in Figure 2 is reduced by a factor of 10.

The behavior of carbon dioxide in the atmosphere includes a variety of phenomena, which could not be represented by a single lifetime; as seen in the blue curve, the $AGWP_{CO2}$ is less than the one for methane because its radiative forcing is smaller; moreover, carbon dioxide in the atmosphere never reaches an asymptotic value because a small fraction of the carbon dioxide emitted is not removed from the atmosphere by natural processes, while the rest of the processes are described by exponential functions with long lifetime.

The ratio of the two curves is the GWP_{CH_4}, a decreasing function with increasing time horizon; when the time horizon approaches the time of release the GWP_{CH_4} tends to 120, which should be interpreted as the radiative forcing of the methane relative to the one of carbon dioxide. From the graph it is deduced that the GWP_{CH_4} values are about 63, 21 and 3, obtained from calculations, for respective time horizons of 20, 100, and 500 years. The IPCC recommends using a time horizon of 100 years.

The methane contents in the atmosphere started to grow since 1750, the year considered as the start of the industrial revolution; at that time, the methane content in the atmosphere was 0.722 ppm; it grew exponentially until about 1980, in the 1990s the rise slowed down and reached the value of 1.893 ppm in 2011, an increment of some 1.171 ppm, i.e. an average increase of 138%. This value is compared with the same temporal increment of carbon dioxide in the atmosphere from 280 ppm in 1750 to the current 395 ppm, an increment of 115 ppm, i.e. an average increase of 36%. From these values, it is deduced that from the year 1750 to now, i.e. 260 years, for which the GWP_{CH_4} is around 10, the increase in the climatic relevance of methane has been 40 times larger than that for carbon dioxide. This proves the relevance of methane as a greenhouse gas.

As in 1750, the atmospheric content of methane was probably in equilibrium and mainly caused by natural sources, it is considered that the noted increment is mainly due to anthropogenic reasons. The cause of the increase has to be attributed to direct atmospheric releases of natural gas during its geological extraction, purification, flaring and venting, liquefaction and transport, as well as storage, manipulation and use of the gas in electricity-generating station and from poor gas combustion. There is much literature, even regulations, on the mass fraction of natural gas leakages from all these operations. Values are quoted [6] from 2% to 10% of natural gas releases when the complete fuel cycle is considered: from the source to the power plant.

When natural gas is used instead of coal or to back up the intermittency and variability of wind/solar photovoltaic systems for load-based electricity generation, the expected climatic effect from the natural gas directly released to atmosphere, also called the *fugitive methane*, has to be added to the corresponding release of carbon dioxide from the natural gas combustion process. To determine the relevance of the radiative forcing of the leaked natural gas, the IPCC [4] has introduced the concept of equivalent carbon dioxide emission (glossary):

"**Equivalent carbon dioxide emission**, the amount of carbon dioxide emission that would cause the same integrated radiative forcing over a given time horizon as an emitted amount of a greenhouse gas or the mixture of greenhouse gases. The equivalent carbon dioxide emission is obtained by multiplying the emission of the greenhouse gas by its global warming potential for the given time horizon".

The use of the equivalent carbon dioxide concept when applied to methane permits to compare the GWP of a given coal station with the one for a gas-fired installation of the same power when gas leakages are included. That relation is obtained from the following algorithm:

$$R_{m/c} = m \left\{ 1 + \psi \frac{M_{CH_4}}{M_{CO_2}} (GWP(t_h)) \right\}, \quad (3)$$

Table 4. Ratio between the greenhouse gases from a gas-fired station including methane leakages and from a coal-fired plant of equal power.

ψ	GWP/t_h		
	120/as.	63/20	21/100
0.02	0.93	0.73	0.57
0.04	**1.37**	0.95	0.65
0.06	**1.80**	**1.18**	0.90

where m is the ratio between the masses of carbon dioxide generated in the combustion of methane and coal per unit of energy generated in the respective electrical power plants, it depends on the quality of the fossil fuels and the efficiency of the plant, the average value of $\frac{1}{2}$ is frequently used in calculations; ψ is the fraction of fugitive methane directly discharged to the atmosphere from leakages in the natural gas cycle; M_{CH_4}/M_{CO_2} is the ratio between the molecular mass of methane and carbon dioxide needed to estimate the methane carbon dioxide equivalent, and $GWP(t_h)$ the global warming potential of methane for time horizon (t_h). In Table 4, estimations are presented for different leakage fractions, the asymptotic and horizon times of 20 and 100 years, corresponding to the GWP (t_h) of 120, 63 and 21.

It is observed from the table that for gas leakages of 2%, the breakeven, although close, is not reached even for the asymptotic value, while for leakages of 4%, the breakeven is close for a time horizon of 20 years. Leakages superior to 6% could not be accepted even for time horizons of 100 years. The results clearly indicate that replacing coal-fired with gas-fired plants does not provide any relevant climate reduction unless gas leakage is reduced to less than 2%.

Likewise, the climatic effect of a gas-fired backup power is obtained by adding the carbon dioxide equivalent of the fugitive methane to the carbon dioxide generated during the fraction of the time that the backup power is needed. In this case, the ratio between the methane/carbon dioxide equivalent due to the fugitive methane and the carbon dioxide release from the combustion of the gas in the backup plant is given by the equation:

$$R_{m/c} = \left\{ \psi \frac{M_{CH_4}}{M_{CO_2}} (GWP(t_h)) \right\}. \quad (4)$$

In Figure 3, estimations are presented for different leakage fractions, the asymptotic and horizon times of 20 and 100 years, corresponding to the $GWP(t_h)$ of 120, 63 and 21.

As in Table 4, it is also observed that for gas leakages of 2%, the breakeven, although close, is not reached even for the asymptotic value of the GWP, while for 4% leakage breakeven is close for the 20-year GWP. It is then concluded that for leakages above 2% and certainly superior to 4% it will be climatically advantageous to backup wind/solar photovoltaic systems with coal-fired instead of gas-fired plants.

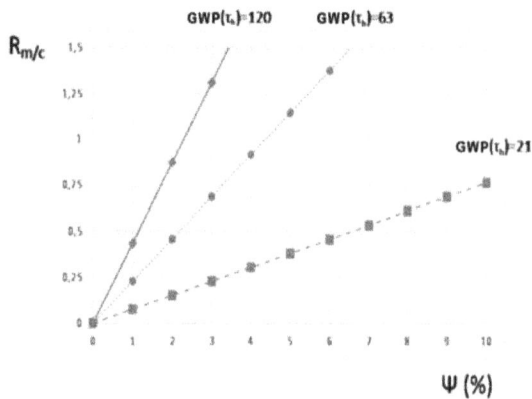

Fig. 3. Ratio between the carbon dioxide equivalent for fugitive methane and the carbon dioxide emitted in a wind/solar photovoltaic system backed by a gas-fired plant.

3 The essential role of nuclear energy in reducing greenhouse gas emissions

Nuclear fission energy is capable of replacing most of the stationary tasks now performed by the combustion of fossil fuels. Other than the generation of electrical energy, it may equally well be used for production of process heat and hydrogen as well as for desalination. However, many environmental organizations and governments oppose the application of nuclear energy. Among the reasons usually given are:

- nuclear energy is not sustainable;
- nuclear energy is not economically viable;
- and nuclear energy is not safe.

3.1 Nuclear energy from fission is sustainable

Today's commercially available uranium-fueled nuclear power plants can provide the world with clean, economical and reliable energy well into the next century on the basis of the already-identified uranium deposits. Furthermore, nuclear reactors operating with fast neutrons are able to fission not only the rare uranium isotope U-235 but also the Pu-239 isotope generated from the transmutation of the abundant uranium isotope U-238. Thus, the deployment of fast-neutron fission reactors transforms uranium into a truly *inexhaustible energy source*, because of their ability to harvest up to one hundred times more energy from the same amount of mined uranium as the commercially available thermal reactors can achieve [7,8].

This fast-neutron fission technology has already been proven, all that is further needed is to develop it to a commercial level and deploy it widely [9]. The amount of depleted uranium that is available and stored at enrichment plants in a number of countries, together with the uranium recoverable from used fuel elements, contains enough energy to power the world for several hundred years without additional mining. Afterwards, mining of small quantities of uranium in future centuries, including extracting uranium from lower-grade ores and, if necessary,

from seawater, could satisfy global energy needs economically for as long as human civilization will endure.

3.2 Nuclear energy from fission is economically viable

Conditions for economic viability of nuclear energy are:

- presence of a *level playing field*, i.e., an open market that is not skewed in favor of some technologies by means of subsidies and/or by a legally imposed priority access for delivery to the electrical grid at a fixed high price;
- standardization of the plants, built in series and supported by a standardized supply chain;
- a long-term governmental energy policy (stable over a time period of several decades) including, among other features, good (unbiased, accurate, evidence-based) public information;
- a stable and streamlined licensing process that is technology-neutral, risk-informed and capable of resolving promptly any safety issues that may arise during construction and operation;
- and gradual introduction of the concept of payment for *external costs*, applied to all energy technologies and based on common standards.

The fact that nuclear energy is economically viable has been shown, among others, by the national energy program in France where the unit price of electricity in a market supplied about 75% by nuclear fission is among the lowest worldwide. An important additional benefit of this reliance on nuclear energy is that per capita emission of greenhouse gases in France is among the lowest for industrial nations worldwide and many times lower than in otherwise similar countries that have no nuclear power plants and that rely on a mix of fossil fuels and *renewables*.

An important aspect of long-term commercial viability of power plants is the future development of their respective fuel costs. Nuclear power plants rank best in this respect because their sensitivity to fuel-cost increases is small as seen in Table 5.

The current temporary abundance (in the USA) of low-cost natural gas may seem to make gas-fired stations appear to be economically attractive. However, this will change because it can be expected that gas prices will rise substantially during the 60+ lifetime of new-build nuclear power plants.

Thus, the fuel supply side of nuclear power reactors eliminates any doubt concerning its sustainability. As to the materials used in the construction of nuclear power

Table 5. Percent sensitivity of generating cost to a 50% increase in fuel price.

Nuclear	IGCC	Coal Steam	CCGT
3	20	22	38

IGCC: integrated gasification combined cycle; CCGT: combine cycle gas turbine. Source: WEO '06/OECD/IEA World Energy Outlook 2006

plants, it is noted that none of them is in short supply (and most are readily recyclable), so that they too do not constitute a sustainability impediment.

3.3 Nuclear energy from fission has a low environmental impact

Numerous scientific comparisons have shown that nuclear fission is among the energy sources that are least polluting and have the lowest overall environmental impact [10]. Operating nuclear power plants do not produce air pollution nor do they emit CO_2. Any CO_2 that is associated with nuclear finds its origin in the mining of uranium and in the production of structural materials necessary for the building of the nuclear plants; small amount of CO_2 are released during the periodic testing of emergency diesel generators and on the use of external power during refuelling outages and maintenance.

Annually, the 435 operating nuclear power plants prevent the emission of more than 2 billion tons of CO_2. By contrast, coal-fired stations emit worldwide about 30 billion tons of CO_2 per year and cause health effects and premature death through air pollution and dispersion of pollutants, including mercury and other poisonous materials [11]. It is to be noted that nuclear power plants emit less radioactive material than do coal-fired stations (uranium and other radioactive isotopes are found naturally in coal ash and soot) [12]. The most severe environmental impact associated with nuclear energy is due to the mining of uranium. However, the need for uranium mining will be reduced after fast reactors have become commercially available, as may be expected within the coming decades.

New methods for efficiently recycling the used fuel will reduce the radioactive hazards as well as the volume of the waste that must be kept isolated from the environment. New technologies have been actively developed to reduce the level of radioactivity of a repository containing this type of waste so that the activity of the waste, after a few centuries, will be comparable to that of the natural uranium deposits that are widely distributed around the world. Furthermore, modern waste isolation technology will equal or exceed the level of isolation originally provided by nature for radioactive ores. In this way the waste will be reduced to a historical time scale of a few hundred years, rather than a geological time scale of hundreds of thousands of years. Furthermore, it is important to note that this waste will be disposed of in an environmentally inert form, i.e., ceramic or vitrified solids that will not start leaching any material into the environment for thousands of years, long after their radioactivity will have dissipated. On the other hand, large amounts of solid and gaseous waste from coal-fired stations (including mercury and heavy metals) will remain poison-ous in perpetuity and they are neither kept well-guarded nor well separated from the environment.

3.4 Nuclear energy from fission meets high safety standards

Nuclear fission is among the safest energy technologies in terms of health effects and fatalities as seen in Figure 4. This

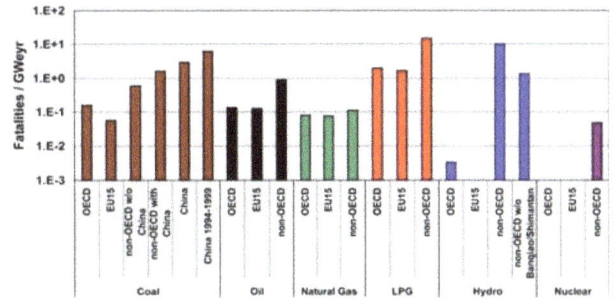

Fig. 4. Comparison of energy-related damage (fatalities per GW/y). Based on historical experience of severe accidents in OECD, non-OECD countries and EU-15 (from Ref. [13]).

is true notwithstanding the three major nuclear accidents that have occurred, namely the 1979 Three Mile Island (TMI) in the USA, the 1986 Chernobyl in Ukraine, and the 2011 Fukushima in Japan. Of these three, only the Chernobyl accident caused a number of fatalities, namely among those persons that were directly exposed to high radiation levels during the urgent initial part of the clean-up operation.

The total number of nuclear-caused fatalities is relatively small (less than one hundred) compared to the number of annual fatalities in the coal and oil/gas industry as seen in Table 6 where there are included the global average values of the mortality rate per billion kWh due to all causes as reported by the World Health Organization (WHO).

Both the accident at Chernobyl and that at Fukushima caused considerable land contamination and required evacuation of the population. However, in both cases the major part of the evacuated areas has/had radiation levels that are lower than the normal background level in many regions around the world, raising the question of how much evacuation was really necessary and for how long. In the case of TMI-2, there was no land contamination, but a short-term evacuation was imposed as a cautionary

Table 6. Mortality rates (deaths per TWh) from energy sources.

Coal global average	100	50% global electricity
Coal China	160	75% China's electricity
Coal USA	15	44% USA electricity
Oil	36	36% global/8% electricity
Natural gas	4	20% of global electricity
Biofuel/biomass	24	21% global energy
Solar (rooftop)	0.44	< 1% global electricity
Wind	0.15	~ 1% global electricity
Hydro-global average	1.4	15% global electricity
Nuclear global average	0.04	17% global electricity

Source: Updated data from: World Health Organization

measure. It should be noted that land contamination is not limited to severe nuclear accidents; it has also occurred following severe accidents in the chemical industry, in which the contaminants were extremely deadly and long lasting (e.g. Bhopal, India; Seveso, Italy).

The radioactive isotopes of iodine (I-131, half-life 8 days) and cesium (Cs-137, half-life 30 years) have dominating importance in accidents in which the containment is breached and radioactivity is released into the environment. The short half-life of I-131 and its biological accumulation in the thyroid requires simple precautions, such as ingesting a small dose of potassium iodine, to prevent its health effects. However, Cs-137 will stay in the environment for a longer time period that is determined by its effective soil removal half-life, i.e., the combination of its radioactive half-life and the rate of removal from the soil surface by natural processes and by adding manure and fertilizers as it has been done in regions contaminated by the Chernobyl releases. This latter process can be accelerated by removal of a thin layer of the top soil in areas where the radiation level exceeds the allowable radiation level, as it is being practiced in soils contaminated by the Fukushima Daiichi accident.

Natural background radiation varies greatly over the world (depending on soil composition and the location's elevation) but higher background has not been found to be correlated with higher rates of cancer in the population. The average background radiation at sea level in much of the world is about three milli-Sievert (mSv) per year whereas that in many regions around the world is considerably higher. As an example, at Ramsar in Iran, the background radiation level is about 138 mSv per year, i.e. about 46 times higher than the average background. Nevertheless, the incidence of cancer in the local population of regions with high background radiation has not been observed to be higher than the normal rate.

The economic damage associated with nuclear accidents can be substantial, as was demonstrated in the above-mentioned three major accidents. This potential for severe economic damage is a strong incentive on the part of the owner/operator of the nuclear power plant to observe extreme caution, observing strictly all safety-related rules and regulations and maintaining a strict safety culture (even without continuous monitoring by the relevant regulatory organization).

As is normal in the evolution of any technology, also the new designs of nuclear power plants incorporated many new safety-related improvements, mainly coming from the worldwide system of analysing, reporting and incorporating operating experience conducted by the World Association of Nuclear Operators (WANO) created after the Chernobyl accident. WANO also conducts periodic external peer reviews of the operational safety of each one of the operating power plants in the system.

The International Atomic Energy Agency (IAEA) produces safety principles, safety requirements and safety guides created by international consensus, to help countries to create their own regulatory regimes, maintains and distributes an Incident Reporting System (IRS) to share operating incidents and an International Nuclear Event Scale, (INES), where events, incidents and accidents are classified. Under the request of governments, the Agency also performs independent evaluations of the operational and safety culture of the requested plant and on the regulatory completeness and practices of the regulatory organization. The Agency is also depositary of the many existing international conventions, of which the *Nuclear Safety Convention* is among the most relevant.

These international activities, together with the national research and advances in technology and regulation, have created a high level of safety assurance for future nuclear power plants and substantial safety improvements in currently operating nuclear stations.

Public opposition to nuclear energy is in part due to fear of radiation caused by recollection of the effects of nuclear weapons used during World War II and by sensationalized media coverage of nuclear incidents. Another cause of the public fear of radiation is the use of the scientifically unsubstantiated linear-no-threshold (LNT) hypothesis in which it is erroneously assumed that the biological effects of nuclear radiation are linear even at very low radiation doses [14].

4 Conclusions

Nuclear power plants are capable of sustainably and reliably supplying the large quantities of clean and economical energy needed to run industrial societies with minimal emission of greenhouse gases.

The world's industrial nations should take the lead in transforming the major part of their electrical energy generating capacity from fossil fuel-based to nuclear fission-based.

Wind/solar photovoltaic systems with gas-fired backup power stations will not be able to reduce the rate of greenhouse-gas emission, even for relatively low atmospheric gas leakage rates.

Distorting the electricity market with subsidies and by legislation to attract intermittent energy technologies into applications for which they are not well suited, is costly, economically wasteful and counterproductive.

Countries that depend on imported natural gas should be aware that they carry full responsibility for their part of the global consequences of the associated atmospheric leakage of methane, including the leakage taking place outside their borders.

Only in specific cases and for some isolated locations without access to an electric grid, may the use of intermittent energy sources for electrical energy generation be economically viable.

References

1. United Nations, Towards sustainable development, in *Report of the World Commission on Environment and Development: Our Common Future* (United Nations, New York, 1987), Part I, Chap. 2
2. D.A. Meneley, Nuclear Energy: The Path Forward, in *CANADA: becoming a sustainable energy powerhouse*, 1st Ed. (Canadian Academy of Engineering, Ottawa, Canada, 2014), Chap. 9

3. The European Nuclear Energy Forum, SWOT Analysis of Energy Technologies, in *2nd Meeting, Prague, 22–23 May 2008* (2008), p. 20

4. IPCC, *Climate Change 2013: The Physical Science Basis* (World Meteorological Organization (WMO), United Nations Environmental Programme (UNEP), Genève, Switzerland, 2013)

5. IPCC 2013, *The Physical Science Basis. Contributions of Working Group I of the Fifth Assessment Report of the Intergovernmental Panel of Climate Change* (Cambridge University Press, Cambridge, New York, USA, 2013), Chap. 8, p. 712, Fig. 8.9

6. NAS, in *Proceedings of the National Academy of Science of the United States of America: Greater Focus Needed in Methane Leakage from Natural Gas Infrastructure, Washington DC, 2014*, (2014), Vol. 109, No. 5

7. B.L. Cohen, Breeder Reactors: A Renewable Energy Source, Am. J. Phys. **51**, 75 (1983)

8. D. Lightfoot et al., Nuclear Fission Fuel is Inexhaustible, in *Climate Change Technology Conference, May 10–12, Ottawa, Canada, 2006* (2006)

9. C.E. Till, Y.I. Chang, *Plentiful Energy: The Story of the Integral Fast Reactor* (CreateSpace 2011)

10. P.A. Kharecha, J.E. Hansen, Prevented Mortality and Greenhouse Gas Emissions from Historical and Projected Nuclear Power, Environ. Sci. Technol. **47**, 4889 (2013)

11. A. Gabbard, Nuclear Resource or Danger, ORNL **26**, 1 (1993)

12. J.M. Cuttler, Remedy for Radiation Fear: Discharge the Politized Science, Dose Response J. **12**, 170 (2014)

13. S. Hirschberg, C. Bauer, P. Burgherr, E. Cazzoli, T. Heck, M. Spada, K. Tryer, Health Effects of Technologies for Power Generation: Contributions from Normal Operation, Severe Accidents and Terrorist Threat, in *Proceedings of PSAM 12 Conference, International Association for Probability Safety Assessment and Management, IAPSAM, Cal. 2014* (2014)

14. E.J. Calabrese, Road to Linearity: Why Linearity at Low Doses Became the Basis for Carcinogenic Risk Assessment, Arch. Toxicol. **83**, 203 (2009)

2

Estimation of the radionuclide inventory in LWR spent fuel assembly structural materials for long-term safety analysis

Stefano Caruso[*]

Radioactive Materials Division, National Cooperative for the Disposal of Radioactive Waste (NAGRA), Hardstrasse 73, 5430 Wettingen, Switzerland

Abstract. The radionuclide inventory of materials irradiated in a reactor depends on the initial material composition, irradiation history and on the magnitude and spectrum of the neutron flux. The material composition of a fuel assembly structure includes various alloys of Zircaloy, Inconel and stainless steel. The existing impurities in these materials are very important for accurate determination of the activation of all nuclides with a view to assessing the radiological consequences of their geological disposal. In fact, the safety assessments of geological repositories require the average and maximum (in the sense of very conservative) inventories of the very long-lived nuclides as input. The purpose of the present work is to describe the methodology applied for determining the activation of these nuclides in fuel assembly structural materials by means of coupled depletion/activation calculations and also to crosscheck the results obtained from two approaches. UO_2 and MOX PWR fuels have been simulated using SCALE/TRITON, simultaneously irradiating the fuel region in POWER mode and the cladding region in FLUX mode and aiming to produce binary macro cross-section libraries by applying accurate local neutron spectra in the cladding region as a function of irradiation history that are suitable for activation calculations. The developed activation libraries have been re-employed in a second run using the ORIGEN-S program for a dedicated activation calculation. The axial variation of the neutron flux along the fuel assembly length has also been considered. The SCALE calculations were performed using a 238-group cross-section library, according to the ENDF/B-VII. The results obtained with the ORIGEN-S activation calculations are compared with the results obtained from TRITON via direct irradiation of the cladding, as allowed by the FLUX mode. It is shown that an agreement on the total calculated activities can be found within 55% for MOX and within 22% for UO_2, whereas the latter is reduced to 9% when more accurate irradiation data are used (core-follow flux data instead of life-average flux data).

1 Introduction

In the context of characterizing spent fuel as technical waste[1], the assessment of the radionuclide inventory not limited to the fuel region but also including the cladding and structural materials is very important because of the build-up of very long-lived nuclides relevant for long-term safety analysis. Moreover, the release of these radionuclides from Zircaloy cladding and structural materials as a result of corrosion processes is much faster than the process of spent fuel dissolution. For this reason, these nuclides need to be treated separately.

An accurate determination of the induced activity can be performed if the activation study relies on knowledge of the real fuel depletion characteristics, such as the neutron flux spectrum in the material investigated. The assessment of the nuclide inventory from the perspective of geological disposal has a double aspect, being related on the one hand to the fuel handling and encapsulation operations (short- to medium-lived nuclides are more relevant for the dose rate contribution) and, on the other hand, to the long-term safety aspects. In fact, the long-lived nuclides, especially those producing the most decay heat, are relevant for the repository safety assessment. All this requires considerable effort in defining and validating a spent fuel depletion/activation methodology that can provide a radionuclide inventory with acceptable accuracy.

---

[*] e-mail: stefano.caruso@nagra.ch
[1] In strict terms the spent fuel is not classified as a waste.

In the present case, the significant heterogeneity of the fuel used in the five Swiss reactors makes this task highly complex. In this work, the SCALE/TRITON depletion sequence [1] and the stand-alone ORIGEN-S code [2] are both used to calculate the induced activity for high burnup fuels. The results obtained with TRITON first and then ORIGEN-S employing activation libraries built with TRITON itself are compared and discussed. Furthermore, several available methodologies are discussed for a more comprehensive but not exhaustive analysis of the topic.

Section 2 discusses the set of spent fuel assemblies to be characterized, including material components and impurities. Section 3 gives an overview of the methodologies used for the activation studies, describing in more detail the depletion/activation codes used and the criteria applied for accounting for the axial variations of the neutron flux. The results and conclusions are presented in Sections 4 and 5 respectively.

2 Spent fuel characteristics

2.1 Representative fuel assembly data

Basically, the assessment of the radiological inventory assumes as a basis the spent nuclear fuel (SNF) that has not been sent for reprocessing and is foreseen for geological disposal. From the available database, and also based on predictions, several SNF categories were generated within the framework of a NAGRA model inventory to be used for long-term safety assessment [3]. An overview of the implemented SNF categories is given in Table 1, according

Table 1. FA (fuel assembly) categorization per NPP, fuel type, average initial enrichment and average burnup to be considered for the repository.

AGT-ISRAM	Owner	Type	^{235}U/Pu [wt.%]	BU [Gwd/t$_{HM}$]
J-B-950001	KKB	PWR / UO$_2$	3.36	35.9
J-B-950002	KKB	PWR / UO$_2$	3.71	43.5
J-B-950003	KKB	PWR / UO$_2$	4.54	52.4
J-B-950004	KKB	PWR / UO$_2$	4.5	55
J-B-950005	KKB	PWR / UO$_2$	4.5	38.3
J-B-950501	KKB	PWR / MOX	0.71/2.73	33.8
J-B-950502	KKB	PWR / MOX	0.26/3.66	36.4
J-B-950503	KKB	PWR / MOX	0.26/3.69	43
J-B-950504	KKB	PWR / MOX	0.27/4.81	53.9
J-B-950505	KKB	PWR / MOX	0.27/4.86	55
J-G-950001	KKG	PWR / UO$_2$	3.5	39.3
J-G-950002	KKG	PWR / UO$_2$	3.46	47
J-G-950003	KKG	PWR / UO$_2$	4.39	56.6
J-G-950004	KKG	PWR / UO$_2$	4.4	55
J-G-950005	KKG	PWR / UO$_2$	4.4	32.9
J-G-950501	KKG	PWR / MOX	0.26/4.78	54.8
J-G-950502	KKG	PWR / MOX	0.25/4.78	55
J-L-950001	KKL	BWR / UO$_2$	1.67	18.3
J-L-950002	KKL	BWR / UO$_2$	2.36	26.3
J-L-950003	KKL	BWR / UO$_2$	2.71	34.5
J-L-950004	KKL	BWR / UO$_2$	3.31	43.9
J-L-950005	KKL	BWR / UO$_2$	4.01	50.6
J-L-950006	KKL	BWR / UO$_2$	4.3	55
J-L-950007	KKL	BWR / UO$_2$	4	32
J-M-95-0001	KKM	BWR / UO$_2$	3.13	40.3
J-M-95-0002	KKM	BWR / UO$_2$	3.67	48.5
J-M-95-0003	KKM	BWR / UO$_2$	4.08	52.1
J-M-95-0004	KKM	BWR / UO$_2$	4.2	55
J-M-95-0006	KKM	BWR / UO$_2$	4.2	30.3

Table 2. Impurities assumed to be present (<1000 ppm) in the FA structural material (in ppm) for the activation calculations [4,5].

	Al	B	C	Ca	Cd	Cl	Co	Cr	Cu	Fe	H	Hf	Mg	Mn	Mo
Zry-4	75	0.5	270	30	0.5	20	1.5	>	50	>	25	100	20	50	50
Steel	0	0	800	0	0	0	500	>	0	>	0	0	0	>	0

	N	Na	Nb	Ni	O	P	Pb	Si	Sn	Th	Ti	U	W	Zr
Zry-4	80	20	100	700	>	30	130	120	>	0.5	50	1.5	100	>
Steel	400	0	0	>	0	450	0	>	0	0.05	>	0.05	0	0

to AGT[2]-ISRAM[3] [4] denomination, the NPP[4], the type of fuel, the average enrichment and burnup. Each one of the 29 categories illustrated in the table will be characterized using the methodology described in this paper. However, the present study is limited to the fuel from the Gösgen NPP, namely the UO$_2$ J-G-950004 and MOX J-G-950502, both with a representative burnup of 55 GWd/t.

2.2 Structural materials and impurities

For the calculation of the induced activity in structural material from irradiated fuel bundles, it is necessary to know exactly the material composition up to the impurities level. For this study, a typical Siemens FA, with 15×15 array, was considered as the reference. The impurity vectors of all the materials involved (e.g. Zircaloy-4, Inconel, steel) were used for the calculation. Table 2 shows the impurities assumed in the Zircaloy and steel, which correspond to the values used in the NAGRA Entsorgungsnachweis[5] project [5,6]. The content of thorium, uranium and cobalt has recently been reviewed on the basis of sample measurements and is reported here. For illustration purposes, the table is limited to an impurities content of less than 1000 ppm.

3 Analysis of LWR fuel assemblies

The accuracy of activation calculations is determined largely by the accuracy and the completeness of the nuclear data associated with the transmutation process (macro cross-section libraries) and decay equations (nuclear data). These are the basic criteria to be used for the qualification of the method.

The methods employed for this activation study are described and discussed in the following sub-sections.

3.1 Methodologies

Several approaches can be used for the assessment of the induced activation in the FA structure. A set of these, mainly based on the SCALE computer code system (SCALE 6.1) that is developed and maintained at ORNL, has been considered in this work and are highlighted here:

- integrated depletion/activation calculation at FA level, using the SCALE/TRITON sequence (fuel depletion by POWER mode and cladding/structure activation by FLUX mode);
- stand-alone ORIGEN-S activation calculation using a self-developed TRITON cladding library (already achieved in point 1);
- development of a neutron activation cross-section library from a defined neutron flux spectrum in cladding/structure (if known) using COUPLE [1] and interfacing the created activation library with ORIGEN-S for the radionuclide activity calculation;
- stand-alone ORIGEN-S activation calculation on the basis of an ORIGEN-ARP standard fuel library (inaccuracy in the neutron flux spectrum).

The first three methods can be considered as the most accurate, since getting the correct neutron spectrum for the cladding and the best cross-sections and decay data available. In particular, the first two are discussed in detail in the following sections. The results for methods 1 and 2 are presented later in Section 4. Methods 3 to 5 are, however, briefly discussed here.

Method 3 is based on the a priori knowledge of the neutron flux spectra in the cladding/structural materials. The spectrum can be calculated by means of dedicated neutron transport calculations, e.g. by employing MCNP [7] to model the reactor core and running the simulation in criticality mode and extracting a representative neutron flux spectrum[6] for the cladding regions. The spectrum can be successively given to COUPLE, which creates binary nuclear data libraries (infinite dilution cross-sections) to be used directly in the depletion code ORIGEN-S for the activation calculations. Method 4 is a less accurate approach consisting of running ORIGEN-S coupled with the ORIGEN-ARP standard fuel libraries. The neutron flux spectrum used is the one in fuel, which is harder than

---

[2]AGT: waste package type.

[3]"Information System for Radioactive Materials (ISRAM)" for the long-term documentation of radioactive wastes and materials in Switzerland [4].

[4]KKB: Beznau; KKG: Gösgen; KKL: Leibstadt; KKM: Mühleberg.

[5]Demonstration of disposal feasibility for spent fuel, vitrified high-level waste and long-lived intermediate-level waste.

[6]i.e. estimating the volume-averaged neutron flux by simulation of particle scoring in "cladding" detector (F4 tally).

the characteristic spectrum in cladding. This introduces a larger uncertainty in the estimation of the inventory. Furthermore, not all ARP libraries are updated with the most recent cross-section libraries. However, this approach has the advantage of being less time-consuming.

It may be worth mentioning other activation codes that can be employed for this analysis. Some of these codes are listed here, with related cross-section databases: FISPACT-2007 [8] with EAF-2007 and the decay data on JEFF3.1, CINDER'90 [9] with ENDF/B-VI and EAF-3 and GRSAKTIV-II [10] with 84-group HAMMER data. The evaluation of these codes is, however, outside the scope of this paper.

3.2 Depletion calculations and neutron activation cross-section libraries

As introduced previously, two methods were employed for this activation study, both based on the SCALE package. The first approach (see point 1 in Sect. 3.1) is an integrated depletion/activation calculation at FA level, based entirely on the TRITON sequence. The second method is based on decoupling the TRITON sequence in a 2-step calculation, where the cladding macro cross-section libraries are first developed by TRITON and later used by ORIGEN-S stand-alone. Because TRITON is at the base of both methodologies, this section is devoted to describing the TRITON depletion model.

The TRITON depletion sequence allows transport and depletion calculations to be performed at fuel assembly level. TRITON consists of a sequence of different modules which are sequential-coupling transport calculations with depletion calculations. The resonance self-shielding is performed using CENTRM to prepare problem-dependent pointwise continuous-energy flux for use in the NEWT multigroup transport solver. CENTRM computes "continuous-energy" neutron spectra using discrete ordinance or other deterministic approximations for the Boltzmann transport equation. TRITON uses the BONAMI module for the unresolved-resonance energy region, performing Bondarenko calculations for the resonance self-shielding correction. Among several processing options, the CENTRM/PCM cross-section processing methodology was applied for the present study, because it is coupled with the most recent neutron libraries (ENDF/B-VII) and also because it can handle heterogeneous structures. The T-DEPL calculation sequence was selected and the fuel region was depleted in POWER mode and the cladding region in FLUX mode.

As shown in Section 2, PWR UO_2 and MOX 15×15 bundles, the type irradiated in the Gösgen (KKG) nuclear power plant, were considered. The UO_2 FA had an initial enrichment of 4.4 wt.% ^{235}U. The MOX fuel is characterized by a 3-region enrichment (high, medium and low Pu-content), giving 0.25 wt.% of ^{235}U and 4.78 wt.% of fissile Pu.

The main fuel characteristics, such as geometry, materials and other reactor-related parameters, were implemented. The control rods have been considered as fully extracted, this being their normal condition for most of their life in the reactor, meaning that the guide tubes are filled with water. Physical boundaries are set to mirror boundaries. The fuel was depleted using core-follow data, based on detailed irradiation history, for a final burnup of 55 GWd/t in both cases. A cross-section of the south-west 1/4 FA, as modeled with TRITON, is presented in Figure 1, where the UO_2 is illustrated on the left and MOX on the right. A 4×4 unit cell coarse mesh structure was used.

Although a 1/8 symmetry is given (in Fig. 1, the rods with the same 1/8 symmetry are illustrated with the same

(a)   (b)

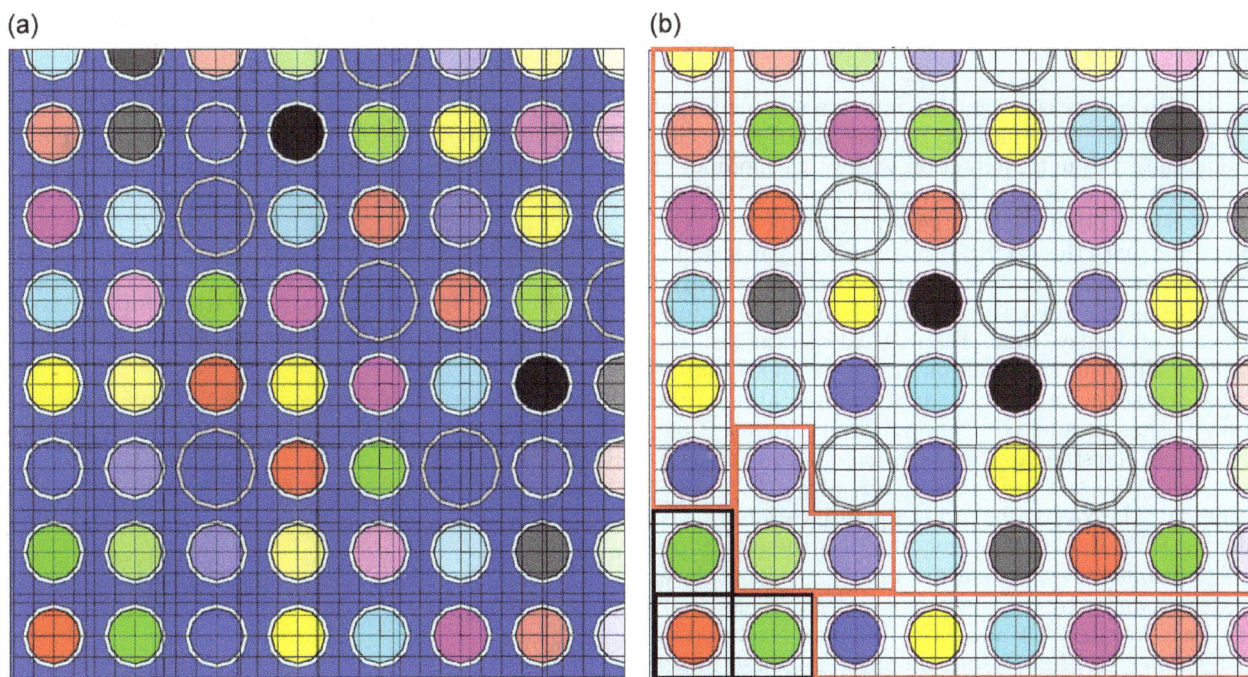

Fig. 1. UO_2 (left) and MOX (right) KKG 1/4 assembly fuel model in TRITON.

color), the model of the FA was represented by 1/4 FA, using mirror boundary conditions. The effect of this model simplification is a reduction of computation time without a significant loss of accuracy.

The assign function, which simplifies the cross-section processing by calculating a particular rod and assigning this one to all other rods, was used for both UO_2 and MOX models. However, because of the 3-region enrichment characterizing the plutonium rods in the MOX case, three different regions were assigned. These regions are defined as low Pu-content fuel (black framed area on Fig. 1 right), medium Pu-content fuel (red framed areas on Fig. 1) and high Pu-content fuel (no framed areas).

TRITON provides the possibility to develop a problem-specific fuel model and, based on the model developed, to create a problem-dependent library. The neutron activation cross-section libraries for the cladding can be produced in the course of a TRITON depletion calculation. In fact, the depletion module allows a simultaneous run in POWER mode for the fuel region and in FLUX mode for the cladding region. The new binary cross-section libraries are produced for each region declared in the depletion module. In this work, the cladding was defined as a unique region and was irradiated according to the neutron flux spectrum calculated in the Zircaloy. The activation library produced in this way is customized only on the neutron spectrum in cladding and is based on the cross-section data ENDF/B-VII, with 238 neutron energy groups.

The multigroup cross-sections are then combined with the neutron flux solution and collapsed using the COUPLE code to create effective one-group cross-sections for use with ORIGEN. Burnup-dependent cross-section libraries for ORIGEN are saved during the TRITON depletion calculation at each depletion step. These libraries are created for each depleted mixture in the analyzed configuration.

3.3 Implementing the activation libraries in ORIGEN-S

The ORIGEN-S code is designed to function as a module of the SCALE code system and obtains problem-specific neutronic data through interaction with other modules of the system, such as the above-mentioned TRITON. ORIGEN-S data libraries can be generated by the TRITON sequence and, with these, ORIGEN-S can be run in a stand-alone configuration. In fact, time-dependent material concentrations are solved using the ORIGEN-S isotope depletion and decay code. For activation studies, the accuracy of the results depends on having an appropriately weighted cross-section library that is representative of the material being irradiated: flux-weighted cross-sections updated from the standard cross-section data on the basis of the real structure of a fuel assembly (FA type-dependent). Moreover, the exact quantification of the impurities is fundamental, being through these isotopes that significant transmutation reactions are taking place; e.g. ^{14}C is produced by (n,p) reactions on ^{14}N, ^{36}Cl as result of (n,γ) reactions on ^{35}Cl, and ^{60}Co, ^{94}Nb also produced by (n,γ) reactions on their stable isotopes. To these, other isotopes abundantly present into the structural materials and having remarkable resonances self-shielding properties

are contributing to the global activation (see (n,γ) reactions of ^{55}Fe, ^{93}Zr and ^{63}Ni). With this approach, any material can be analyzed under specific spectral irradiation conditions. It is worth noting that ORIGEN-S is able to utilize multi-energy-group neutron flux and cross-sections in any group structure. However, the 238 multigroup space and energy cross-sections need to be collapsed in a spectrum-averaged (one-group) cross-section in order to solve the activation equation.

In order to employ ORIGEN-S stand-alone, the total neutron flux intensity must be given in the input as a function of irradiation time. The values for total flux can be extracted accordingly from the TRITON output, which gives values for any defined material region (e.g. fuel, cladding). Here, the average values 5.04×10^{14} n/cm^{2}s for UO_2 and 5.55×10^{14} n/cm^{2}s for MOX were used respectively.

3.4 Multi-region flux activation calculations

A significant limitation of the current approach is the assumption of a two-dimensional model, which ignores the very important disuniformity of the axial neutron flux. The cladding in the extremities and the structural materials of the top and bottom of the FA are irradiated with a lower neutron flux than in the central position. As a consequence, the use of a constant axial flux introduces unacceptable inaccuracy. To overcome this, a neutron flux region-dependent factor was implemented. The factor was used to normalize the mass of material to the corresponding neutron flux for a specific region. The FA is divided into four main regions, each one characterized by an average representative neutron flux (see Tab. 3) coming from the determined extrapolation distance of the neutron flux along the full axial length of the core (see also Ref. [11]). The scaling factors are directly employed as a mass weighting factor for each material region of the FA, so that the mass of the material is normalized to the neutron flux. The employment of these weighting factors on the irradiated mass is equivalent to the application of a reduction factor on the neutron flux itself. This approach has the advantage of performing the simulation in one single run.

4 Results and comparison

The results of fuel activation calculations for the modeled fuel assemblies described above are given in Table 4, as

Table 3. Neutron flux regions in FA [11].

Reactor type	PWR
Fuel type	Westinghouse
Region of FA	Flux scaling factor
Top end fitting	10%
Gas plenum	20%
Fueled region	100%
Bottom end fitting	20%

Table 4. Isotopic activities (Bq/t_{HM}) in cladding for PWR UO$_2$ and MOX fuel (cooling time 60 days).

Nuclides	PWR (KKG) UO$_2$ - 4.4% enrich. / 55 GWd/t						PWR (KKG) MOX - 0.25/4.78% enrich. / 55 GWd/t					
	TRITON Core-Foll	Origen-S Flux-avg	Origen-S Core-Foll	Origen-S Flux-avg	Orig/Trit Flux-avg	Orig/Trit Core-Foll	TRITON Core-Foll	Origen-S Flux-avg	Origen-S Core-Foll	Origen-S Flux-avg	Orig/Trit Flux-avg	Orig/Trit Core-Foll
	Hot Cld	Hot Cld	Hot Cld	Cold cld	Hot Cld	Hot Cld	Hot Cld	Hot Cld	Hot Cld	Cold cld	Hot Cld	Hot Cld
ag108m	6.26E+03	6.95E+03	5.50E+03	1.30E+04	1.11	0.88	3.71E+03	4.45E+03	4.36E+03	4.98E+03	1.20	1.17
am241	2.79E+07	2.36E+07	2.42E+07	2.23E+07	0.84	0.86	2.69E+07	3.06E+07	3.25E+07	3.92E+07	1.14	1.21
am242m	1.07E+06	1.03E+06	1.03E+06	8.25E+05	0.96	0.97	1.20E+06	1.47E+06	1.54E+06	1.85E+06	1.23	1.28
am243	7.60E+06	1.09E+07	8.14E+06	1.35E+07	1.43	1.07	1.87E+06	2.29E+06	2.20E+06	5.89E+06	1.22	1.18
ar39	3.79E+06	3.97E+06	3.40E+06	4.80E+06	1.05	0.90	3.19E+06	4.09E+06	4.05E+06	3.87E+06	1.29	1.27
ba133	2.78E+03	3.36E+03	1.84E+03	5.43E+03	1.21	0.66	1.17E+03	1.44E+03	1.41E+03	2.42E+03	1.24	1.21
ba137m	8.78E+09	9.94E+09	8.29E+09	1.30E+10	1.13	0.94	4.62E+09	5.66E+09	5.54E+09	8.53E+09	1.22	1.20
c14	2.89E+10	3.27E+10	2.81E+10	3.92E+10	1.13	0.97	1.26E+10	1.52E+10	1.50E+10	1.94E+10	1.21	1.20
cd113m	8.43E+06	8.26E+06	8.22E+06	7.92E+06	0.98	0.97	1.40E+07	1.99E+07	1.99E+07	1.55E+07	1.42	1.42
cl36	6.42E+08	7.25E+08	6.28E+08	8.80E+08	1.13	0.98	2.49E+08	2.96E+08	2.92E+08	4.09E+08	1.19	1.18
cm242	8.18E+09	9.73E+09	8.19E+09	1.11E+10	1.19	1.00	4.18E+09	5.14E+09	4.92E+09	9.39E+09	1.23	1.18
cm243	5.01E+06	6.31E+06	4.65E+06	7.73E+06	1.26	0.93	2.05E+06	2.38E+06	2.25E+06	4.91E+06	1.16	1.10
cm244	1.63E+09	2.79E+09	1.72E+09	3.76E+09	1.71	1.05	2.47E+08	3.22E+08	3.05E+08	9.88E+08	1.30	1.23
cm245	3.05E+05	5.53E+05	3.07E+05	7.27E+05	1.82	1.01	4.58E+04	6.05E+04	5.66E+04	1.94E+05	1.32	1.24
cm246	7.57E+04	1.47E+05	6.58E+04	2.66E+05	1.94	0.87	4.53E+03	5.26E+03	4.86E+03	2.50E+04	1.16	1.07
co60	4.35E+13	4.81E+13	4.25E+13	5.44E+13	1.11	0.98	2.54E+13	3.17E+13	3.11E+13	3.61E+13	1.25	1.22
cs134	1.60E+10	2.16E+10	1.61E+10	3.19E+10	1.35	1.01	7.61E+09	1.02E+10	9.77E+09	1.72E+10	1.35	1.29
cs135	3.21E+04	3.62E+04	2.99E+04	4.05E+04	1.13	0.93	2.73E+04	3.53E+04	3.55E+04	4.12E+04	1.30	1.30
cs137	9.27E+09	1.05E+10	8.76E+09	1.38E+10	1.13	0.94	4.88E+09	5.97E+09	5.86E+09	9.01E+09	1.22	1.20
eu152	1.40E+05	1.45E+05	1.47E+05	1.26E+05	1.04	1.05	2.48E+05	3.29E+05	3.60E+05	2.79E+05	1.33	1.45
eu154	8.03E+08	9.07E+08	7.37E+08	1.92E+09	1.13	0.92	4.83E+08	6.14E+08	5.97E+08	9.45E+08	1.27	1.24
eu155	4.54E+08	4.81E+08	3.97E+08	5.89E+08	1.06	0.87	2.33E+08	2.87E+08	2.79E+08	4.66E+08	1.23	1.20
fe55	1.76E+14	2.00E+14	1.78E+14	2.31E+14	1.13	1.01	8.58E+13	1.07E+14	1.05E+14	1.34E+14	1.25	1.22
h3	3.33E+10	3.45E+10	2.97E+10	3.31E+10	1.04	0.89	3.19E+10	4.08E+10	4.02E+10	3.85E+10	1.28	1.26
ho166m	5.37E+01	6.59E+01	4.31E+01	1.11E+02	1.23	0.80	2.57E+01	3.32E+01	3.20E+01	5.18E+01	1.29	1.25
i129	3.31E+03	3.67E+03	3.09E+03	4.84E+03	1.11	0.93	1.82E+03	2.23E+03	2.19E+03	3.26E+03	1.22	1.20
kr85	6.87E+08	7.72E+08	6.44E+08	9.66E+08	1.12	0.94	3.74E+08	4.57E+08	4.48E+08	6.35E+08	1.22	1.20
nb93m	4.69E+11	5.12E+11	4.42E+11	1.66E+09	1.09	0.94	3.83E+11	4.90E+11	4.83E+11	4.61E+11	1.29	1.26
nb94	2.43E+10	2.65E+10	2.30E+10	2.81E+10	1.09	0.95	1.76E+10	2.22E+10	2.20E+10	2.29E+10	1.26	1.25
ni59	7.44E+10	8.28E+10	7.33E+10	9.82E+10	1.11	0.99	3.14E+10	3.77E+10	3.72E+10	4.99E+10	1.20	1.19
ni63	1.04E+13	1.17E+13	1.01E+13	1.44E+13	1.13	0.98	4.03E+12	4.80E+12	4.74E+12	6.60E+12	1.19	1.18
np237	3.11E+03	3.16E+03	2.84E+03	2.93E+03	1.02	0.91	2.74E+03	3.48E+03	3.45E+03	3.43E+03	1.27	1.26

Table 4. (continued).

Nuclides	PWR (KKG) UO₂ - 4.4% enrich. / 55 GWd/t						PWR (KKG) MOX - 0.25/4.78% enrich. / 55 GWd/t					
	TRITON Core-Foll Hot Cld	Origen-S Flux-avg Hot Cld	Origen-S Core-Foll Hot Cld	Origen-S Flux-avg Cold cld	Orig/Trit Flux-avg Hot Cld	Orig/Trit Core-Foll Hot Cld	TRITON Core-Foll Hot Cld	Origen-S Flux-avg Hot Cld	Origen-S Core-Foll Hot Cld	Origen-S Flux-avg Cold cld	Orig/Trit Flux-avg Hot Cld	Orig/Trit Core-Foll Hot Cld
pd107	1.92E+04	2.18E+04	1.83E+04	2.96E+04	1.14	0.95	9.72E+03	1.19E+04	1.17E+04	1.90E+04	1.22	1.20
pm146	5.36E+03	3.71E+03	3.53E+03	5.79E+03	0.69	0.66	6.38E+03	8.19E+03	8.06E+03	7.84E+03	1.28	1.26
pm147	1.16E+10	1.18E+10	1.12E+10	1.39E+10	1.02	0.96	7.32E+09	8.95E+09	8.73E+09	1.23E+10	1.22	1.19
pu238	1.15E+08	1.35E+08	1.03E+08	1.50E+08	1.18	0.90	5.79E+07	6.68E+07	6.71E+07	9.99E+07	1.15	1.16
pu239	3.41E+07	2.88E+07	2.98E+07	3.03E+07	0.84	0.87	4.36E+07	6.41E+07	6.40E+07	5.96E+07	1.47	1.47
pu240	4.31E+07	2.73E+07	2.81E+07	3.10E+07	0.63	0.65	3.67E+07	4.62E+07	4.60E+07	4.47E+07	1.26	1.25
pu241	2.11E+10	2.06E+10	2.08E+10	2.15E+10	0.98	0.98	2.07E+10	2.70E+10	2.66E+10	3.49E+10	1.31	1.29
pu242	4.62E+05	5.44E+05	4.69E+05	7.39E+05	1.18	1.02	1.93E+05	2.20E+05	2.14E+05	4.42E+05	1.14	1.11
sb125	8.19E+13	8.91E+13	7.96E+13	9.19E+13	1.09	0.97	6.37E+13	8.20E+13	7.99E+13	8.21E+13	1.29	1.25
se79	7.09E+03	7.84E+03	6.59E+03	1.04E+04	1.11	0.93	3.97E+03	4.86E+03	4.77E+03	6.71E+03	1.22	1.20
sm151	3.02E+07	3.34E+07	3.08E+07	3.71E+07	1.10	1.02	3.08E+07	4.31E+07	4.26E+07	4.90E+07	1.40	1.38
sn121m	1.23E+11	1.32E+11	1.15E+11	1.27E+11	1.08	0.93	9.69E+10	1.22E+11	1.21E+11	1.19E+11	1.26	1.24
sn126	4.49E+04	4.95E+04	4.17E+04	7.06E+04	1.10	0.93	2.75E+04	3.39E+04	3.33E+04	4.46E+04	1.24	1.21
sr90	4.48E+09	4.94E+09	4.11E+09	6.07E+09	1.10	0.92	2.65E+09	3.24E+09	3.18E+09	4.28E+09	1.22	1.20
tc99	1.80E+08	1.84E+08	1.62E+08	2.02E+08	1.02	0.90	1.40E+08	1.67E+08	1.65E+08	1.67E+08	1.19	1.18
te125m	1.92E+13	2.07E+13	1.85E+13	2.11E+13	1.08	0.96	1.48E+13	1.88E+13	1.85E+13	1.88E+13	1.27	1.25
u234	3.34E+05	3.59E+05	3.12E+05	3.73E+05	1.08	0.93	2.18E+05	2.66E+05	2.61E+05	2.97E+05	1.22	1.20
u235	6.24E+01	5.91E+01	6.21E+01	5.32E+01	0.95	0.99	8.76E+01	1.11E+02	1.11E+02	9.03E+01	1.26	1.27
u236	1.14E+03	1.09E+03	1.04E+03	1.09E+03	0.95	0.91	8.09E+02	9.23E+02	9.19E+02	9.93E+02	1.14	1.14
u238	4.47E+03	4.49E+03	4.65E+03	4.13E+03	1.00	1.04	3.73E+03	4.70E+03	4.72E+03	4.38E+03	1.26	1.26
y90	4.56E+09	5.01E+09	4.17E+09	6.14E+09	1.10	0.92	2.71E+09	3.35E+09	3.27E+09	4.39E+09	1.24	1.21
zr93	8.91E+09	9.84E+09	8.44E+09	8.38E+09	1.10	0.95	5.90E+09	7.37E+09	7.28E+09	5.98E+09	1.25	1.23
Total	**4.56E+15**	**5.54E+15**	**4.98E+15**	**4.60E+15**	**1.22**	**1.09**	**3.75E+15**	**5.79E+15**	**5.11E+15**	**4.65E+15**	**1.55**	**1.37**

UO2 PWR, 4.4 % enrich., 55 GWd/t burnup

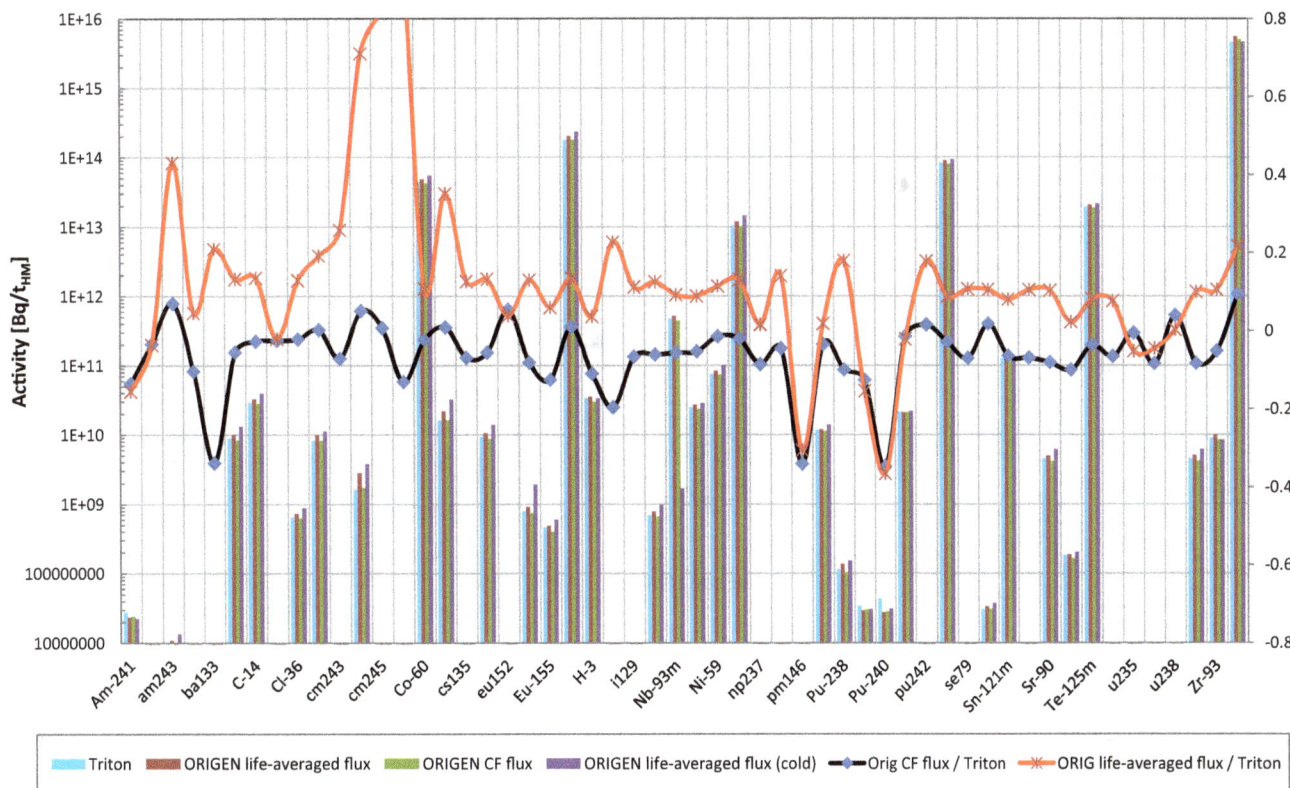

Fig. 2. Specific activities and deviations between TRITON and ORIGEN-S for the UO$_2$ PWR case (4.4% enrichment, burnup = 55 GWd/t, CT = 60 days).

specific activity (Bq/t$_{HM}$) normalized to 1 ton of heavy metal, for the cases of KKG UO$_2$ and MOX PWR with 55 GWd/t burnup. Furthermore, the results are given for cladding in fuel rods (TRITON macro cross-section libraries for hot cladding) as well as for water rod cladding (TRITON macro cross-section libraries for cold cladding), aiming to show the influence of different spectral irradiation conditions on the final results. The calculations with ORIGEN-S were carried out using first a life-averaged flux value and then quasi-core-follow flux values (respectively *Flux-avg* and *Core-Foll* in Tab. 4). The calculation produced results for 328 nuclides[7], including activation products, fission products and actinides. However, Table 4 reports only a restricted set of nuclides relevant for long-term safety analysis. All values refer to a cooling time of 60 days, chosen to filter out the very short-lived isotopes.

The TRITON values in the table refer to the fuel rod cladding and are calculated using core-follow irradiation data. These results are compared against:

– the activities calculated for the hot cladding with ORIGEN-S using a life-averaged flux value (same time mesh as the following point);

– the activities calculated for hot cladding, but employing core-follow flux data;
– the activities calculated for the cold cladding, based on life-averaged flux.

The following observations are derived from this study:

– a tendency to overpredict the activity in the ORIGEN-S calculation, more pronounced for the MOX case. This can be attributed to the approximation of the irradiation data used for ORIGEN-S calculations;
– in fact, the ORIGEN-S core-follow calculation agrees much better with the TRITON one (9% for UO$_2$ and 37% for MOX) than the ORIGEN-S life-averaged one (22% for UO$_2$ and 55% for MOX);
– there are isotopes that are very sensitive to irradiation history (e.g. ^{246}Cm activity for the UO$_2$ fuel which shows a factor 1.94). The build-up of curium, in fact, involves a sequence of neutron captures. Thus, uncertainties in the determination of this element build-up as a result of uncertainties propagation of its precursors;
– around 20% higher global activity between the ORIGEN calculation based on the hot cladding (fuel rod) library and the cold cladding (water rod) library, using the same irradiation conditions.

It is worth mentioning that the total activity given in the last row of the table refers to the full set of nuclides generated. Furthermore, the decay time windows due to outage were

---

[7]The set of 328 nuclides includes all relevant nuclides needed for long-term safety assessment, starting at time of emplacement into the geological disposal.

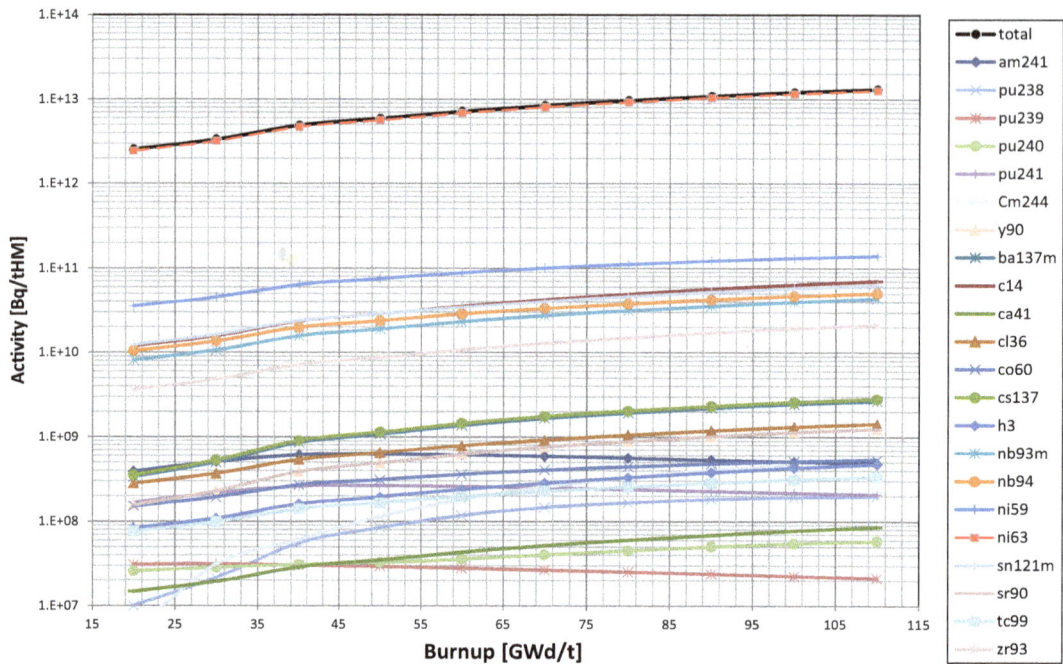

Fig. 3. Actinide, activation product and fission product activities as a function of fuel burnup, decay-corrected up to 100 years.

considered only by the TRITON calculations. Figure 2 illustrates the specific activities, per individual nuclide, and deviations between TRITON and ORIGEN-S, carried out employing different levels of irradiation data: core-follow flux (CF) and life-average flux. The values of the deviations (continuous lines) are given on the right axis. The case represented is a UO_2 PWR fuel assembly with 4.4% enrichment, 55 GWd/t burnup and cooling time of 60 days.

A sensitivity analysis aimed at investigating the relationship between burnup and induced activity in the structural material was also carried out. The build-up of isotopes as a function of burnup is illustrated graphically in Figure 3 for fission and activation products and actinides. The values refer to the case of a UO_2 fuel assembly, as previously described, and were calculated using ORIGEN-S on the basis of a macro cross-section library built with TRITON (hot cladding) by depleting the fuel up to 110 GWd/t burnup (using life-averaged flux), using steps of 10 GWd/t. All values reported are decay-corrected up to 100 years, in order to treat only the more long-term relevant nuclides. It can be observed that:

– the total activity build-up is a linear function with the burnup, as expected for activation products;
– the global activity is dominated by the activation of nickel, reaching 1.2×10^{13} Bq/t_{HM} at 110 GWd/t (see the contribution of [63]Ni to the total activity in Fig. 2);
– the fission products show linear behavior for high burnup (after 40 GWd/t);
– there is a factor 5 difference from the total activity at 20 GWd/t and 110 GWd/t;
– the build-up of actinides is not linear. The activity is dominated by [241]Am ($\sim 6 \times 10^8$ Bq/t_{HM}) and, for very high burnup, by [244]Cm ($< 5 \times 10^8$ Bq/t_{HM}).

5 Conclusions and further developments

The objective of activation studies of Zircaloy cladding and structural material from UO_2 and MOX spent fuel assemblies is to establish an approach that will serve as a sound basis for the assessment of all spent fuel to be disposed of in the high-level waste geological repository in Switzerland. The large heterogeneity of the fuel used in the five Swiss reactors makes this task highly complex so that different methodologies need to be investigated.

Among the different approaches discussed in this work, the author focused on the development of macro cross-section libraries customized on the Zircaloy cladding which ensure the employment of a more accurate neutron spectrum for the activation calculation. The library can be built using the SCALE/TRITON sequence and the activation analysis can be carried out with ORIGEN-S on the basis of this library. The computational time needed by SCALE/TRITON calculations is nevertheless quite long, but the time devoted to the ORIGEN-S activation calculations is extremely short. This makes the initial effort of developing the libraries worthwhile, being the use of ORIGEN not constrained by the material composition, which can be modified at will but always in agreement with the neutron spectral conditions. The axial neutron flux dependence was also taken into account by using flux weighting factors applied to the mass of the irradiated components.

A tendency to overpredict the activity by ORIGEN-S as compared to TRITON was observed, even more clearly for the MOX case (55% more in the total activity). This is due to the dependence on the accuracy of the employed irradiation history. This dependence was investigated by employing irradiation data at different levels (life-averaged against core-follow), showing that the more accurate the irradiation history is, the smaller the gap is between the results. Different spectral irradiation conditions were also investigated (cold cladding against hot cladding), the cold one being 20% lower in total activity. Furthermore, a sensitivity analysis was carried out to investigate the activity build-up as a function of the burnup at individual isotopic level. As expected, the global activity, dominated by the activation of nickel, shows a linear behavior with burnup, with the exception of the small contribution from actinides.

The study is limited to simulation of UO_2 and MOX PWR spent fuel assemblies; a validation against measured data has not yet been performed but is still desirable. An international benchmark would be also desirable, as platform to infer the quality of these results and future works. Additional effort will be needed to include all the spent fuel types irradiated in the Swiss NPPs and foreseen for geological disposal. Particularly interesting will be the case of BWR, where a 3D model is needed to account for the neutron spectra inhomogeneity along the FA axial profile. The scope of the investigation could be even extended: the method as illustrated in this work is mainly focused on the determination of the induced activity in fuel cladding, but it could be directed to other relevant reactor components, e.g. control and safety rods and/or thimble plugs.

References

1. ORNL (Oak Ridge National Laboratory), SCALE: a comprehensive modeling and simulation suite for nuclear safety analysis and design, ORNL/TM-2005/39, vs. 6, 2011
2. I.C. Gauld, O.W. Hermann, R.M. Westfall, *ORIGEN scale system module to calculate fuel depletion, actinide transmutation, fission product buildup and decay, and associated radiation terms* (Oak Ridge National Laboratory, Oak Ridge, Tennessee, 2009) ORNL/TM 2005/39, Version 6, Vol. II, Sect. F7
3. NAGRA (National Cooperative for the Disposal of Radioactive Waste), Model inventory for radioactive materials, MIRAM 14. NAGRA Technical Report NTB 14-04, Wettingen, Switzerland, 2014
4. H. Maxeiner, M. Vespa, B. Volmert, M. Pantelias, S. Caruso, T. Müller, Development of the inventory for existing and future radioactive wastes in Switzerland: ISRAM & MIRAM, ATW Int. J. Nucl. Power **58**, 625 (2013)
5. NAGRA (National Cooperative for the Disposal of Radioactive Waste), Model radioactive waste inventory for reprocessing waste and spent fuel, NAGRA Technical report NTB 01-01, Wettingen, Switzerland, 2002
6. NAGRA (National Cooperative for the Disposal of Radioactive Waste), Entsorgungsprogramm 2008 der Entsorgungspflichtigen, NAGRA Technical report NTB 08-01, Wettingen, Switzerland, 2008
7. X-5 Monte Carlo Team, MCNP - A general Monte Carlo N-particle transport code, Version 5; Vol. I: Overview and theory, Technical report LA-UR-03-1987, Los Alamos National Laboratory, 2005
8. R.A. Forrest, FISPACT-2007: User manual, Technical report, UKAEA FUS 534, EURATOM/UKAEA Fusion Association, 2007
9. W.B. Wilson et al., A manual for CINDER'90 version 07.4 codes and data, Technical report LA-UR-07-8412, Los Alamos National Laboratory, 2008
10. U. Hesse, K. Hummelsheim, GRSAKTIV-II: Ein Programmsystem zur Berechnung der Aktivierung von Brennelement- und Core-Bauteilen in Vielgruppendarstellung, Technical report GRS-A-3002, Gesellschaft für Anlagen- und Reaktorsicherheit, 2001
11. A.T. Luksic, B.D. Reid, Using the ORIGEN-2 computer code for near core activation calculations, in *Proceedings of the third international conference on high level radioactive waste management, ANS Las Vegas, NV (USA) 1992* (1992)

3

Characterization of the ion-amorphization process and thermal annealing effects on third generation SiC fibers and 6H-SiC

Juan Huguet-Garcia[1*], Aurélien Jankowiak[1], Sandrine Miro[2], Renaud Podor[3], Estelle Meslin[4], Lionel Thomé[5], Yves Serruys[2], and Jean-Marc Costantini[1]

[1] CEA, DEN, Service de Recherches Métallurgiques Appliquées, 91191 Gif-sur-Yvette, France
[2] CEA, DEN, Service de Recherches en Métallurgie Physique, Laboratoire JANNUS, 91191 Gif-sur-Yvette, France
[3] ICSM-UMR5257 CEA/CNRS/UM2/ENSCM, Site de Marcoule, bâtiment 426, BP 17171, 30207 Bagnols-sur-Cèze, France
[4] CEA, DEN, Service de Recherches en Métallurgie Physique, 91191 Gif-sur-Yvette, France
[5] CSNSM, CNRS-IN2P3, Université Paris-sud, 91405 Orsay, France

Abstract. The objective of the present work is to study the irradiation effects on third generation SiC fibers which fulfill the minimum requisites for nuclear applications, i.e. Hi-Nicalon type S, hereafter HNS, and Tyranno SA3, hereafter TSA3. With this purpose, these fibers have been ion-irradiated with 4 MeV Au ions at room temperature and increasing fluences. Irradiation effects have been characterized in terms of micro-Raman Spectroscopy and Transmission Electron Microscopy and compared to the response of the as-irradiated model material, i.e. 6H-SiC single crystals. It is reported that ion-irradiation induces amorphization in SiC fibers. Ion-amorphization kinetics between these fibers and 6H-SiC single crystals are similar despite their different microstructures and polytypes with a critical amorphization dose of $\sim 3 \times 10^{14} \text{cm}^{-2}$ ($\sim 0.6\,\text{dpa}$) at room temperature. Also, thermally annealing-induced cracking is studied via in situ Environmental Scanning Electron Microscopy. The temperatures at which the first cracks appear as well as the crack density growth rate increase with increasing heating rates. The activation energy of the cracking process yields 1.05 eV in agreement with recrystallization activation energies of ion-amorphized samples.

1 Introduction

Future nuclear applications include the deployment of the so-called Generation IV fission and fusion reactors, which are devised to operate at higher temperatures and to higher exposition doses than nowadays nuclear reactors. One of the critical issues to the success of future nuclear applications is to develop high performance structural materials with good thermal and radiation stability, neutron transparency and chemical compatibility [1].

Structural materials for nuclear applications are exposed to high temperatures, aqueous corrosive environments and severe mechanical loadings while exposed to neutron and ion irradiation. Its exposure to incident energetic particles displaces numerous atoms from the lattice sites inducing material degradation. Such degradation is the main threat to the safe operation of core internal structures and is manifested in several forms: radiation hardening and embrittlement, phase instabilities from radiation-induced

or enhanced precipitation, irradiation creep and volumetric swelling [2]. As can be observed in Figure 1, nominal temperatures and displacement doses can reach up to 1100 °C and 200 dpa depending on the nuclear reactor design. As a consequence, conventional nuclear materials, mostly metallic alloys, do not meet the requirements to operate neither under nominal nor accidental conditions.

Nuclear grade Silicon Carbide based composites – made of third generation SiC fibers densified via chemical vapor infiltration (CVI) with a SiC matrix; SiC_f/SiC_m – are among the most promising structural materials for fission and fusion future nuclear applications [3]. However, several remaining uncertainties place SiC_f/SiC_m in a position that requires further research and development, notably the radiation behavior of the fiber reinforcement which is crucial for the composite radiation stability.

The objective of the present work is to study the irradiation effects on third generation SiC fibers which fulfill the minimum requisites for nuclear applications, i.e. Hi-Nicalon type S, hereafter HNS, and Tyranno SA3, hereafter TSA3. With this purpose, these fibers have been ion-irradiated at room temperature to different doses under

---

*e-mail: juan.huguet-garcia@cea.fr

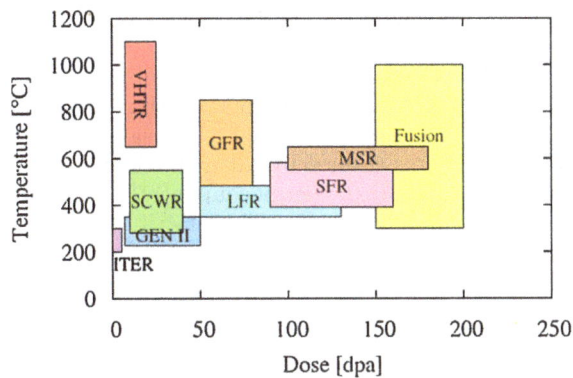

Fig. 1. Nominal operating temperatures and displacement doses for structural materials in different nuclear applications. The acronyms are defined in the Nomenclature section (adapted from Ref. [2]).

Table 1. Main characteristics of third generation SiC fibers.

Fiber	Tyranno SA3	Hi-Nicalon type S
Producer [6]	Ube Industries	Nippon Carbon
Diameter (μm) [6]	7.5	12
Density (g cm^{-3}) [6]	3.1	3.05
C/Si ratio[a] [7]	1.03–1.2	1.07
Composition [6]	68Si+32C +0.6Al	69Si+31C+0.2O
Grain Size (nm)[b] [5]	141–210	26–36

[a] Values correspond to the edge and core of the fiber respectively.
[b] Min. and max. Feret diameters.

elastic energy loss regimes to simulate neutron damage. The irradiation effects have been characterized in terms of micro-Raman Spectroscopy (μRS), Transmission Electron Microscopy (TEM) and Environmental Scanning Electron Microscopy (E-SEM) and compared to the response of the as-irradiated model material, i.e. 6H-SiC single crystals.

2 Materials and methods

2.1 6H-SiC single crystals and third generation SiC fibers

6H-SiC single crystals of $246\,\mu$m thickness were machined from N-doped (0001)-oriented 6H-SiC single crystal wafers grown by CREE Research using a modified Lely method. Crystals were of n-type with a net doping density (n_D–n_A) of $10^{17}\,$cm^{-3}. All samples were polished to achieve a microelectronics "epiready" quality.

Main characteristics of HNS and TSA3 fibers are summarized in Table 1. Figure 2 shows TEM images of the microstructures of both fibers. Both fibers consist in highly faulted 3C-SiC grains and intergranular pockets of turbostratic C visible as white zones in Figure 2. Stacking Faults (SFs) in SiC grains are clearly observed for both fibers as striped patterns inside the grains. Stacking fault linear density yields $0.29\pm0.1\,$nm^{-1} for HNS fibers and $0.18\pm0.1\,$nm^{-1} for TSA3 fibers. It has been determined by counting the number of stripes per unit length in the perpendicular direction using ImageJ [4] image analysis software. Also, mean maximum and minimum Feret diameters – which correspond to the shortest and the longest distances between any two points along the grain boundary (GB) – were determined. These values yield, respectively, 26 and 36 nm for the HNS fibers and 141 and 210 nm for the TSA3 fibers [5].

2.2 Ion-irradiation

Different 6H-SiC single crystals and SiC fibers were irradiated at room temperature (RT) with 4 MeV Au^{2+} to 5×10^{12}, 10^{13}, 5×10^{13}, 10^{14}, 2×10^{14}, 3×10^{14}, $10^{15}\,$cm^{-2} at JANNUS-Orsay facility and to $2\times10^{15}\,$cm^{-2} at JANNUS-Saclay facility [8]. To evaluate the irradiation

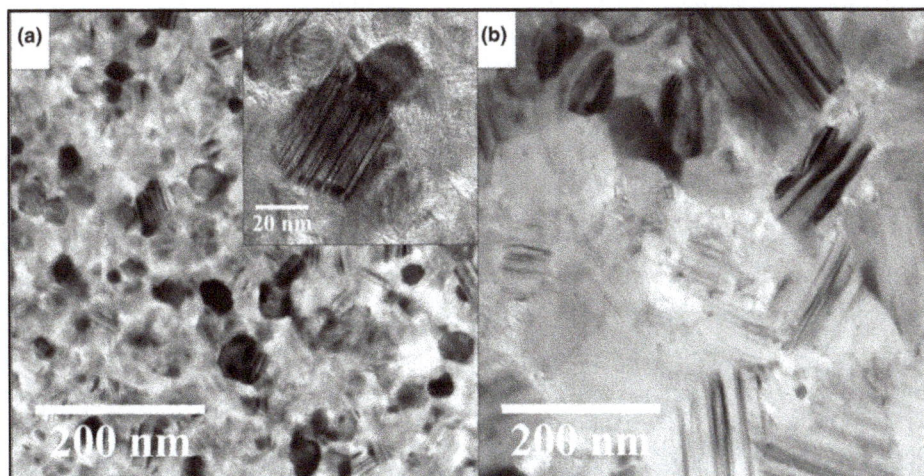

Fig. 2. TEM images of the as-received (a) HNS and (b) TSA3 fibers. Stripped patterns inside the grains indicate the high density of stacking faults in both samples (reproduced from Ref. [5]).

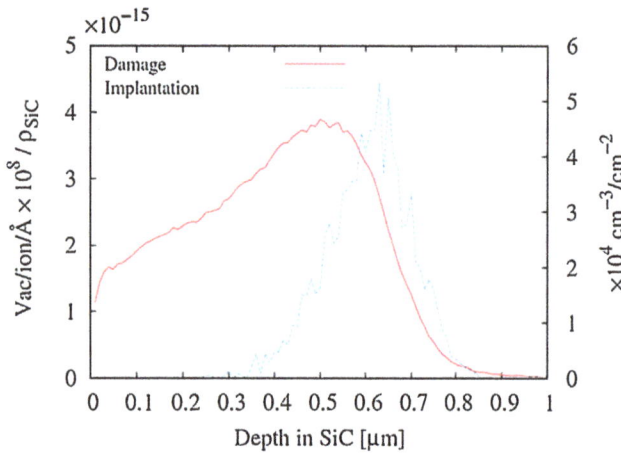

Fig. 3. Damage and implantation profiles for 4 MeV Au in SiC. Fluence-dpa estimation can be obtained by direct multiplication of the y-axis per the ion fluence.

damage, ion-fluences have been converted to dpa with equation (1):

$$dpa = \frac{\frac{Vac}{ion\ A} \times 10^8}{\rho_{SiC}[atoms\ cm^{-3}]} \times \varphi[ions\ cm^{-2}] \qquad (1)$$

where φ is the ion fluence, ρ_{SiC} the theoretical density of SiC ($3.21\,g\,cm^{-3}$) and $\frac{Vac}{ion\ A}$ the vacancy per ion ratio given by SRIM-2010 calculations [9]. Figure 3 shows the vacancy per ion ratio and the implantation profiles as a function of the SiC depth. SRIM calculations have been performed with full damage cascades. Threshold displacement energies for C and Si sublattices were set to 20 and 35 eV respectively [10].

2.3 Micro-Raman Spectroscopy (μRS)

Irradiated samples were characterized at JANNUS-Saclay facility by surface μRS at RT using an Invia Reflex Renishaw (Renishaw plc, Gloucestershire, UK) spectrometer. The 532 nm line of a frequency-doubled Nd-YAG laser was focused on a $0.5\,\mu m^2$ spot and collected through a $100\times$ objective. The laser output power was kept around 2 mW to avoid sample heating.

2.4 Transmission (TEM) and Environmental Scanning Electron Microscopy (E-SEM)

Thin foils for TEM observations were prepared using the Focused Ion Beam (FIB) technique. The specimens were extracted from the samples irradiated to $2 \times 10^{15}\,cm^{-2}$ using a Helios Nanolab 650 (FEI Co., Hillsboro, OR, USA) equipped with electron and Ga ion beams. The specimen preparation procedure is described elsewhere [5]. TEM observations were conducted in a conventional CM20 TWIN-FEI (Philips, Amsterdam, Netherlands) operated at 200 kV equipped with a LaB6 crystal as electron source and a Gatan (Gatan Inc, Warrendale, PA, USA) heating specimen holder (25–1000 °C)

with manual temperature control. The CCD camera used to take pictures is a Gatan Orius 200.

The E-SEM observation was conducted in a FEI QUANTA 200 ESEM FEG equipped with a heating plate (25–1500 °C), operated at 30 kV. Precise sample temperature measurement is ensured by a homemade sample holder containing a Pt-Pt-Rh10 thermocouple [11]. H_2O pressure was kept constant at 120 Pa. The 6H-SiC samples were quickly heated up to 900 °C to then set the heating rate to values ranging from 1 to 30 °C/min for each test.

3 Results and discussion

3.1 Third generation SiC fibers microstructure and Raman spectra

μRS is a powerful characterization technique based on the inelastic scattering of light due to its interaction with the material atomic bonds and the electron cloud providing a chemical fingerprint of the analyzed material. SiC is known to have numerous stable stoichiometric solid crystalline phases, so-called polytypes, the cubic (3C-SiC) and the hexagonal (6H-SiC) being the most common ones [12]. Raman peak parameters such as intensity, bandwidth and wavenumber provide useful information related to the phase distribution and chemical bonding of SiC and SiC fibers [13]. Table 2 gathers the characteristic Raman peak wavenumber for 3C- and 6H-SiC polytypes.

Figure 4 shows the collected Raman spectra for the as-received samples. For the 6H-SiC spectrum, group-theoretical analysis indicates that the Raman-active modes of the wurtzite structure (C_{6v} symmetry for hexagonal polytypes) are the A_1, E_1 and E_2 modes. In turn, A_1 and E_1 phonon modes are split into longitudinal (LO) and transverse (TO) optical modes. Also, the high quality of the sample allows the observation of second order Raman bands as several weaker peaks located at $500\,cm^{-1}$ and between $1400-1850\,cm^{-1}$.

Raman spectra collected from as-received TSA3 and HNS fibers differ notably from the single crystal one. Their polycrystalline microstructure and the intergranular free C shown in Figure 2 induce the apparition of several peaks related to their chemical fingerprint. Peaks located between the $700\,cm^{-1}$ and $1000\,cm^{-1}$ are related to the cubic SiC polytype. Satellite peaks around $766\,cm^{-1}$ are attributed to disordered SiC consisting of a combination of simple polytype domains and nearly periodically distributed stacking faults [13,14]. This explanation is consistent with the high SF density observed in Figure 2.

High-intensity peaks located between $1200\,cm^{-1}$ and $1800\,cm^{-1}$ are attributed to the intergranular free C despite the little free C content of both fibers. The high contribution of these peaks to the spectra is due to the high Raman cross-section of C−C bonds which is up to ten times higher than the Si−C bonds [15]. Regarding the C chemical fingerprint, the G peak centered around $1581\,cm^{-1}$ is related to graphitic structures as a result of the sp^2 stretching modes of C bonds and the D peak centered around $1331\,cm^{-1}$, according to Colomban et al. [13], should be attributed to vibrations

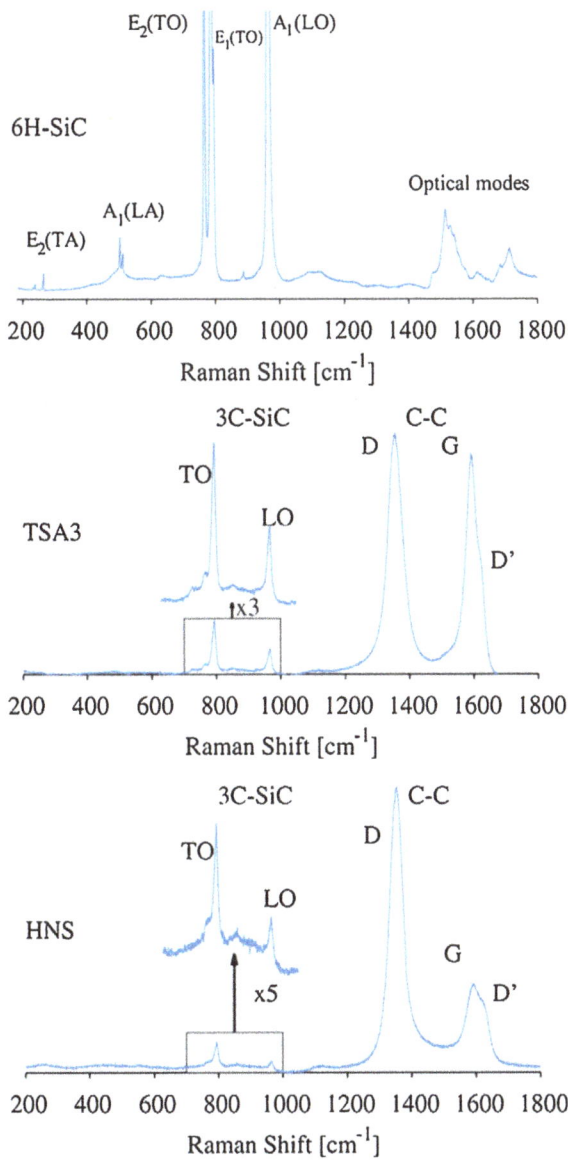

Fig. 4. Surface Raman spectra for as-received 6H-SiC single crystal and third generation SiC fibers (adapted from Ref. [5]).

involving sp^3–$sp^{2/3}$ bonds. Finally, the shouldering appearing on the G band in both fibers, D', results from the folding of the graphite dispersion branch corresponding to G at Γ point.

There is a remarkable difference in the G peak intensity between TSA3 and HNS fibers. It has been stated that the G over D peak intensity ratio is proportional to the in-plane graphitic crystallite size [17]. Therefore, the smaller size of the intergranular free C pockets of HNS takes account for such difference.

3.2 Ion-irradiation-induced amorphization

During service as nuclear structural material, SiC composites will be subjected to neutron and ion-irradiation. When an energetic incident particle elastically interacts with a lattice atom, there is a kinetic energy exchange between them. If this transmitted energy is higher than the threshold displacement energy of the knocked lattice atom, it will be ejected from its equilibrium position giving birth to a Frenkel pair: a vacancy and an interstitial atom. In turn, if the kinetic energy transfer is high enough, the displaced atom may have enough kinetic energy to displace not only one but many atoms of the lattice, which, in turn, will cause other displacement processes giving birth to displacement cascade. The number of surviving defects after the thermal recombination of the displacement cascade may pile up dealing to the degradation of the exposed material [18].

Ion-irradiation has been widely used by the nuclear materials community to simulate neutron damage due to the tunability of the radiation parameters (dose, dose rate, temperature) and the similarity of the defect production in terms of displacement cascade creation [19].

In this work, the samples have been irradiated to increasing fluences at RT with 4 MeV Au ions in order to simulate neutron damage. Figure 5 shows the evolution of the Raman spectra as a function of the irradiation dose. As can be observed, ion-irradiation induces sequential broadening of the Si–C bond related peaks until they combine in a unique low-intensity broad peak. Also, ion-irradiation induces the appearance of new low-intensity broad peaks at ~500 cm^{-1} and ~1400 cm^{-1}. These changes with dose in the Raman spectra are the consequence of the increasing damage of the crystal lattice and are usually attributed to the dissociation of the Si–C bonds and the creation of Si–Si and C–C homonuclear bonds [20], in agreement with EXAFS [21] or EELS [22] data and theoretical analyses [23]. However, some authors have pointed out that changes

Table 2. Raman shift for 3C- and 6H-SiC [16].

Polytype	X = q/qB	Raman shift [cm^{-1}]			
		Planar acoustic	Planar optic	Axial acoustic	Axial optic
		TA	TO	LA	LO
3C-SiC	0	-	796	-	972
	0	-	797	-	965
6H-SiC	2/6	145,150	789	-	-
	4/6	236,241		504,514	889
	6/6	266	767	-	-

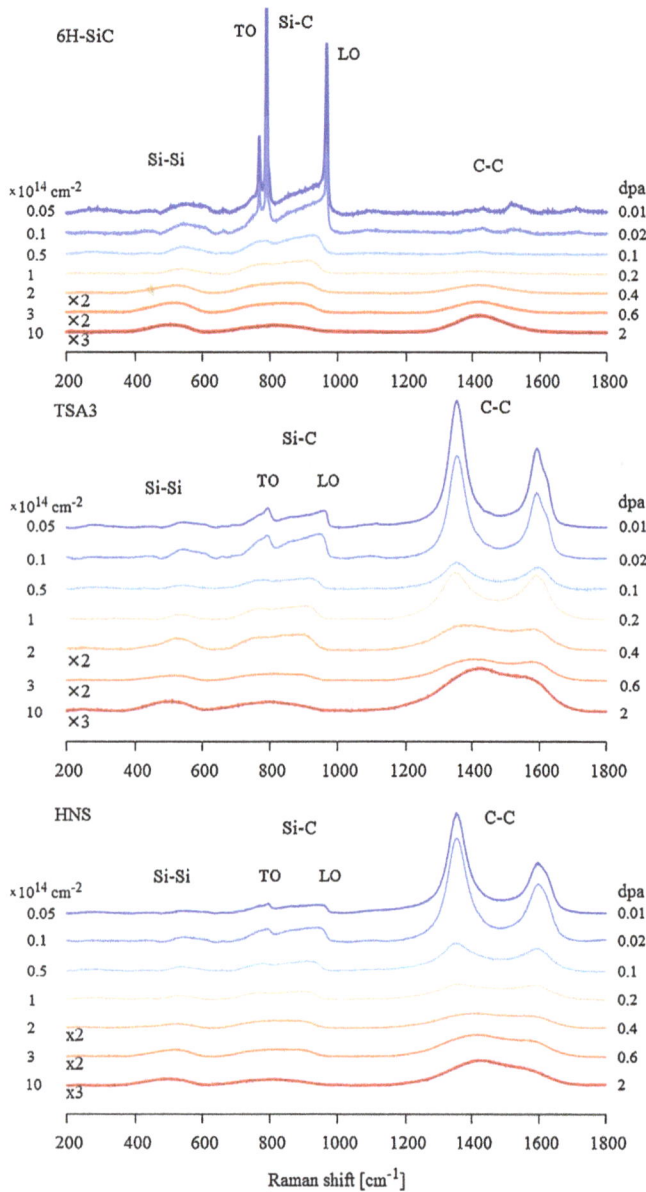

Fig. 5. Surface Raman spectra for ion-irradiated 6H-SiC single crystal and third generation SiC fibers.

intensity broad peaks at $\sim 800\,\mathrm{cm}^{-1}$ characteristic of amorphous SiC. As can be observed in Figure 6, complete amorphization of the ion-irradiated layer is confirmed by TEM imaging and electron diffraction of samples irradiated to $4\,\mathrm{dpa}\,(2 \times 10^{15}\,\mathrm{cm}^{-2})$. SAED patterns of these zones are composed of diffuse concentric rings.

Ion-amorphization kinetics for 6H-SiC single crystals has been previously studied by μRS in terms of the total disorder parameter and the chemical disorder. The former is defined as $(1\text{-}A/A_{\mathrm{cryst}})$ corresponding to the total area A under the principal first-order lines normalized to the value A_{cryst} of the crystalline material. The latter is defined as the ratio of C–C homonuclear bonds to Si–C bonds and denoted as $\chi_{(\mathrm{C\text{-}C})}$, ranging from zero for perfect short-range order to unity for random short-range disorder. Short-range order describes the degree of the chemical state with respect to the local arrangement of atoms, which can be partially preserved even when the LRO is completely lost [20,25].

In our work, the use of these parameters to study the ion-amorphization of SiC fibers is limited by two factors. First, the Si-C signal increases at low doses, hence invalidating A/A_{norm} as an indicative of the total disorder evolution, and secondly, the enormous impact of the free C of the as-received fibers in their Raman spectra, hence invalidating $\chi_{(\mathrm{C\text{-}C})}$ as a good indicative of the short-range order evolution. In order to overcome these limitations, chemical disorder has been calculated as the ratio of Si–Si homonuclear bonds to Si–C bonds ($\chi_{(\mathrm{Si\text{-}Si})}$) under the assumption that the intensity of the Raman peaks is proportional to the concentration of the related atomic bond [20].

Figure 7 shows the $\chi_{(\mathrm{Si\text{-}Si})}$ evolution as a function of the dose for the three samples. Data has been fitted with a multistep damage accumulation (MSDA) model given by equation (2):

$$f_d = \sum_{i=1}^{n} \left[\left(f_{d,i}^{sat} - f_{d,i-1}^{sat} \right) \left(1 - e^{-\sigma_i(\varphi - \varphi_{i-1})} \right) \right] \quad (2)$$

where n is the number of steps in damage accumulation, $f_{d,i}^{sat}$ the level of damage saturation for the step i, σ_1 the damage cross-section for the step i, and ϕ and ϕ_{i-1} the dose and the saturation dose of the ith step [26].

MSDA model assumes that damage accumulation is a sequence of distinct transformations of the current structure of the irradiated material and that reduces to a direct impact (DI) model meaning that amorphization is achieved in a single cascade [26]. Table 3 gathers the best-fit (non-linear least-squares Marquardt-Levenberg algorithm) parameters for $n=2$ of the $\chi_{(\mathrm{Si\text{-}Si})}$ evolution with dose.

MSDA parameters for 6H-SiC amorphization kinetics are consistent with previous reported ones based in RBS and μRS data [25,27] hence confirming $\chi_{(\mathrm{Si\text{-}Si})}$ as a relevant indicative for the amorphization level of the sample.

According to the MSDA parameters, there is a significant difference in the first stage of the amorphization process between SiC fibers and 6H-SiC. However, this difference may arise from the difficulty to treat the Raman spectra of SiC fibers due to their C signal so it cannot be directly attributed to a prompt amorphization. More experimental data is needed to confirm this hypothesis.

in the Raman spectra in SiC for moderated irradiation damage do not necessarily imply the formation of Si and C homonuclear bonds. For instance, the abrupt end of the broad band observed near the $950\,\mathrm{cm}^{-1}$ for samples irradiated to $10^{14}\,\mathrm{cm}^{-2}$ in Figure 5 can be attributed to a release of the Brillouin zone-center Raman selection rules due to a loss of translation symmetry caused by minor and local damage without amorphization [24]. It is worth to highlight that in SiC fibers irradiation at low doses increases the intensity of the Si–C related peak despite its randomization. As commented, there is a remarkable influence of the free C in the SiC fibers Raman spectra due to the high Raman cross-section of C–C bonds. Under irradiation, the rupture of these bonds will imply the drop of its cross-section allowing the SiC Raman signal to emerge over the free C one. Finally, the spectra show similar low-

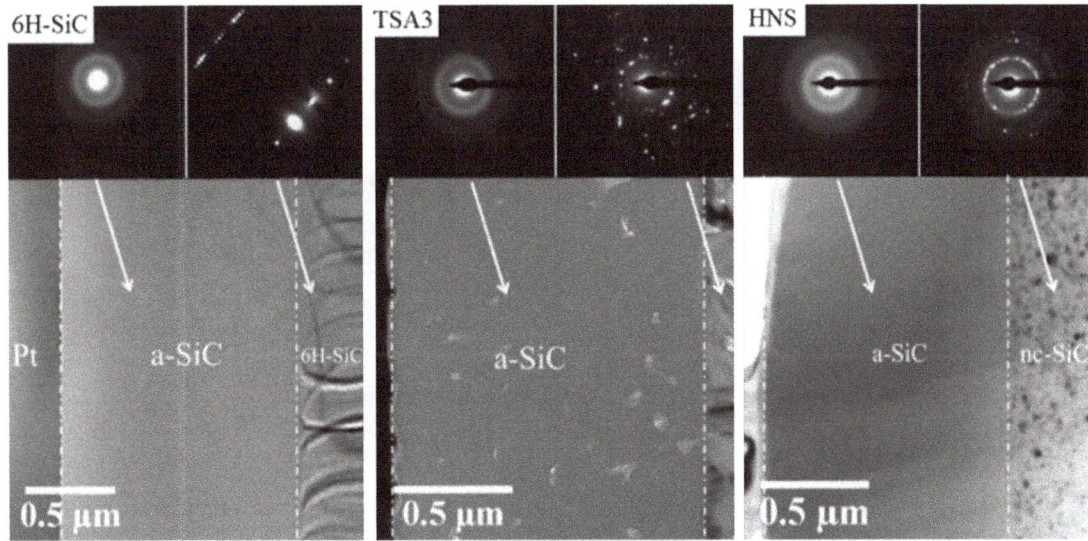

Fig. 6. TEM images and SAED patterns obtained from the irradiated 6H-SiC and SiC fibers with 4 MeV Au^{3+} ions at RT to 2×10^{15} cm^{-2}. The concentric and diffuse rings in SAED patterns indicate that the irradiated layer is completely amorphous (a-SiC). nc-SiC: nanocrystalline SiC (adapted from Ref. [5]).

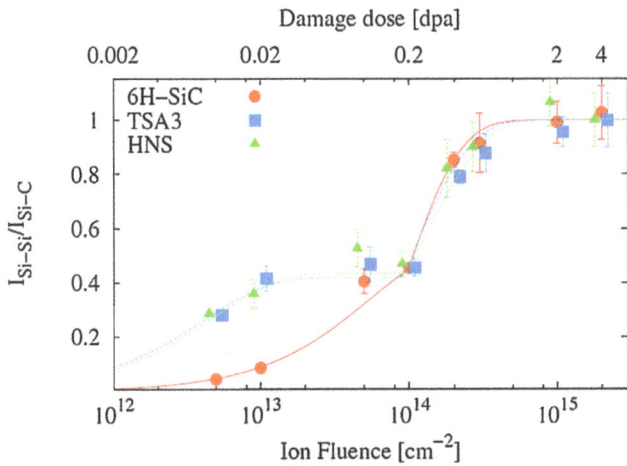

Fig. 7. Intensity of the Raman peaks associated to homonuclear Si–Si bonds normalized to the intensity of the Raman peaks associated to Si–C bonds. Experimental data is horizontally offset for the sake of clarity and fitted with the MSDA model ($n = 2$) presented in equation (2).

Table 3. Best-fit MSDA parameters for $n = 2$ (two-step) of the $\chi_{(Si-Si)}$ evolution with dose.

| Sample | $n=2$ | | | |
| | $i=1$ | | $i=2$ | |
	f_d^{sat}	σ_1[a]	f_d^{sat}	σ_2[a]
6H-SiC	0.58	0.54	1	0.82
TSA3	0.45	0.046	1	0.94
HNS	0.46	0.049	1	1.18

[a]Cross-sections in $\times 10^{-14}$ cm^{-2} units.

On the other hand, all irradiated samples show an inflexion point around 10^{14} cm^{-2} (0.2 dpa) and reach the saturation value over 3×10^{14} cm^{-2} (0.6 dpa). Therefore, it can be asserted that the three samples present a two-step amorphization process regardless of their different polytype, composition and microstructure.

It is widely accepted that GBs act as point defect sinks [28]. However, the grain size must be optimized because a small grain size has two opposing effects on the free energy of an irradiated material. For instance, a smaller grain size hinders intragranular point defects accumulation which, in turn, decreases the free energy resulting from irradiation-induced defects. However, a smaller grain size also may increase the free energy resulting from the increase on the GB density which can favor the path toward an amorphous phase [29]. The microstructure influence of the behavior of SiC under irradiation is controversial as both experimental and computational studies can be found concerning whether grain refinement enhances or reduces SiC radiation resistance [30–33]. The similar ion-amorphization doses of 6H-SiC, TSA3 and HNS suggest that the microstructure of these fibers is not refined enough to show significant enhanced or reduced radiation resistance – not even for the HNS fibers which grain sizes are around 20 nm.

3.3 In situ E-SEM thermal annealing

Radiation-induced amorphization is detrimental for the use of SiC under nuclear environments at low temperatures as it causes the degradation of the material's physico-chemical properties [34]. Even though amorphous SiC (a-SiC) is known to be highly stable, irradiation-induced damage in SiC can be recovered and the a-SiC layer recrystallized by thermal annealing at high temperatures [25,35]. However,

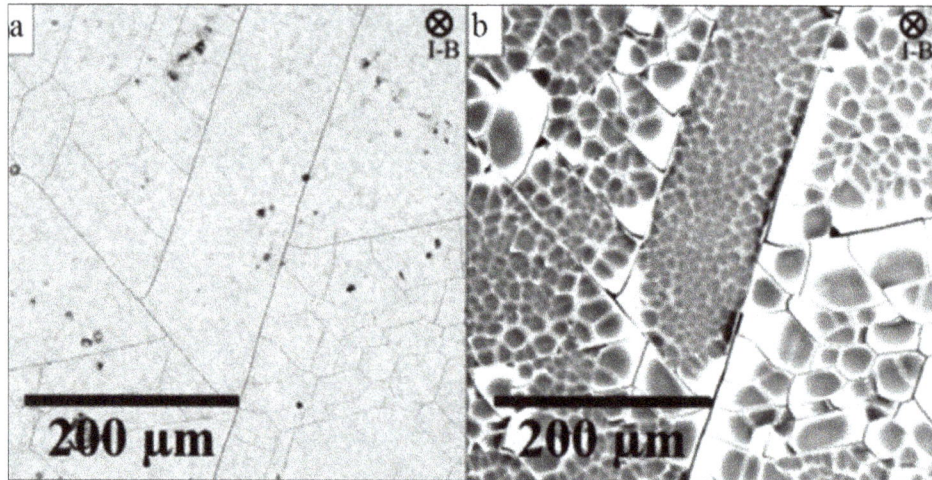

Fig. 8. Mechanical failure evolution of the SiC ion-amorphized layer during thermal annealing: (a) cracks appear along the cleavage planes and eventually lead to (b) exfoliation (adapted from Ref. [37]).

it has been reported that thermal annealing has an undesirable side effect. As shown in Figure 8, it induces mechanical failure of the ion-amorphized layers in single crystals SiC [25,36] and in HNS fibers [37]. However, little information concerning thermal annealing-induced mechanical failure is available for SiC. It has been reported that thermal stresses – arising from a mismatch between the coefficient of thermal expansion of the irradiated layer and the pristine substrate – are not responsible for the mechanical failure [37] and recrystallization-related stresses have been pointed out as the cracking and delamination cause [36,37].

In order to provide further information on how recrystallization is related to mechanical failure, several thermal annealing tests on ion-amorphized 6H-SiC single crystals have been conducted and observed at different temperature ramps via in situ E-SEM.

Figure 9 shows the evolution of the linear crack density as a function of time for different heating rates. As it can be observed, crack density reaches similar saturation values independently of the heating rate whereas cracking kinetics are heating rate-dependent. For instance, both the temperatures at which cracking is triggered and the crack density growth rate increase with increasing heating rates.

Cracking kinetics appears to be thermally activated phenomenon. In order to obtain the characteristic activation energy (E_a) of the process, the experimental data have been assumed to obey an Arrhenius law. For instance, Figure 10 shows the log-plot of two characteristic features of the cracking phenomenon: the inverse of the time necessary to reach the 50% of the cracking density ($t_{50\%}$) as a function of the inverse of the sample temperature at time $t_{50\%}$, denoted as $T_{50\%}$. These two parameters have been successfully applied for the study of the recrystallization temperature of tungsten as a function of the heating rate and allow to get rid of the time dependency of the test [38]. Linear fit to the log-plot yields an E_a of 1.05 eV. This value falls in the range of recrystallization activation energies

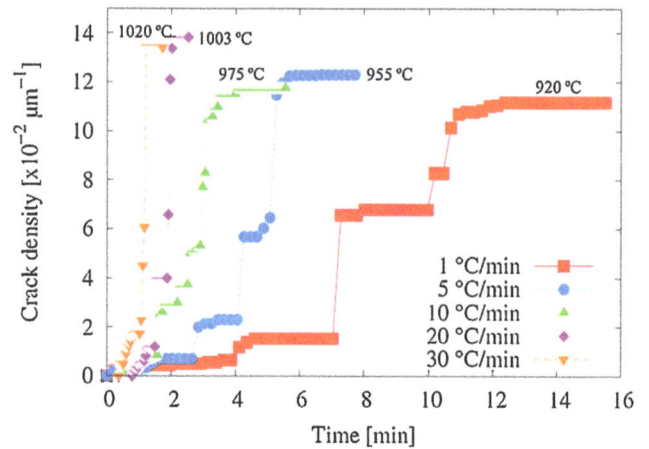

Fig. 9. Crack density evolution during the in situ thermal annealing for different temperature ramps. Values near the curves refer to the temperature at which the first crack was observed during the respective test.

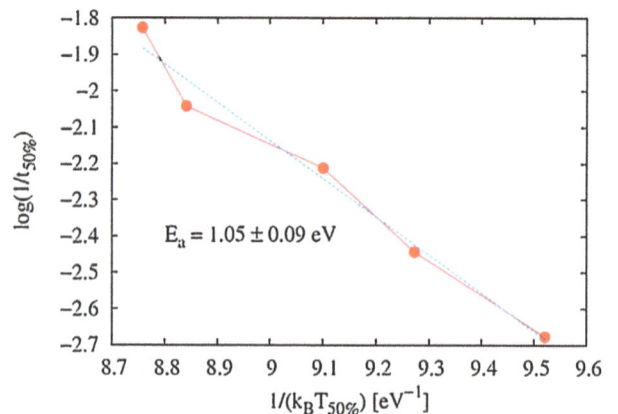

Fig. 10. Log-plot of the inverse of the time necessary to reach the 50% of the cracking density ($t_{50\%}$) as a function of the inverse of temperature at this moment ($T_{50\%}$). E_a is the activation energy for the cracking phenomenon.

found by isothermal annealing of ion-amorphized SiC, i.e. 0.36–0.65[25] to 2.1[36] eV, sustaining that recrystallization-related stresses are the underlying mechanism which induced mechanical failure.

4 Conclusions

In this work, ion-amorphization of SiC fibers has been studied in terms of surface μRS and TEM imaging and compared to the model material, i.e. 6H-SiC. It is reported that SiC fibers, HNS and TSA3, and 6H-SiC display a similar ion-amorphization process despite their different SiC polytypes and microstructures. Critical amorphization dose yields $\sim 3 \times 10^{14}$ cm^{-2} (~ 0.6 dpa) for 4 MeV Au ions at RT.

Also, the kinetics of thermally annealing-induced cracking is studied via in situ E-SEM observations. It is reported that the temperatures at which the first cracks appear as well as the pace of crack density growth increase with increasing heating rates. The activation energy of the cracking process yields 1.05 eV in agreement with recrystallization activation energies of ion-amorphized samples. This observation supports recrystallization as the stress source causing the mechanical failure of the annealed samples.

The authors would like to thank JANNUS staffs for their technical support during irradiations and EMIR network for funding the irradiation time. Also we are grateful to B. Arnal and D. Troadec for FIB sample preparation and T. Vandenberghe for TEM observations.

Nomenclature

μRS	micro-Raman Spectroscopy
CVI	Chemical Vapor Infiltration
dpa	displacements per atom
DTA	Dose To Amorphization
E-SEM	Environmental Scanning Electron Microscope
GB	Grain Boundary
GENII	Generation II (current nuclear reactors)
GFR	Gas Fast Reactor
HNS	Hi-Nicalon type S
LFR	Lead Fast Reactor
MSDA	Multistep Damage Accumulation
MSR	Molten Salt Reactor
RBS	Rutherford Backscattering Spectrometry
SAED	Selected Area Electron Diffraction
SCWR	Super Critical Water Reactor
SFR	Sodium Fast Reactor
TEM	Transmission Electron Microscope
TSA3	Tyranno SA3
VHTR	Very High Temperature Reactor

References

1. P. Yvon, F. Carré, Structural materials challenges for advanced reactor systems, J. Nucl. Mater. **385**, 217 (2009)
2. S.J. Zinkle, J.T. Busby, Structural materials for fission & fusion energy, Mater. Today **12**, 12 (2009)
3. A. Iveković, S. Novak, G. Dražić, D. Blagoeva, S.G. de Vicente, Current status and prospects of SiCf/SiC for fusion structural applications, J. Eur. Ceram. Soc. **33**, 1577 (2013)
4. C.A. Schneider, W.S. Rasband, K.W. Eliceiri, NIH Image to ImageJ: 25 years of image analysis, Nat. Methods **9**, 671 (2012)
5. J. Huguet-Garcia, A. Jankowiak, S. Miro, D. Gosset, Y. Serruys, J.-M. Costantini, Study of the Ion-irradiation behavior of advanced SiC fibers by Raman Spectroscopy and Transmission Electron Microscopy, J. Am. Ceram. Soc. **98**, 675 (2015)
6. A.R. Bunsell, A. Piant, A review of the development of three generations of small diameter silicon carbide fibres, J. Mater. Sci. **41**, 823 (2006)
7. C. Sauder, J. Lamon, Tensile creep behavior of SiC-based fibers with a low oxygen content, J. Am. Ceram. Soc. **90**, 1146 (2007)
8. Y. Serruys, P. Trocellier, S. Miro, E. Bordas, M.O. Ruault, O. Kaïtasov, S. Henry, O. Leseigneur, T. Bonnaillie, S. Pellegrino, S. Vaubaillon, D. Uriot, JANNUS: a multi-irradiation platform for experimental validation at the scale of the atomistic modelling, J. Nucl. Mater. **386-388**, 967 (2009)
9. J.F. Ziegler, M.D. Ziegler, J.P. Biersack, SRIM–The stopping and range of ions in matter (2010), Nucl. Instrum. Methods Phys. Res. B **268**, 1818 (2010)
10. R. Devanathan, W.J. Weber, F. Gao, Atomic scale simulation of defect production in irradiated 3C-SiC, J. Appl. Phys. **90**, 2303 (2001)
11. R. Podor, D. Pailhon, J. Ravaux, H.-P. Brau, Development of an integrated thermocouple for the accurate sample temperature measurement during high temperature Environmental Scanning Electron Microscope (HT-ESEM) experiments, Microscopy and Microanalysis **21**, 307 (2015)
12. F. Bechstedt, P. Käckell, A. Zywietz, K. Karch, B. Adolph, K. Tenelsen, J. Furthmüller, Polytypism and properties of silicon carbide, Phys. Status Solidi **202**, 35 (1997)
13. P. Colomban, G. Gouadec, L. Mazerolles, Raman analysis of materials corrosion: the example of SiC fibers, Mater. Corros. **53**, 306 (2002)
14. G. Gouadec, P. Colomban, Raman Spectroscopy of nano-materials: how spectra relate to disorder, particle size and mechanical properties, Prog. Cryst. Growth Charact. Mater. **53**, 1 (2007)
15. M. Havel, P. Colomban, Raman and Rayleigh mapping of corrosion and mechanical aging in SiC fibres, Compos. Sci. Technol. **65**, 353 (2005)
16. S. Nakashima, H. Harima, Raman investigation of SiC polytypes, Phys. Status Solidi **162**, 39 (1997)
17. L.G. Cançado, K. Takai, T. Enoki, M. Endo, Y.A. Kim, H. Mizusaki, A. Jorio, L.N. Coelho, R. Magalhães-Paniago, M.A. Pimenta, General equation for the determination of the crystallite size L[sub a] of nanographite by Raman spectroscopy, Appl. Phys. Lett. **88**, 163106 (2006)
18. S. Zinkle, Radiation-induced effects on microstructure, Compr. Nucl. Mater. **1**, 65 (2012)
19. G.S. Was, R.S. Averback, Radiation damage using ion beams, Compr. Nucl. Mater. **1**, 195 (2012)
20. S. Sorieul, J.-M. Costantini, L. Gosmain, L. Thomé, J.-J. Grob, Raman spectroscopy study of heavy-ion-irradiated α-SiC, J. Phys.: Condens. Matter **18**, 5235 (2006)
21. W. Bolse, Formation and development of disordered networks in Si-based ceramics under ion bombardment, Nucl. Instrum. Methods Phys. Res. B **141**, 133 (1998)

22. M. Ishimaru, A. Hirata, M. Naito, I.-T. Bae, Y. Zhang, W.J. Weber, Direct observations of thermally induced structural changes in amorphous silicon carbide, J. Appl. Phys. **104**, 033503 (2008)

23. M. Ishimaru, I.-T. Bae, Y. Hirotsu, S. Matsumura, K.E. Sickafus, Structural relaxation of amorphous silicon carbide, Phys. Rev. Lett. **89**, 055502 (2002)

24. F. Linez, A. Canizares, A. Gentils, G. Guimbretiere, P. Simon, M.-F. Barthe, Determination of the disorder profile in an ion-implanted silicon carbide single crystal by Raman spectroscopy, J. Raman Spectrosc. **43**, 939 (2012)

25. S. Miro, J.-M. Costantini, J. Huguet-Garcia, L. Thomé, Recrystallization of hexagonal silicon carbide after gold ion irradiation and thermal annealing, Philos. Mag. **94**, 3898 (2014)

26. J. Jagielski, L. Thomé, Damage accumulation in ion-irradiated ceramics, Vacuum **81**, 1352 (2007)

27. X. Kerbiriou, J.-M. Costantini, M. Sauzay, S. Sorieul, L. Thomé, J. Jagielski, J.-J. Grob, Amorphization and dynamic annealing of hexagonal SiC upon heavy-ion irradiation: effects on swelling and mechanical properties, J. Appl. Phys. **105**, 073513 (2009)

28. W.G. Wolfer, Fundamental properties of defects in metals, Compr. Nucl. Mater. **1**, 1 (2012)

29. T.D. Shen, Radiation tolerance in a nanostructure: is smaller better?, Nucl. Instrum. Methods Phys. Res. B **266**, 921 (2008)

30. W. Jiang, H. Wang, I. Kim, I.-T. Bae, G. Li, P. Nachimuthu, Z. Zhu, Y. Zhang, W. Weber, Response of nanocrystalline 3C silicon carbide to heavy-ion irradiation, Phys. Rev. B **80**, 161301 (2009)

31. W. Jiang, H. Wang, I. Kim, Y. Zhang, W.J. Weberb, Amorphization of nanocrystalline 3C-SiC irradiated with Si ions, J. Mater. Res. **25**, 2341 (2010)

32. L. Jamison, P. Xu, K. Sridharan, T. Allen, Radiation resistance of nanocrystalline silicon carbide, in *Advances in materials science for environmental and nuclear technology II: ceramic transactions*, edited by S.K. Sundaram, K. Fox, T. Ohji, E. Hoffman (John Wiley & Sons, Inc., Hoboken, NJ, USA, 2011), Vol. 227

33. L. Jamison, M.-J. Zheng, S. Shannon, T. Allen, D. Morgan, I. Szlufarska, Experimental and ab initio study of enhanced resistance to amorphization of nanocrystalline silicon carbide under electron irradiation, J. Nucl. Mater. **445**, 181 (2014)

34. Y. Katoh, L.L. Snead, I. Szlufarska, W.J. Weber, Radiation effects in SiC for nuclear structural applications, Curr. Opin. Solid State Mater. Sci. **16**, 143 (2012)

35. S. Miro, J.-M. Costantini, S. Sorieul, L. Gosmain, L. Thomé, Recrystallization of amorphous ion-implanted silicon carbide after thermal annealing, Philos. Mag. Lett. **92**, 633 (2012)

36. A. Höfgen, V. Heera, F. Eichhorn, W. Skorupa, Annealing and recrystallization of amorphous silicon carbide produced by ion implantation, J. Appl. Phys. **84**, 4769 (1998)

37. J. Huguet-Garcia, A. Jankowiak, S. Miro, R. Podor, E. Meslin, Y. Serruys, J.-M. Costantini, In situ E-SEM and TEM observations of the thermal annealing effects on ion-amorphized 6H-SiC single crystals and nanophased SiC fibers, Phys. Status Solidi **252**, 149 (2015)

38. C.J.M. Denissen, J. Liebe, M. van Rijswick, Recrystallisation temperature of tungsten as a function of the heating ramp, Int. J. Refract. Met. Hard Mater. **24**, 321 (2006)

4

Modelling of powder die compaction for press cycle optimization

Jean-Philippe Bayle[1,*], Vincent Reynaud[2], François Gobin[1], Christophe Brenneis[1], Eric Tronche[1], Cécile Ferry[1], and Vincent Royet[1]

[1] CEA, DEN, DTEC, SDTC, 30207 Bagnols/Cèze, France
[2] Champalle Company, 151 rue Ampère, ZI Les Bruyères, 01960 Peronnas, France

Abstract. A new electromechanical press for fuel pellet manufacturing was built last year in partnership between CEA-Marcoule and Champalle[Alcen]. This press was developed to shape pellets in a hot cell via remote handling. It has been qualified to show its robustness and to optimize the compaction cycle, thus obtaining a better sintered pellet profile and limiting damage. We will show you how 400 annular pellets have been produced with good geometry's parameters, based on press settings management. These results are according to a good phenomenological pressing knowledge with Finite Element Modeling calculation. Therefore, during die pressing, a modification in the punch displacement sequence induces fluctuation in the axial distribution of frictional forces. The green pellet stress and density gradients are based on these frictional forces between powder and tool, and between grains in the powder, influencing the shape of the pellet after sintering. The pellet shape and diameter tolerances must be minimized to avoid the need for grinding operations. To find the best parameters for the press settings, which enable optimization, FEM calculations were used and different compaction models compared to give the best calculation/physical trial comparisons. These simulations were then used to predict the impact of different parameters when there is a change in the type of powder and the pellet size, or when the behavior of the press changes during the compaction time. In 2016, it is planned to set up the press in a glove box for UO_2 manufacturing qualification based on our simulation methodology, before actual hot cell trials in the future.

1 Introduction

The electronuclear closed fuel cycle chosen by France plans the reprocessing of spent fuel and will enable natural uranium resource saving, as well as a reduction in the volume of wastes and their toxicity compared with the choice of direct storage (once-through cycle). The nuclear waste from spent fuel is classified depending on its activity and half-life. The High Activity (HA) waste represents more than 95% of the total radioactivity of French nuclear waste. The liquid extraction process called PUREX enables the Minor Actinides (MAs) to be separated from the Fission Products (FP) in HA waste. The advanced management of the MAs is a goal for the transmutation envisaged in fourth generation reactors or in specially-dedicated reactors. Two approaches to MA transmutation in fast breeder reactors (FBRs) are envisaged, i.e. homogeneous and heterogeneous recycling. The heterogeneous mode consists in concentrating the MAs in special assemblies located in the periphery of the reactor core. The neutronic impact on the core limits the introduction of a higher quantity of MAs, restricted to 10 to

20%. Materials including Americium (Am) located around the reactor core can be of target type if the MA supports an inert matrix, or else part of a Minor Actinide Bearing Blanket (MABB) if the MAs are directly incorporated into fertile UO_2 fuels.

2 Context

The manufacturing of fuel pellets incorporating minor actinides by remote handling in hot cells requires simple, effective operations and robust technologies. Rejects must be minimized, which is harder with higher and higher actinide concentrations. The process of pellet shaping is well known from the literature [1–4]. It is generally carried out by uniaxial cold compaction in die to obtain green pellets (rough pellets from the pressing) with a density of about 65% of the theoretical density (th.d). This shaping is then followed by a sintering operation which enables the density to reach 95% of the th.d. At present, the pressing technology used in Atalante hot cells (Marcoule, France) is based on a manual process with a radial opening die, compared to the conventional process of a floating die

---
* e-mail: jean-philippe.bayle@cea.fr.

where a downward movement of the die occurs, enabling the ejection of the pellet. Another process with a fixed die enables pellet ejection by the lower punch which pushes with a pressure support from the upper punch. Damages can be present after the ejection stage if the pressure from the two punches is not coordinated, and these are generally revealed during the sintering stage. They can be worsened by the radiological behavior of the pellet, depending on its composition, and by the manufacturing process. Different defect types occur for sintered pellets, in particular cracks, end-capping and spalling [5]. Cracks can form down the sides of pellets and be longitudinal or lateral, or happen in the ends and sometimes cause "end capping" in the top or bottom of the pellets. Spalling can be found on the sides or the ends. The green pellets can have defects which depend essentially on the level of support pressure during die ejection. Other sources of damage can also be identified in the process of powder shaping [6]. First, the introduction of secondary phases composed of hard inclusions or air pockets leads to an excessive relaxation during ejection, with spalling occurring on the pellets, and to different wear patterns on the internal walls of the die and thus to blocked pellet sliding and to shearing. Secondly, inappropriate press settings for compression level, pressing time, or punch accompanying pressure during ejection can cause damage.

The mechanical stress distribution within pellets during the ejection step influences the surface defects. The mechanical stress induced by the die can be high, in particular at the corner of the die, where the springback occurs during the pellet ejection. The stress concentrations are accentuated by springback, which corresponds to the volume expansion of the pellet by relaxation of stress during ejection. Some authors have used digital simulation to estimate the mechanical stresses in pellets during this step. Aydin and Briscoe [1] attempted to determine the residual stress distributions in cylindrical pellets. Their study showed that axial residual tensile stress appears at the extremities of the pellet from the axial stress relaxation stage in die (decompression in die). These stresses are due to the friction forces between the die and the pellet, which block the axial springback when the pressure is released. In their study, neither the pellet slide and release phase nor the interactions with the edge of the die were taken into account, as the radial walls of the die were artificially removed. Jonsen and Haggblad [7] took into account the compaction and the ejection with the real kinematics of ejection. The distribution of the residual stress consolidated by measurements of neutron diffraction shows that the pellet edges are submitted to axial compression over a thin layer (200–400 μm), and the part below this layer undergoes traction over a thicker zone (600 μm). From these two studies, it is known that residual stresses after ejection are strongly influenced by the tool shapes and kinematics of ejection. In this context, an ejection performed by a radial die opening is expected to be less damaging. Therefore, this mode of ejection was used for the manufacturing of the minor actinide fuel pellets considered in this study.

Another issue is that minor actinide fuel pellet grinding after sintering must be minimized in order to limit highly radioactive dust. Consequently, geometrical tolerance for the diameter needs to be rather wide, $\pm 50\,\mu$m around

nominal values (8–10 mm). Pellet geometrical dimension mastery is necessary in order to obtain "net shape" pellets. It is well known that the pressing stage is critical for the shape of the pellet after sintering. For instance, when uniaxial compaction is performed green densities decrease along the height of the compact from the extremity which was in contact with the moving punch. After sintering, the shrinkage follows the density gradient and a conical shaped pellet is formed. With two mobile punches, a double-conical (hourglass) shaped pellet is obtained. In die compression, the heterogeneous density is due to the friction forces between the powder and the wall of the die, as well as the friction between the grains of the powder [1,8]. These friction effects have been extensively studied for perfectly cylindrical dies, but never investigated for a specially shaped die. More particularly, the diametrical profile of the die could be designed in order to counterbalance the effect of friction.

3 Objectives

The density gradients obtained in the compact depends on various parameters such as the tool quality, the powder behavior, the compaction cycles, the lubrication type, etc. Because the powders used for nuclear fuel manufacturing are precious, pellet damage must be minimized and a net-shaped pellet is necessary because it does not require grinding. The main objective of this study was to be able to anticipate the demanding manufacturing factors, which can influence the press settings before the production cycle, and then during the manufacturing, to be able to have the shortest possible response time to correct parameters to ensure finished products with stable quality. Consequently, the study firstly concerned the optimization of the fuel manufacturing cycles of an innovative nuclearized press for nuclear fuel manufacturing in a hostile and restricted environment. To meet this need, a capability study of the press is described, with on the first press regulation results in the inactive conditions of a mock-up. An annular geometry pellet with compulsory manufacturing tolerances is taken into account. From the results of the study, simulations are proposed on the basis of previous simulations where the model parameters of the compaction were characterized for various powders. We can thus act on the cycle compaction parameters of the press, on the model parameters of each powder, and on certain friction coefficients depending on the lubricant type.

4 Materials and methods

4.1 Alumina powder (Al$_2$O$_3$_ T195)

Alumina powder was used in this study. Its behavior is known from the literature [4], and it is widely used in the compaction field. Alumina powder was used to guarantee the conformity of the measurement and calculation results which could be compared with those from unpublished works [3]. Furthermore, it will be used to carry out

Table 1. Characteristics of Al_2O_3 powder.

Powder	Supplier	Morphology	Size (μm)	Bulk density (g.cm^{-3})	Theoretical density (g.cm^{-3})	E_{Th} Theoretical Young's modulus (GPa)	ν_{Th} Theoretical Poisson's ratio
Al_2O_3	Ceraquest	Spherical	50–200	1.24	3.970	530	0.22

qualification trials for a new nuclearized press currently undergoing testing. The particles are spherical, 50 to 200 μm in diameter. These spheres in turn are composed of 1–10 μm grains [9]. Main characteristics of studied Al_2O_3 powder are summarized in Table 1.

4.2 New nuclear press description and characteristics

One of the fuel manufacturing processes originates in the conventional process of the powder metallurgy industry and enables pellet shaping in dies, followed by sintering. The shaping of the Minor Actinide Bearing Blanket (MABB) pellets is currently done manually in hot cells. Manufacturing Automation and a better control of the shaping parameters were tested during this study, in order to prepare the way for a new automatic nuclear press under a collaboration set up between the CEA and Champalle[Alcen]. The minimization of criticality risks is an important goal for MABB pellet manufacturing, and is the main reason why the press is being built to operate without oil, and is completely electromechanical. It is a uniaxial automatic mono-punch simple effect press, with a displacement-piloted die. Its capacity is 10 tons, the maximum height is limited to 1.2 m and the production rate is one to five cylindrical annular pellets per minute. Installing the apparatus in an existing hot cell for nuclear fuel production required a modular design and simulation studies, which were carried out using 3D software to show the entry of all modules through the airlock. The objective was to validate the modular units' ability to be assembled, dismantled and maintained by remote handling techniques. The 30 separate units making up the press had to go through a 240 mm diameter air-lock to enter the hot cell. To be sure the remote handling scenarios were appropriate, virtual reality simulation studies were carried out, taking into account force feedback and inter-connectability between the different units [10,11]. In parallel, different radiological software checked that the press components' radiological dimensioning would ensure radiation resistance during operation in a hostile environment. A mock-up simulating the future hot cell and equipped with the real remote handling systems has been built in the CEA/ Marcoule HERA facility technological platform, in order to physically test press unit assembly by remote handling, and the apparatus operations. The press, adapted to nuclear conditions, is patented. The press is a uniaxial mono-punch press, with a single compaction cycle. The upper punch and die are mobile at different velocities and the lower punch is fixed. The die is used for the ejection step with an upper punch pressure support. The hot cell press location imposed

the use of an existing hot cell, without modifications or external motors being possible. A transfer of the module units through the 240 mm diameter of the Lacalhene Leaktight Transfer Double Door had to be carried out. To minimize the criticality impact and because hydrogenated liquids are prohibited in hot cell, we replaced hydraulic energy by electric energy. This is the main reason why the choice was made of electric motors with transmission systems with a minimum gap, combining rotary and translatory mechanisms for the upper punch and the die. To decrease the height needed, the die motorization was placed to one side and the effort transmitted via a toggle joint to the die plate. The press production rate is about four pellets per minute and its pressure capacity is 10 tons. The base structure has one lower plate. This plate is fixed to a circular rail built into the hot cell floor. The press can therefore be rotated in order to enable access to any of the five main parts as required. The first part includes the rigid frame of the press, consisting of the lower and upper plates connected by four guide columns. The plates support respectively the motors of the die and of the upper punch. The lower plate holds the fixed lower punch equipped with a displacement sensor. Between these two plates, the upper punch and the die plates (parts 2 and 3) slide up and down. Plate displacements are monitored by sensors, and the mobile upper punch is also fitted with a force sensor. The powder load system and displacement motor of the filling shoe are set up on the mobile die plate. The filling shoe is moved laterally by an electric motor and a rack system. The powder load system has a tippable powder transfer jar which can be completely connected using remote handling. The press was patented under a CEA and Champalle common patent [12]. The nuclear press has enabled the manufacturing of Al_2O_3 anular pellets with a 10 mm die diameter in CEA Marcoule mock-up. The Al_2O_3 powder was used, with zinc stearate lubrication in the mass measured at 2%.

4.3 Optimization cycle background

The use of the press with slave die displacement (equivalent to a double effect cycle) can enable cycle optimization and operating, in order to reduce the difference between the minimum and maximum pellet diameters. An optimal operating cycle enabling uniform stress distribution throughout the pellet means making the applied and transmitted forces equivalent. The difference between these forces is called Δ. To influence Δ, several parameters were varied in the compaction cycle. Figure 1 shows the upper, applied and lower transmitted punch forces, die and upper

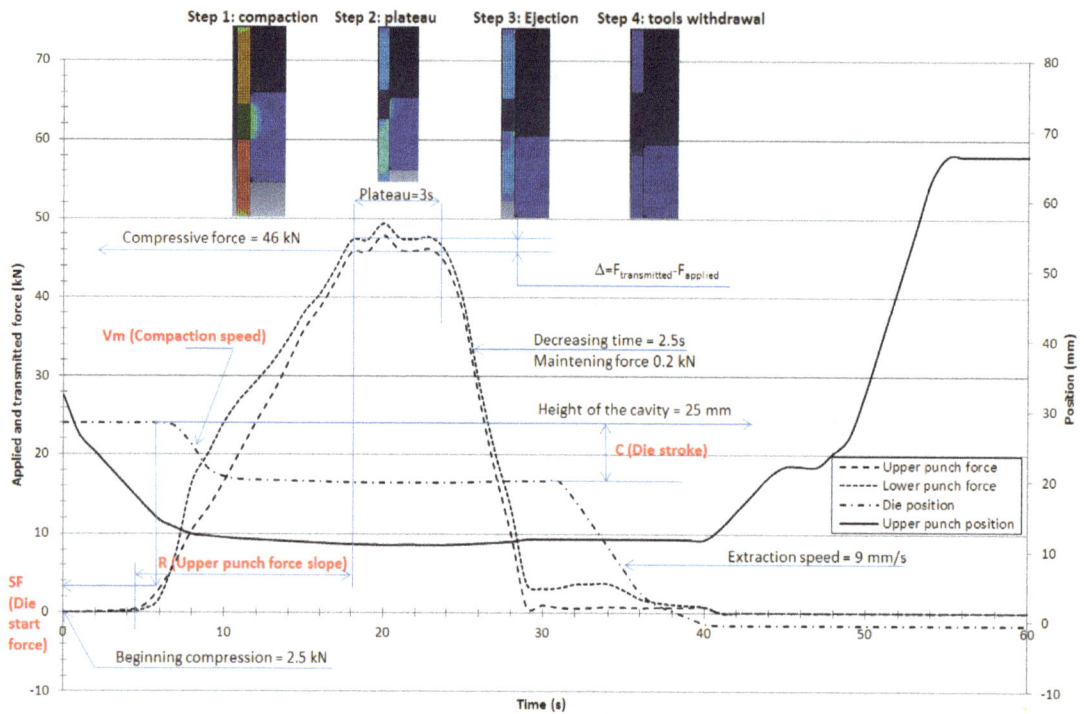

Fig. 1. Upper, applied and lower transmitted punch forces, die and upper punch displacements depending on time, Von Mises stresses during step calculations (1 to 4), corresponding to the compaction cycle.

Table 2. Regulation parameters of the cycle press, SF, R, C, V_m.

Parameters	Symbol	Value	Unit
Die start force	SF	3.5	kN
Upper punch force slope	R	5	s
Die stroke	C	6	mm
Die compaction speed	V_m	7	mm/s

punch displacements depending on time. The compaction cycle settings for a given powder thus require an optimization of the press setting parameters.

For the Al_2O_3 powder studied, in order to obtain a geometrical tolerance of ±0.012 mm for a diameter sintered to 9.015 mm, the chosen parameters are indicated in Table 2. For R, it is the time to increase from 3.5 to 46 kN. Parameter C is equal to the difference between the position of the height of the powder column (25 mm) and the position of the compression start point (19 mm) which enables a green pellet height to be compacted to 11 mm.

Constants are: the force at the beginning of compression is set (punch) at 2.5 kN; the compressive force (punch) is 46 kN; time to the compression plateau (punch) is 3 s; the decreasing slope (punch) is set at 2.5; the maintaining force (punch) at 0.2 kN; the position of the height of the cavity (die) at 25 mm; the extraction speed is set (die) at 9 mm/s. For the Al_2O_3 powder studied, we obtained a geometrical

tolerance of ±0.012 mm for a diameter sintered to 9.015 mm. The die diameter was of 10.000 mm. These optimal settings meant the best pellet quality was obtained, with a lubricant inside the powder and with a good flowing powder. To summarize, to minimize Δ, you must find a compromise between V_m and R in order to reduce the friction index depending on the flow index (powder behavior) and the friction coefficient (powder and die friction) [9].

5 Modelling

5.1 Model description

Roscoe et al. of Cambridge University first established general relationships of soil behavior based on the theory of elastoplasticity with strain hardening, in the field now described by Cam-Clay (CC) Model. These models are based on four main elements: the study of isotropic compression tests, the concept of critical state, a force relation-dilatancy and the rule of normality for plastic strain. In the CC model, the elliptical load surface (plastic potential), in isodensity, is defined in the plan of invariants (P, Q) by the expression below [2,4,8,13]:

$$f(P, Q, \rho(\varepsilon_V^P)) = (Q/M)^2 + P(P - P_C) = 0,$$

$- P = (\sigma_{Applied} + 2\sigma_{Radial})/3$, hydrostatic stress (MPa),
$- Q = |\sigma_{Applied} - \sigma_{Radial}|$, deviatoric stress (MPa),

$- M = \phi\,(\beta,\ \mu,\ \sigma_{\text{Applied}},\ \sigma_{\text{Transmitted}},\ \sigma_{\text{Radial}})$, critical state stress (de-densification/densification),

$- P_C(\varepsilon_V^P) = \left[(P_0 + Cohe) \times e^{-k \cdot \varepsilon_V^P}\right] - Cohe$, consolidation pressure (MPa),

- *Cohe*, powder cohesion pressure (MPa),
- P_0, the initial consolidation pressure (MPa),
- ε_V^P is the plastic volumetric strain with $(\rho = \rho_0 * e^{-\varepsilon_V^P}$, ρ_0 is the initial density),
- $k = {}^{(1+e_0)}/_{(Lambda+Kappa)} = \varphi(\sigma_Z, M, \beta, \rho_c, \rho_{ref})$,
  - $e_0 = (1 - \rho_{ref})/\rho_{ref}$, void ratio with $(\rho_{ref} = \rho_{real}/\rho_{theo})$,
  - $\sigma_Z\,(h_{Sensor})$, axial stress at height of the radial sensor (Janssen model),
  - $\beta = \sigma_{Radial}/\sigma_Z\,(h_{Sensor})$, Flow index,
  - $\rho_c = \rho_{ch} \times \exp\,(-3\sigma_Z/(1-2v)E)$,
  - Lambda = plastic contribution,
  - Kappa = elastic contribution (takes to the oedometric test),
  - E, v, Young modulus and Poisson coefficients depending on β, ε_{Vol}, ε_{Diam}, σ.

In this model, we can choose to use the plastic strain, or density, we preferred to pilot model by the strain hardening variable. The plastic flow occurs when the state of stress meets the condition $f = 0$.

5.2 Model parameter identifications

To determine the σ_{Applied}, $\sigma_{\text{Transmitted}}$ and σ_{Radial}, we used instrumented INSTRON® press with upper, lower and radial sensors (strain gauges include in the carbide die). The Jansen model enables the calculation of the axial stress at the level height of the pellet, where the radial strain is measured. Then, we calculated the flow index β (friction in the powder) with the ratio between the radial and axial stresses to the level of the sensor. Also, we calculated with these measures Q and P. Then, we identified elastic (E, v, Kappa) and plastic coefficients (M, k and Lambda) [14]. Finally, we calculated P_c and the behavior between P and Q depending on the volumetric plastic strain, or density. It is possible to determine Kappa and Lambda without k formulation with the isotropic (oedometric) compression tests. In these, the powder is compacted in a die and then changes in powder height H are drawn up as a function of the applied pressure P. Next, the void ratio is drawn up as a function of the P logarithm with: $e = n/(1 - n)$, where $n = 1 - \rho/\rho_{\text{theo}}$ is the powder porosity. The isotropic compression test results give curves $e = f(\ln\sigma)$ which can be considered as lines, a blank consolidation curve, called the Lambda curve, which describes the load during the test and an unloading-reloading curve, called the kappa curve, which describes the non-linear elastic behavior during the test. Another method proposed by Abaqus® consists to take into account the tabulation of the curve Pc depending on ε_V^P based on oedometric test [15]. For CC model, we identified coefficients for two powders, Al_2O_3 reference powder (atomized powder) and Ceria powder (microsphere powder) synthesized by WAR process [9].

5.3 Another model

During the calculations with CC model, we observed convergence problems during the first calculations, because of the raw curve considerations stemming from the press data acquisition concerning the upper punch load evolution of force as well as the die and needle displacements. This problem was solved by separating compaction and accompaniment into several steps, so as to soften the slope changes. Another problem of convergence comes from the CC model itself, because it cannot represent a tensile stress (no section of the load surface corresponding to the negative hydrostatic pressures). There is thus a 10% failure of convergence in elastic return. Furthermore, when you draw up Q depending on P, we observed that the first part of the load surface corresponds to a softening ellipse. Rather than implementing this special feature in the initial model like previous Cast3 m® study [4,16], we opted for a better adapted Drucker-Prager type model. A Drucker-Prager Cap model (DPC) was tried and compared to CC Model. DPC takes into account the powder cohesion, the linear elasticity or non-linear porous elasticity. It used two main yield surface segments, a linearly pressure dependent DPC shear failure surface: $F_S = t - p\,\tan\beta - d = 0$.

The cap yield surface: $F_C = \sqrt{(P - P_a)^2 + [Rt/(1 + \alpha - \alpha/\cos\beta)]^2} + R(d + P_a\tan\beta) = 0$ and the transition surface: $F_t = \sqrt{(P - P_a)^2 + [t - (1 - \alpha/\cos\beta)(d + P_a\tan\beta)]^2} - \alpha(d + P_a\tan\beta) = 0$.

All parameters are given in references [2,15].

5.4 Finite element simulations

The geometrical model is an axisymmetric 2D type. It is established based on the powder column, the die and the lower and upper punch. The upper punch and the die are mobile. A connector (equation between two nodes) was used to ensure the speed ratio between the upper punch and the die (punch with rigid connection for piloting via a reference node).

The punch mesh is relatively large and uniform. That of the powder is also uniform, and a little finer. On the other hand, that of the die is much more refined, in particular at the rounded corners in touch with the powder where the stress concentrations are situated, and where the generation of residual stress can be high during the pellet ejection springback. It is the sensitive point which must be handled carefully to avoid generating problems of convergence during the calculation.

During the simulation, the uniaxial simple effect cycle of shaping with a floating die is composed of a succession of discrete stages, each run in a succession of iterations. At the beginning of the calculation, the die is considered to be full of powder, with the upper punch in contact with the powder. At this stage, there is the first step which consists in powder compaction with the upper punch at the speed of 14 mm/s while exercising a push with the die in the same

Fig. 2. Press in the mock-up, with conveyor and tubular container of 37 pellets.

direction as the upper punch but at a more moderate speed, i.e. 10 mm/s. The step is finished when a plateau of a few seconds is reached at 47 kN (600 MPa). The second step consists in pellet ejection by a vertical die withdrawal and the preservation of a support pressure fixed at 11 kN (150 MPa). During this stage, pellet radial springback takes place. The third step consists in withdrawal of the upper punch and complete pellet freeing, when the pellet axial springback occurs. The final step involves the sintering process, and creates shrinkage depending on the density gradients generated during the shaping.

For the press contact elements, the model used is based on the non-penetration of the two bodies in contact. Abaqus® uses the Lagrange multipliers method. The algorithm imposes the non-penetration condition on the resolution system by adding unknowns to the system. This greatly increases calculation time. The friction is defined as a Coulomb friction. The Coulomb coefficient taken into account in the calculation is equal to 0.094.

As indicated in reference [13], we used a simple sintering model based on thermal strain to one dimension $\alpha \Delta T = \varepsilon$th, with $\alpha = (\rho_c / 95\% \rho_{th})^{1/3} - 1$. To summarize, for each meshing element of the powder, the green density ρ_c was calculated with the Cam-Clay model as well as the corresponding α coefficient. Next, the thermal dilation model of the green density map, the α coefficient map and a temperature level $(\Delta T = 1)$ were entered. Shrinkage was thus calculated. A subroutine was developed in a Python language in the Abaqus® code to take the sintering step into account [17].

6 Comparisons and discussions

To highlight the best comparison between experimentations and simulations, we have chosen to take into account the capability study realized with the electromechanical

press. Capability machine is the ability of the apparatus to reach the required input performance. This takes into account the statistical process control and permits a measurement of whether the machine can respect the interval tolerances (defined by the top and bottom targets) given in the specifications. The apparatus concerned in this study is the nuclear press described in previous chapter and the sintered input dimensions of the pellet are given bellow by: diameter = 8.45 ± 0.09 mm, hole diameter = 2.2 mm and, height < 12 mm. The results of optimization study have been used to calculate size of a new tool. The proportional law is possible for the small gap and the new calculate diameter is 9.370 mm with the high tolerance fixed to ±0.005 mm. The needle diameter has not been changed. Two new tools have been built, one with needle and one without needle. We decided to shape 400 pellets with each tool. Only hole pellet results are presented in this study. The compaction cycle during 400 annular pellet productions has been realized [17]. The powder volume depends on the weight of the pellet (2.180 g) and the bulk powder density. The volume of the powder necessary to make 400 pellets is 0.689 L. For information, the capacity of the jar is 0.751 L and 0.374 L between the jar and the powder column.

The pellets were shaped in continuous compaction, and a pathway system was built to keep the order and the direction of the pellet. This order was monitored to check the press variations (drift) and direction, and to see the side where the upper punch applied the force. All the compaction cycles were recorded in the press database software. The pellets were put into glass tubes containing 37 pellets (Fig. 2). After compaction, each green pellet was measured by laser profilometer (height, and diameter corresponding to height) and weighed with precision scales. A chronological number was written on the side directly in contact with the upper punch. All the pellets (100 per batch) were then placed in an alumina crucible and sintered in a furnace under air. The sintering conditions were

Fig. 3. Sintered pellets in the alumina container (100 pellets).

1600 °C, with 4 °C/min for the heating up, for a duration set at 4 hours, followed by 2 °C/min for the cooling (Fig. 3). The same measurements were carried out on pellets after sintering (height, diameter and weight) [18].

As shown in the Figure 4, the average diameter of the pellets is 8.510 mm, the maximum and minimum diameters are respectively 8.533 mm and 8.487 mm. The project objective was reached, but the diameter of the die must be reduced because the average diameter is still too high. We found out that the distribution is not centred and the asymmetric coefficient is 0.572. The average is 8.508 mm. The maximum is 8.533 and the minimum is 8.490. The standard deviation is 0.0068 and the variance is 4.75×10^{-5}. The Alfa coefficient of the confidence gap is 0.05. The Cp capability process is 5.03. The performance process coefficients Pp and Ppk are respectively 4.35 and 1.52 and we must conclude that the process is very capable [19].

Table 3 summarizes objectives and results of all studies, the die dimensions of each die calculated with proportionality law.

To better understand the results, the curves in Figure 5 show different experimental and calculation results. It shows the evolution of the pellet height depending on the diameter. The optimization and capability study conclusions are indicated. For each study, you have the green and sintered pellet diameters and the die diameters obtained by the application of the proportionality law (data shown in

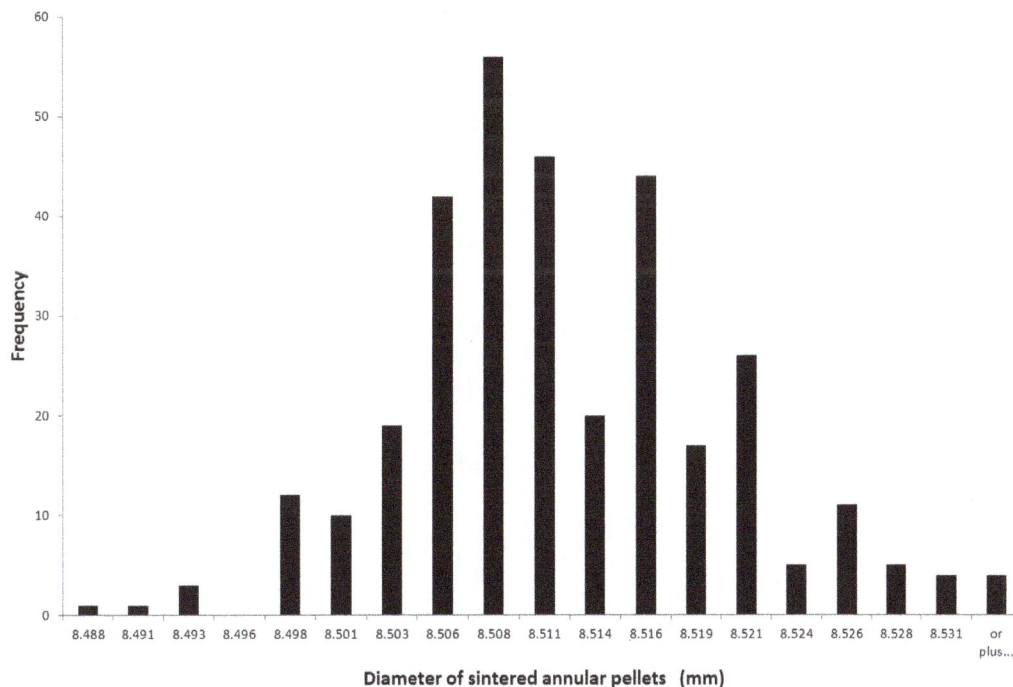

Fig. 4. Histogram of the sintered pellet diameter.

Table 3. Comparison between objective and result diameters, compared to trial number 308.

	Φ_{die} (mm)	Φ_{sintered} (mm)
Optimization study result	10.000	9.015 ± 0.012
Capability study objective		8.450 ± 0.090
Capability study result	9.370 ± 0.005	8.510 ± 0.023

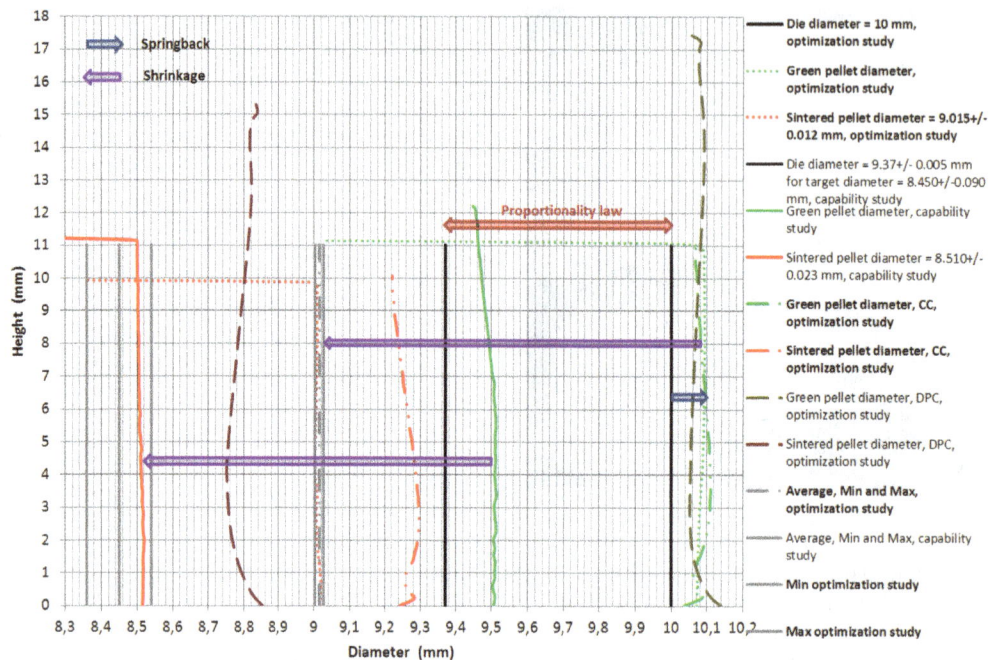

Fig. 5. Comparison between optimization and capability studies, experimental and calculated results.

green for green pellets and red for sintered pellets). We can see the springback between the die and green pellet, as well as shrinkage between green and sintered pellets. Finally, the calculated green and sintered pellet diameters with CC and with DPC models, used for an optimization study without hole and carried out with the Abaqus® software are shown. The calculation results show that the model parameters must be optimized. DPC behavior is better than that of CC, as the shape of the sintered pellet is conical. The model behavior at the base of the pellet does not suit the requirements. The height of the sintered pellet must be modified, and the sintered densities are weak. The sintering is too high and must be reduced.

7 Conclusions and perspectives

Producing tomorrow's fuel pellets in a hot cell will require the use of new press technology, due to the nuclear constraints and very strict shape criteria. This publication describes an optimization study on the pressing cycle of the CEA-Champalle electromechanical press, an apparatus which is compact, modular and nuclearized for hot cell operation. It is known that the pressing cycle influences the density gradients in green pellets. In order to predict the diameter tolerance of pellets after sintering, a compaction modelling and simulation program was also undertaken. The density card of the pellet enabled the shrinkage to be calculated and compared to experimental results. With lubricant in the powder and new pellet diameter and tolerance, the capability of the press to manufacture 400 pellets with hole was studied in a mock-up with remote handling. Results showed that the die diameter calculated and the press cycle set enabled pellets to be shaped with satisfactory tolerances. Simulation and associated model-

ling are present in a main goal to be able to anticipate the demanding manufacturing factors, which can influence the press settings before the production cycle, and then during the manufacturing, to be able to have the shortest possible response time to correct parameters to ensure finished products with stable quality. Research into powder compaction behaviour will continue, in order to obtain an improved model response with a new powder and using a discrete element method to better take into account the behaviour between aggregates [20,21].

We would like to thank the Simulia/Abaqus team for their support and in particular C. Geney. Our thanks also to Champalle and all the Process Cycle Advanced Technology Laboratory team.

References

1. I. Aydin, J. Briscoe, Dimensional variation of die pressed ceramic green compacts, comparison of a FEM with experiment, J. Eur. Ceram. Soc. **17**, 1201 (1997)
2. P.R. Brewin, O. Coube, P. Doremus, J.H. Tweed, *Modelling of powder die compaction*, Springer Engineering Materials and Processes (Springer-Verlag, London, 2008), p. 57, §4.2.2, p. 59 §4.2.3
3. P. Pizette, C.L. Martin, G. Delette, P. Sornay, F. Sans, Compaction of aggregated ceramic powders: From contact laws to fracture and yield surfaces, Powder Technol. **198**, 240 (2010)
4. P. Pizette, C.L. Martin, G. Delette et al., J. Eur. Ceram. Soc. **33**, 975 (2013)
5. G. Kerboul, Étude de l'endommagement des produits céramiques crus par émission acoustique, Thèse INSA Lyon, 1992
6. D.D. Zenger, H. Cai, *Handbook of the common cracks in green P/M compacts* (Powder Metallurgy Reserch Center, WPI, 1997)

7. P. Jonsen, A. Haggblad, Modelling and numerical investigation of the residual stress in a green metal powder body, Powder Technol. **155**, 196 (2005)
8. G. Delette, P. Sornay, J. Blancher, A Finite Element modelling of the pressing of nuclear oxide powders to predict the shape of LWR fuel pellet after die compaction and sintering, in *AIEA Technical Committee, Brussels, 20–24 October 2003* (2003)
9. J.-P. Bayle, Minor actinide bearing blanket manufacturing press and associated material studies for compaction cycle optimization, in *NuMat 2014 Nuclear Materials conference, 27–30 October 2014 Clearwater Beach, Florida* (2014)
10. J.-P. Bayle, *Electromechanical press for nuclear compaction in hot cell* (WNE, Paris, 2014)
11. J.-P. Bayle, *Minor actinide bearing blanket manufacturing press* (Hotlab, Baden, 2014)
12. J.-P. Bayle, WO2015/181121A1, Brevet CEA/Champalle, Presse pour mettre en forme des pastilles dans un environnement restreint et hostile et procédé d'assemblage de la presse
13. J.-P. Bayle, Finite element modeling and experiments for shaping nuclear powder pellets, Procedia Chem. **7**, 444 (2012)
14. C. Dellis et al., PRECAD, A Computer-Assisted Design and Modelling Tool for Powder Precision Moulting, in *HIP'96 Proceeding of the international conference on Hot Isostatic Pressing, 20–22 May 96 Andover, Massachusetts* (1996) pp. 75–78
15. Abaqus® User manual, Vs 6.11 Analysis User's, Manual Volume III: Materials, section 22.3.1, 22.3.2, 22.3.4
16. Cast3 m® User manual, Modèle non linéaire, T. Charras, Edition 2011
17. J.-P. Bayle, Modelling of powder die compaction for press cycle optimization, in *TopFuel 2015, Sept. Zurich* (2015)
18. O. Gillia, Modélisation phénoménologique du comportement des matériaux frittants et simulation numérique du frittage industriel de carbure cémenté et d'alumine, Thèse INPG, 2000
19. F. Desnoyer, Mémento sur la notion de capabilité, TI, ag1775, versus 10/01/2004
20. E. Remy, J. Eur. Ceram. Soc. **32**, 3199 (2012)
21. P. Parant, *Study and modelling of compaction of metal oxide microspheres into pellets* (E-MRS, Warsaw, Poland, 2014)

5

Sensitivity analysis of minor actinides transmutation to physical and technological parameters

Timothée Kooyman[*] and Laurent Buiron

CEA Cadarache, DEN/DER/SPRC/LEDC, Bat. 230, 13108 Saint-Paul-lez-Durance, France

Abstract. Minor actinides transmutation is one of the three main axis defined by the 2006 French law for management of nuclear waste, along with long-term storage and use of a deep geological repository. Transmutation options for critical systems can be divided in two different approaches: (a) homogeneous transmutation, in which minor actinides are mixed with the fuel. This exhibits the drawback of "polluting" the entire fuel cycle with minor actinides and also has an important impact on core reactivity coefficients such as Doppler Effect or sodium void worth for fast reactors when the minor actinides fraction increases above 3 to 5% depending on the core; (b) heterogeneous transmutation, in which minor actinides are inserted into transmutation targets which can be located in the center or in the periphery of the core. This presents the advantage of decoupling the management of the minor actinides from the conventional fuel and not impacting the core reactivity coefficients. In both cases, the design and analyses of potential transmutation systems have been carried out in the frame of Gen IV fast reactor using a "perturbation" approach in which nominal power reactor parameters are modified to accommodate the loading of minor actinides. However, when designing such a transmutation strategy, parameters from all steps of the fuel cycle must be taken into account, such as spent fuel heat load, gamma or neutron sources or fabrication feasibility. Considering a multi-recycling strategy of minor actinides, an analysis of relevant estimators necessary to fully analyze a transmutation strategy has been performed in this work and a sensitivity analysis of these estimators to a broad choice of reactors and fuel cycle parameters has been carried out. No threshold or percolation effects were observed. Saturation of transmutation rate with regards to several parameters has been observed, namely the minor actinides volume fraction and the irradiation time. Estimators of interest that have been derived from this approach include the maximum neutron source and decay heat load acceptable at reprocessing and fabrication steps, which influence among other things the total minor actinides inventory, the overall complexity of the cycle and the size of the geological repository. Based on this analysis, a new methodology to assess transmutation strategies is proposed.

1 Introduction

Minor actinides transmutation represents a potential solution to decrease the amount and hazards caused by nuclear. It can be achieved by subjecting minor actinides nuclei to a neutron flux. Minor actinides transmutation can take two forms, either the minor actinide nuclei undergoes fission and yields fission products which are shorter lived or captures a neutron and is transmuted into another heavy nuclide. The main minor actinides that are produced in nuclear reactors are:

- ^{237}Np, produced by neutron capture on ^{235}U in light-water reactors, decaying to ^{233}Pa with a half-life of

2.14×10^6 years. It is also produced by $(n, 2n)$ reactions on ^{238}U in fast reactors and from ^{241}Am decay;
- ^{241}Am, produced by decay of ^{241}Pu and decaying to ^{237}Np with a half-life of 432 years;
- ^{243}Am, produced by neutron capture on ^{242}Pu and decaying to ^{239}Pu with a half-life of 7370 years;
- ^{244}Cm which is produced by capture on ^{243}Am which decays to ^{240}Pu with a half-life of 18.1 years and which is mainly found in MOX fuels.

When plutonium is recovered from the spent fuel by reprocessing and then reused, only minor actinides and fission products remain in the final waste, along with the small uranium and plutonium losses from the reprocessing step. In this case, both long-term radiotoxicity and final spent fuel repository design constraints are dominated by minor actinides, as the fission products contribution become negligible after a few hundred years.

---

* e-mail: `timothee.kooyman@cea.fr`

Minor actinides transmutation consequently appears as a potential strategy to minimize the fraction and mass of MA in the waste and reduce the spent fuel burden. As such, it was included in the 2006 French law on nuclear waste management as a research option to deal with nuclear wastes management. In the asymptotic case of a complete multi-recycling of all minor actinides, the only waste would be the associated reprocessing losses, which can be as low as 0.01% [1] of the reprocessed mass, thus dividing by a factor up to 1000 the impact of minor actinides.

Minor actinides transmutation has been studied for several decades and many concepts have been discussed so far. We will only focus here on transmutation in critical reactors. Studies have been made on transmutation in thermal [2] and fast reactors [3], either in dedicated [4] or industrial reactors with various types of fuel, coolant and minor actinides isotopic vector. Several experiments have also been carried in various reactors such as the SUPER-FACT experiment in the PHENIX reactor [5] or more recently the METAPHIX experiments in the same reactor.

Fast reactors exhibit an advantage for transmutation compared to thermal systems as they have a higher neutron excess and as they produce less minor actinides from capture on plutonium isotopes. So far, transmutation options for such reactors can be divided in two different approaches. In the homogeneous approach, minor actinides are loaded in the core in fractions higher than in the natural fraction of minor actinides present in the fuel at equilibrium (below 0.5% depending on the spectrum). For minor actinides content above 2–5%, depending on the core design, reactivity coefficients such as Doppler feedback and coolant void worth are negatively impacted, which has an impact on safety performances of the core (see Refs. [3,6]). If we consider reference core design for sodium cooled reactors, minor actinides fraction in the fuel is limited to 2.5–3% to keep acceptable reactivity coefficients [7].

Additionally, this approach exhibits the drawback of "polluting" the entire fuel cycle with minor actinides, thus increasing the cost of every step of the fuel cycle [8].

In the heterogeneous approach, minor actinides are inserted into transmutation targets which can be located in the center or in the periphery of the core. This presents the advantage of dissociating the management of the minor actinides from the conventional fuel and not impacting the core reactivity coefficients.

A transmutation strategy can have various objectives: the goal can be to limit the minor actinides inventory while operating a nuclear reactor fleet, or to transmute the minor actinides stockpile originating from the current operations of LWRs. The interest of the use of a given reactor type for transmutation purposes must be evaluated bearing in mind this final objective. Preliminary questions such as the use of dedicated reactors and reprocessing facilities must also be solved before designing a complete transmutation strategy.

We considered here that the main goal of transmutation issues was to minimize the volume and burden in terms of repository size and radiotoxicity of the waste associated with nuclear energy. Consequently, we made the hypothesis of a closed cycle with plutonium multi-recycling. Only transmutation in fast reactors spectrum was studied here, as a fast spectrum appears to be more suited to

transmutation. The results are detailed here for transmutation in homogeneous mode which shows the best performances but the conclusions are quite similar for transmutation in heterogeneous mode.

2 Scope of the study

Most of the work related to transmutation has been carried out seeking for an efficient transmutation system, that is to say a reactor design which exhibits high minor actinides consumption rate with "acceptable" safety parameters. The common approach to this problem was to start from an existing core and modify it to accommodate the loading of a given fraction of minor actinides, as it is proposed for instance in [9], where the core geometry is modified to decrease the sodium void worth, permitting a subsequent addition of minor actinides in the reactor.

A drawback of this approach is that it focuses solely on the reactor side of the transmutation process while additional constraints on the strategy related to fuel cycle must also be taken into account. Indeed, minor actinides bearing fuels typically lead to complications at the fabrication stage and have more stringent mechanical requirements due to an increased helium production in the fuel. Minor actinides bearing fuel handling and reprocessing are also more complicated, due to their important decay heat and neutron source. Consequently, they also require enhanced radioprotection shielding during the fabrication process.

The aim of this work was to implement a low-level approach to the transmutation concept. First, a global study of the reactor parameters which may have an impact on transmutation has been carried out. Then, fuel cycle considerations and constraints were taken into account to evaluate their effect both on transmutation and on the reactor parameters. From the results, it was then possible to identify a set of relevant parameters which encompassed both reactor and fuel cycle constraints and to outline a global methodology for the design of a comprehensive transmutation strategy.

3 Methodology

A simplified approach of reactor physics was used to evaluate the impact of various parameters on transmutation performances. In order to assess their effect on fuel cycle aspects, a simplified equilibrium algorithm described in Figure 1 was used. The list of parameters which were studied for the reactor part is given in Table 1. The plutonium fraction in the fuel is set at 20%.

A moderating material was added to the cell in some cases to evaluate the effect of a degraded spectrum, even in unrealistic quantities. Even if this material is denominated "moderator" in this paper for simplicity of language, it was not added in the medium as a design feature but as solution to explore a wide range of potentially available spectrum. Variation ranges of the various materials were voluntarily taken as extreme in order to correctly evaluate all possible configurations. As such and due to the simplicity of the model considered, the results cannot be directly transposed to reach a

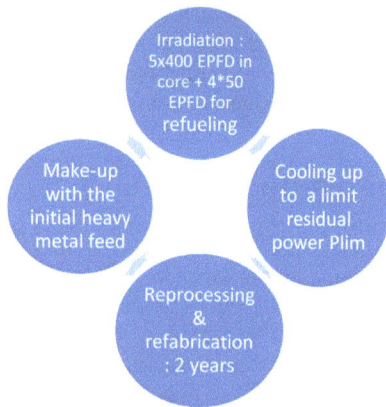

Fig. 1. Algorithm used for multi-recycling simulation.

$$\tau_i(X) = 1 - \frac{final\ content\ in\ nuclide\ X}{initial\ content\ in\ nuclide\ X}$$

$$\tau_f(X) = \frac{fraction\ of\ nuclide\ X\ that\ has\ fissioned}{fraction\ of\ nuclide\ X\ that\ has\ disappeared}.$$

The core was initially loaded with a given volume fraction of minor actinide with the vector given in Table 2. This vector is deemed representative of what should be available in France by 2035. The mass of fuel which had undergone fission was replaced with the equivalent mass of initial feed to keep a constant mass of fuel. The algorithm was stopped once the difference between the ^{241}Am fraction in the fuel at the beginning of two consecutive cycles was below 0.5%.

It has to be noted here that for several nuclides, the main one being ^{241}Am, the transmutation rate can be defined either between the beginning and the end of irradiation or between the start of irradiation and the end of reprocessing, as there will be a production of ^{241}Am during reprocessing due to ^{241}Pu decay. We referred to the first one as the "irradiation transmutation rate" and the second one as the "cycle transmutation rate".

For decay heat calculations, the isotopes used are given in Table 3. Fission products contribution to the total residual power was neglected, as their contribution to the residual power is only 10% after 5 years cooling for MOX fuels irradiated in fast reactors, which is the minimal delay which was considered. The power density was calculated for an assembly of 175 kg of heavy nuclides.

conclusion regarding full-core model. For instance, a full-core model would incorporate information about the geometrical design of the assembly, which feasibility depends on the fuel/coolant/moderator chosen for instance. However, the model used is enough to give a broad understanding of the effects of neutron spectrum variation and broad design parameters such as fuel or coolant fraction in the core, or power.

To simulate the irradiation, 33 group cross-sections were calculated using a homogeneous cell model and the ECCO cell code [10]. These cross-sections were then collapsed into one group cross-sections for depletion calculation which were carried out with a constant flux and a depletion chain ranging from ^{234}U to ^{252}Cf. Cooling down was simulated using the same depletion calculation without flux. Minimal cooling time was set at 5 years and no limits were considered for upper cooling time. The effect of the following parameters was studied:

– maximum allowable decay heat at reprocessing;
– maximum allowable neutron source at reprocessing;
– manufacturing time.

The impact of each parameter on the transmutation performances was estimated using the transmutation rate, defined as the fraction of a nuclide having disappeared either by capture or by fission, and the fission rate, which is the ratio of nuclides which have undergone fission over the number of nuclides having disappeared.

4 Results from reactor analysis

The impact of each parameter discussed in Table 1 on the fission rate and the transmutation was assessed in order to pinpoint the relevant ones. Representative volume fractions of a typical fast reactor were taken with 40% of fuel, 40% of coolant and 20% of structures.

4.1 Effect of the fuel type and fraction

The decrease in the transmutation rate concomitant with the increase in the fission rate when going from oxide to metal, as

Table 1. Reactor parameters.

Physical/technological parameter	Variation range
Fuel type and fraction	Oxide/Nitride/Carbide/Metal between 20 to 50%
Coolant type and fraction	Sodium/LBE[a]/Helium between 20 to 50%
Moderating material type if any	None/ZrH$_2$/Be/MgO
Fraction of MA in the fuel (MA/total heavy metals)	1 to 50%
Fraction of moderator	0 to 20%
Irradiation time	300 to 10,000 EPFD
Flux level	10^{13} to 10^{15} n/cm^2/s
Composition of the MA feed	Am/Am + Cm + Np

[a]Lead-Bismuth Eutectic.

Table 2. Minor actinides isotopic vector.

Element	Mass fraction (%)
^{237}Np	16.87
^{241}Am	60.62
^{242}Am	0.24
^{243}Am	15.7
^{242}Cm	0.02
^{243}Cm	0.07
^{244}Cm	5.14
^{245}Cm	1.26
^{246}Cm	0.08

Table 4. Effect of the fuel type for a cell with 6.5% Am, averaged over the results for the three coolant types.

Fuel type	Transmutation rate (%)	Fission rate (%)
Oxide	67.5	11.5
Carbide	62	14.5
Nitride	60.8	15.5
Metal	57.7	16.3

Table 5. Effect of the coolant material type for a cell with 6.5% Am, averaged over the results for the four fuel types.

Coolant type	Transmutation rate (%)	Fission rate (%)
Helium	66	11.9
LBE	61	14.6
Sodium	62.5	14.9

seen in Table 4, is explained by the modification of the spectrum in the cell. With a metal alloy fuel, the harder spectrum leads to a lower capture cross-section for the minor actinides, which decreases the total absorption cross-section and thus the transmutation rate while increasing the fraction of fissions for the same irradiation time of 1000 EPFD.

One can note that the range of variations of both rates is limited to a few percent, which leads to the conclusion that the choice of the fuel will be mainly dictated by thermo-mechanical constraints pertaining to the residence time, flux level and reactor technology rather than by solely neutronic considerations. The increase of the fuel fraction slightly hardens the spectrum thus slightly decreases transmutation rate by a few percent.

4.2 Effect of the coolant type and fraction

Similarly to the previous case, we can observe in Table 5 here that a small change in the spectrum due to the use of a lighter or heavier coolant has a small effect on the fission and transmutation rate, but once again, this is limited to a few percent so it is concluded that coolant choice will more likely be driven by safety constraints and technological considerations rather than by neutronic aspects.

4.3 Effect of the neutron spectrum

Figure 2 shows the effect of the various moderator materials on the transmutation rate for a cell with 30% fuel, 30%

coolant and between 40 and 20% of structures. It is clear that the hydrogenated moderator ZrH_2, which has a more efficient moderating effect, is the most effective to slow down the neutrons. However, its use in reactor is difficult mainly due to dissociation issues that were not taken into account here. The two other moderators are less efficient and their impact on the transmutation rate is consequently smaller. In each case, the impact on fission rate is inversely proportional to the impact on the transmutation rate. This is explained by the change in the spectrum which has already been discussed before.

However, this highlights a potential use of the moderator to accelerate transmutation kinetic. Using moderator material increases the total absorption cross-section and thus the transmutation rate while decreasing the fission cross-section. This leads to an increase in the production of curium and heavier minor actinides. However, using moderator appears as a possible solution to tune the production of curium with regards to cycle constraints in order to maximize the transmutation rate. This will be discussed in the next parts. It should also be noted that addition of moderating material may lead to damaging power peaking issues [11].

Table 3. Isotopes used for residual power calculations and neutron source calculations.

Isotope	^{242}Cm	^{244}Cm	^{241}Am	^{238}Pu
Power density (W/g)	121.4	2.84	0.11	0.57
Isotope	^{244}Cm	^{245}Cm	^{248}Cm	^{252}Cf
Neutron emission (10^7 n/s/g)	1.4	1	4.4	2.1×10^5

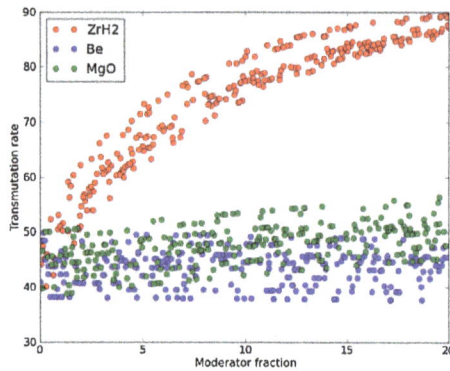

Fig. 2. Transmutation rate versus moderator fraction.

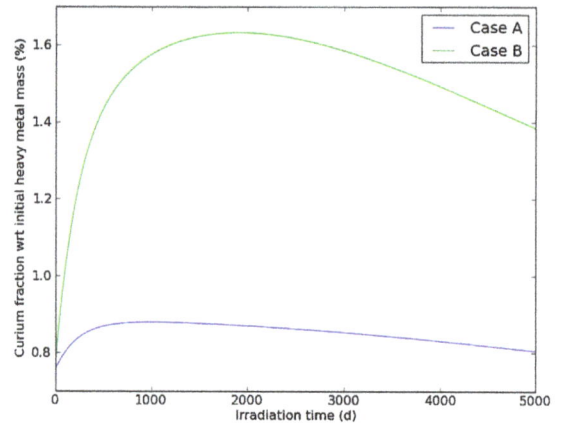

Fig. 4. Illustration of the curium peak for case A and B (detailed below in Tab. 6).

4.4 Effect of the MA isotopic vector and volume fraction

Increasing the minor actinides volume fraction in the fuel has two effects which are opposite. On the one hand, an increase in the MA fraction leads to a harder spectrum, which decreases the transmutation rate. On the other hand, the increase in the loaded fraction of minor actinides also increases the transmutation rate by displacing the fuel isotopic vector further way from its equilibrium value. The first effect is predominant at high fraction and the second one is more visible at low fraction. This can be seen on Figure 3, which shows the transmutation rate at 1000 EPFD versus the fraction of moderator and the fraction of minor actinides for 10^{15} n/cm^2/s flux with 40% fuel at 20% Pu fraction and 20% structures. One can see that for a constant moderator volume fraction, the transmutation rate first increases with the minor actinides fraction and then decreases. This is seen with all kind of coolant/fuel combination, all moderator material and with Am only or all minor actinides. This means that there is an optimal value for MA fraction loaded in the fuel, which depends also on the moderator fraction. In our calculations, no impact of the minor actinides vector on the transmutation rate or fission was found. However, in a "true" reactor, this vector will have an impact on reactivity and safety coefficients of the reactor.

It should also be noted that the position of this optimum may not be adequate with regards to the minor actinides inventory management. Indeed, it corresponds to relatively low minor actinides fraction.

4.5 Effect of flux level and irradiation time

At the first order, transmutation rates variation with regards both to flux level and irradiation time goes as $1 - e^{-\varphi T}$ so an increase in any of these two parameters will lead to an increase in the transmutation rate without any impact on the fission rate, which is verified by our calculations.

Consequently, there is an interest in using the highest possible flux level to accelerate the transmutation process. For irradiation time, the reasoning is similar but appropriate care should be taken with regards to the so-called "curium peak", which can be seen on Figure 4.

This peak is due to the competition between the production of curium from capture on americium isotopes and the destruction of these curium nuclei by fission or capture. At beginning of irradiation, the americium fraction is high which leads to a high production rate of curium with a low consumption rate as the curium fraction is still low. The height of this peak is proportional to the ratio of absorption cross-sections of Cm and Am isotopes. In a fast flux, this ratio is lower than in an epithermal flux, thus explaining why the peak appears to be lower in case A on Figure 4, which corresponds to a fast spectrum than in case B which corresponds to a more degraded spectrum. Both cases were introduced as "extremal" spectrum that can be found in a fast reactor, either with a very energetic spectrum (case A) or a very degraded spectrum (case B). It should also be noted that evolution kinetic of the curium fraction depends on the absorption cross-sections of Cm and Am, which explains the difference observed in terms of evolution on Figure 4.

From the previous analysis, we can conclude that the most important parameters in terms of reactor design for transmutation purposes are the amount of minor actinides loaded in the core and the spectrum. The other parameters studied have an impact which is small compared to the

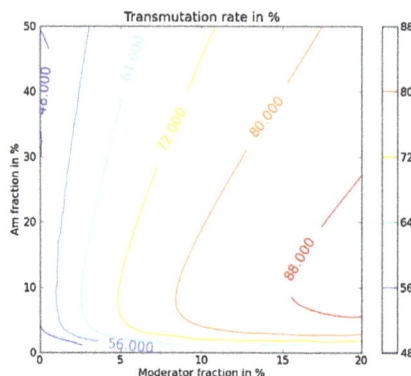

Fig. 3. Transmutation rate versus fraction of moderator.

Table 6. Details of the cases used for the fuel cycle calculations.

Parameter	Case A	Case B
Fuel	Metal	Oxide
Coolant	Na	Na
Mod	None	ZrH_2
Fuel fraction (%)	40	30
Mod fraction (%)	None	20
Coolant fraction (%)	40	20
Pu fraction in fuel (%)	20	30
MA fraction in fuel (%)	8	12

impact of the two previous parameters, and thus they can be neglected in a first step of optimization.

5 Results from the fuel cycle analysis

5.1 Methodology

Considering the results obtained in the previous part, two limit cases were used for the fuel cycle parameters analysis and it is assumed that any intermediate case in terms of spectrum leads to intermediate results. These cases are described in Table 6. They correspond to the optimum in moderator and minor actinides fraction visible on Figure 3. Case A corresponds to an asymptotic case with a very fast spectrum while case B corresponds to a very degraded spectrum for a typical fast reactor. In both cases, a loading with the MA vector described in Table 2 was considered. A manufacturing time of 2 years was also taken into account for the calculations concerning the residual power and neutron source. Sensitivity to the manufacturing time was also assessed. A final point which was discussed is the total

inventory in minor actinides in the fuel cycle, which is proportional to the cooling time.

5.2 Evolution of decay heat and neutron source

The evolution of decay heat and neutron source is plotted in Figure 5 for both cases. Several comments can be made here. The first point that should be made here is that the use of a moderator material to shift the spectrum leads to an increase in the Cm production and thus in the decay heat in neutron source. Second, the sharp decrease in the decay heat is due to the decay of ^{242}Cm with a period of 163 days. Once ^{242}Cm has disappeared, the decay heat is dominated by ^{244}Cm and ^{238}Pu which are longer-lived (respectively 18.1 years and 87 years). A consequence of this is that the feasibility of reprocessing with significant quantities of curium must be demonstrated in order to consider transmutation otherwise prohibitively long cooling delays will have to be considered. For case A, this leads to value around 25 W/kg and for case B this leads to a limit around 50 W/kg.

Additionally, given the very high decay heat level in the first year of cooling, manipulation of such spent fuel assembly will be more complicated than with regular fuel (standard MOX fuel). It should be noted here that the fission products contribution at short timescales was neglected here so Figure 5 is actually underestimating decay heat in the first 5 years of cooling. A sensitivity analysis showed that the main contributor to the decay heat after 5 years was ^{243}Am which yields ^{244}Cm by neutron capture. At shorter timescale, decay heat is dominated by ^{242}Cm contribution which comes from ^{241}Am.

Neutron source being essentially driven by ^{244}Cm contribution, its timescale is different from decay heat load and its decrease slower. For comparison, a typical UOx spent fuel discharged at 47.5 GWd/t has a typical heat load of around 4 W/kg and a neutron source of 2000 n/s/g after 4 years of cooling. One can consequently conclude from this short analysis that an increase in the reprocessing limit both in terms of decay heat and neutron source will be necessary in order to avoid large cooling times and minor actinides in-cycle inventories.

5.3 Impact on the transmutation performances

In the next paragraphs, we will consider case B, which is the most penalizing case in terms of neutron source and decay heat due to the "moderated" spectrum. An important point that can be made about Figure 6 is the difference between the irradiation transmutation rate and the cycle transmutation rate, which is explained by:

– the increased fraction of ^{241}Pu in the fuel due to the spectrum shift;
– the longer cooling time due to the higher decay heat.

One can also see that there is an interest to maximize the allowable decay heat for reprocessing in order to increase the cycle transmutation rate of ^{241}Am. The saturation effect observed around 35 W/kg is due to the hypothesis

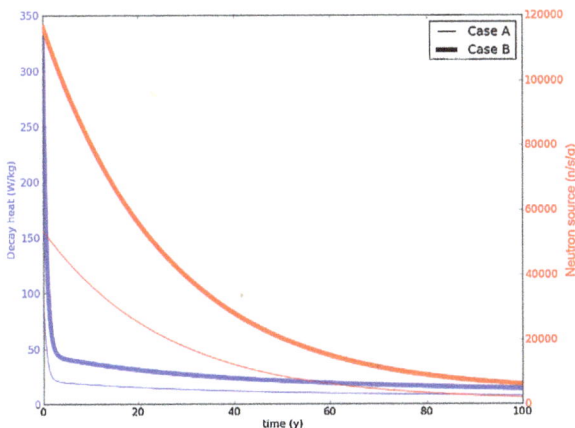

Fig. 5. Evolution of decay heat and neutron source versus cooling time.

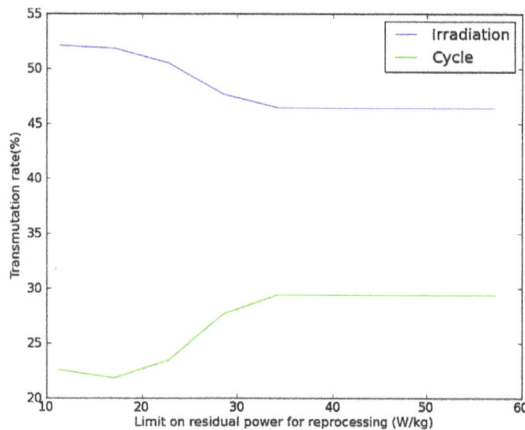

Fig. 6. Transmutation rate of ^{241}Am for irradiation and cycle versus limit on decay heat during reprocessing.

that the minimal cooling time is 5 years. The same behavior can be observed for neutron source. The small increase at low decay heat in the cycle transmutation rate is explained by the very long cooling which leads to decay of a significant fraction of ^{241}Am ($T_{1/2} = 432$ years). Minimizing the cooling time has also the interest of both minimizing the time necessary to reach an equilibrium situation and the total inventory of minor actinide.

However, more active fuel at reprocessing increases the losses during treatment and consequently increases the amount of long-lived wastes coming from spent fuel reprocessing. It also leads to the production of more active wastes, which necessitate more waste packages for final storage. Work is still ongoing to quantify the loss level and identify the optimum solution of the problem.

This phenomenon is not seen in case A as the use of a fast spectrum leads to a lower production of curium isotopes and a lower equilibrium fraction of ^{241}Pu. Consequently, the cooling times are lower by a factor three compared to case B and the minimal value of 5 years is reached. However, the transmutation rate during irradiation is also divided by three and the cycle transmutation rate by two.

5.4 Impact of the manufacturing time

It was also found that the manufacturing time has a non-negligible impact on the total time necessary to reach an equilibrium situation. Indeed, contrary to the cooling time which decreases with the number of cycles, manufacturing is the same for each cycle. An increase of 2 years in fabrication leads then to an increase of 2 years of each cycle. This effect is more visible in case A where residual heat load is lower so cooling time is also lower and the fabrication time share in the entire reprocessing time is higher.

5.5 Outline of an optimization methodology

The conclusion from this fuel cycle parameters analysis is then that optimization of a global transmutation strategy

requires both the optimization of the irradiation and reprocessing parts. Indeed, the good performances obtained from a given reactor can be cancelled by non-adapted reprocessing specifications and delayed reprocessing times. More specifically, the irradiation time and spectrum should be tuned so as to maintain the cooling time within acceptable boundaries while keeping acceptable transmutation performances.

A new methodology of reactor design is currently being developed to take into account these results to settle a multi-recycling transmutation strategy. Bearing in mind technological limitations such as maximal residence fuel time, it is aimed at selecting the best spectrum that ensures an efficient transmutation while allowing reprocessing within acceptable limits. In a second step, a core image will be designed to obtain a relevant spectrum as close as possible to the optimal one while keeping adequate safety parameters.

6 Conclusions

An analysis of the various parameters influencing the performances of a transmutation strategy was carried out including parameters from the reactor and the cycle. The neutron spectrum and the volume fraction of minor actinides in the core were found to be the most relevant parameters for the core, while the cooling time through the limitations on decay heat and neutron source for reprocessing was identified as a critical parameter for the fuel cycle part. It was shown that an optimization of a transmutation strategy required considering at the same time parameters from the cycle and the reactor.

Further work is ongoing to add a third component to this analysis, namely the waste and final repository aspect. Indeed, a goal of transmutation strategy is to reduce the size of the final deep geological repository. The decay heat and alpha activity of the waste are the main constraints impacting the number of waste packages to be stored and thus the volume occupied by the repository. In a second time, a full implementation of the optimization methodology will be carried out taking into account the three sides of the problem along with additional options such as heterogeneous recycling.

References

1. D. Warin, C. Rostaing, Recent progress in Advanced Actinide recycling process, in *Informal Exchange Meeting on partitioning and transmutation, San Francisco, 2010* (NEA, 2010)
2. NEA, *Minor actinides burning in thermal reactors* (NEA, Paris, 2013)
3. NEA, *Homogeneous versus heterogeneous recycling of transuranics in fast nuclear reactors* (NEA, Paris, 2012)
4. J. Tommasi, S. Massara, S. Pillon, M. Rome, Minor actinides destruction in dedicated reactors, in *Informal Exchange Meeting on partitioning and transmutation, Mol, 1998* (NEA, 1998)
5. C. Prunier, F. Boussard, L. Koch, M. Coquerelle, Some specific aspects of homogeneous Am and Np based fuels

transmutation through the outcomes of the superfact experiment in PHENIX fast reactor, in *GLOBAL 1993, Seattle* (1993)

6. J. Tommasi, M. Delpech, J.P. Grouiller, A. Zaetta, Long-lived waste transmutation in reactors, Nucl. Technol. **1111**, 133 (1995)

7. L. Buiron et al., Transmutation abilities of the SFR low void effect core concept CFV 3600 MWth, in *ICAPP 2012, Chicago* (2012)

8. C. Chabert, C. Coquelet-Pascal, A. Saturnin, G. Mathonniere, B. Boullis, D. Warin, L.V.D. Durpel, M. Caron-Charles, C. Garzenne, Technical and economic assessment of different options for minor actinides transmutation: the French case, in *GLOBAL 2011, Tokyo* (2011)

9. K. Kawashima, K. Sugino, S. Ohki, T. Okubo, Design study of a low sodium void reactivity core to accomodate degraded TRU fuel, Nucl. Technol. **3**, 270 (2013)

10. G. Rimpault, The ERANOS code and data system for fast reactor neutronic analyses, in *PHYSOR, Seoul* (2002)

11. T. Wakabayashi, Improvement of core performance by introduction of moderators in a blanket region of fast reactors, Sci. Technol. Nucl. Ins. **103**, 879634 (2013)

6

Heterogeneous world model and collaborative scenarios of transition to globally sustainable nuclear energy systems

Vladimir Kuznetsov[*] and Galina Fesenko

International Atomic Energy Agency, Vienna International Centre, PO Box 100, 1400 Vienna, Austria

Abstract. The International Atomic Energy Agency's International Project on Innovative Nuclear Reactors and Fuel Cycles (INPRO) is to help ensure that nuclear energy is available to contribute to meeting global energy needs of the 21st century in a sustainable manner. The INPRO task titled "Global scenarios" is to develop global and regional nuclear energy scenarios that lead to a global vision of sustainable nuclear energy in the 21st century. Results of multiple studies show that the criteria for developing sustainable nuclear energy cannot be met without innovations in reactor and nuclear fuel cycle technologies. Combining different reactor types and associated fuel chains creates a multiplicity of nuclear energy system arrangements potentially contributing to global sustainability of nuclear energy. In this, cooperation among countries having different policy regarding fuel cycle back end would be essential to bring sustainability benefits from innovations in technology to all interested users. INPRO has developed heterogeneous global model to capture countries' different policies regarding the back end of the nuclear fuel cycle in regional and global scenarios of nuclear energy evolution and applied in a number of studies performed by participants of the project. This paper will highlight the model and major conclusions obtained in the studies.

1 Introduction

The International Atomic Energy Agency's (IAEA's) International Project on Innovative Nuclear Reactors and Fuel Cycles (INPRO) has the objective of helping to ensure that nuclear energy is available to contribute to meeting global energy needs of the 21st century in a sustainable manner. The INPRO task titled "Global scenarios" has the objective to develop, based on scientific and technical analysis, global and regional nuclear energy scenarios that lead to a global vision of sustainable nuclear energy in the 21st century [1–5].

Existing nuclear energy systems, which are almost entirely based on thermal reactors operating in a once-through cycle, will continue to be the main contributor to nuclear energy production for at least several more decades. However, results of multiple national and international studies show that the criteria for developing sustainable nuclear energy cannot be achieved without major innovations in reactor and nuclear fuel cycle technologies.

New reactors, nuclear fuels and fuel cycle technologies are under development and demonstration worldwide. Combining different reactor types and associated fuel chains creates a multiplicity of nuclear energy system arrangements potentially contributing to global sustainability of nuclear energy. In this, cooperation among countries having different policy regarding fuel cycle back end would be essential to bring sustainability benefits from innovations in technology to all interested users. It is becoming increasingly clear that national strategies will have to be harmonized with regional and global nuclear power architectures to make national nuclear energy systems more sustainable.

INPRO is a part of the integrated services of the IAEA provided to Member States considering initial development or expansion of nuclear energy programmes. To provide such countries with better understanding of the options available to achieve sustainable nuclear energy, INPRO has developed an internationally verified analytical framework for assessing transition scenarios to future sustainable nuclear energy systems (hereafter, the framework) and applied in a number of studies performed by participants of the project.

The economic studies carried out by INPRO have shown that investments in Research, Design & Demonstration (RD&D) for innovative technologies, such as fast reactors and a closed nuclear fuel cycle, are huge and provide reasonable pay-back times only in the case of a foreseen large scale deployment of such technologies. Not all of the countries interested in nuclear energy could and would afford such

*e-mail: V.Kuznetsov@iaea.org

investments. Then, benefits associated with innovative technologies can be amplified, and may also be brought to many interested users through mutually beneficial cooperation among countries in fuel cycle back end.

Reflecting upon this finding, the INPRO collaborative project on Global Architecture of Innovative Nuclear Energy Systems based on Thermal and Fast Reactors Including a Closed Fuel Cycle (GAINS) has developed heterogeneous global model to capture countries' different policies regarding the back end of the nuclear fuel cycle and to analyze cooperation options available thereof. The heterogeneous model may involve certain degrees of cooperation between groups of non-personified, non-geographical countries (synergistic case) or it may involve no cooperation (non-synergistic case). The heterogeneous world model is included in the framework to consider specific fuel cycle development strategies that different countries may pursue and examine a potential for mutually beneficial cooperation.

Synergies among the various existing and innovative nuclear energy technologies and options to amplify them through collaboration among countries in fuel cycle back end are being further examined in the INPRO collaborative project on Synergistic Nuclear Energy Regional Group Interactions Evaluated for Sustainability (SYNERGIES). This project is still ongoing; it is to be finalized in 2015.

2 INPRO collaborative project on global architecture of innovative nuclear energy systems with thermal and fast reactors and a closed nuclear fuel cycle (GAINS)

The INPRO collaborative project, GAINS addressed technical and highlighted some institutional issues to develop a global architecture for sustainable nuclear energy in the 21st century, and it also outlined plausible transitions to such architecture.

Sixteen participants from different regions of the world – Belgium, Canada, China, Czech Republic, France, India, Italy, Japan, Republic of Korea, Russian Federation, Slovakia, Spain, Ukraine, USA, European Commission (EC), plus Argentina as an observer, carried out coordinated investigations and contributed to the GAINS final report [1].

GAINS has developed an international analytical framework for assessing transition scenarios to future sustainable nuclear energy systems and conducted sample analyses, including [1]:

- A common methodological approach, including basic principles, assumptions, and boundary conditions.
- Storylines for nuclear power evolution and long-term nuclear energy demand scenarios based on IAEA Member States' high and low estimates for nuclear power demand until 2050, and expected trends until 2100 based on forecasts of international energy organizations.
- A heterogeneous world model comprised of groups of non-personified countries with different policies regarding the nuclear fuel cycle back end.

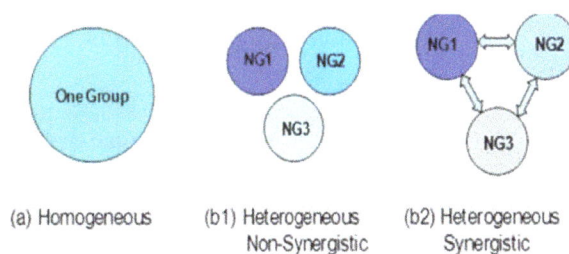

Fig. 1. Possible world models for fuel cycle analysis.

- Metrics and tools for the sustainability assessment of scenarios for a dynamic nuclear energy system, including a set of key indicators and evaluation parameters.
- An international database with best-estimate characteristics of existing and advanced nuclear reactors and associated nuclear fuel cycles required for material flow and economic analysis; this database extends other IAEA databases and takes into account preferences of different countries.

All previous studies of global nuclear energy scenarios, even those done region-wise [2], used the so-called homogeneous world model, wherein all countries in the world or a region were assumed to pursue the same policy regarding nuclear reactors and nuclear fuel cycle and use the same facilities at a given time. Different from that, GAINS has introduced a model of the heterogeneous world comprising different nuclear strategy groups of countries non-personified, non-geographical (NG) based on the spent nuclear fuel management strategy being pursued for the back end of the nuclear fuel cycle (Fig. 1).

For the purpose of GAINS analysis, three country groups (NGs) were defined as follows: NG1 recycles spent nuclear fuel and pursues a fast reactor programme; NG2 directly disposes of spent fuel or sends it for reprocessing to NG1; and NG3 sends spent nuclear fuel to NG1 or NG2. The methodology applied in the analysis does not assign individual countries to groups, but allocates a fraction of future global nuclear energy generation to each group as a function of time to explore "what if" scenarios. For the GAINS studies the NG1:NG2:NG3 ratio was fixed at 40:40:20 allowing further sensitivity analysis to variations of the NG fractions. In this, two alternative scenarios of nuclear power growth were considered in GAINS ending at 2500 GW(e) and 5000 GW(e) by the century end.

The GAINS metrics is presented in Table 1. It reflects sustainability areas related to power production, nuclear material resources, discharged fuel, radioactive waste and minor actinides, fuel cycle services, system safety and costs, and investment.

Innovative reactors expected to have a major impact on the future nuclear energy system architecture include advanced light water reactors (ALWRs), advanced heavy water reactors (AHWRs), high temperature reactors (HTRs), fast reactors (FRs), and potentially, accelerator driven systems (ADSs) and/or molten salt reactors (MSRs). Combining the different reactor types and associated fuel chains creates a multiplicity of nuclear energy system arrangements aimed at solving specific goals,

Table 1. GAINS key indicators and evaluation parameters [1].

No.	Key indicators and Evaluation Parameters			Resource Sustainability	Waste Management and Environmental Stressors	Safety	Proliferation Resistance and Physical Protection	Economics	Infrastructure
	Color coding indicative of relative uncertainty level in estimating specific quantitative values for future NES (can vary based on a particular scenario)		Low						
			Medium-low						
			Medium-high						
			High						
Power Production									
KI-1	Nuclear power production capacity by reactor type								X
EP-1.1	(a) Commissioning and (b) decommissioning rates				X				X
Nuclear Material Resources									
KI-2	Average net energy produced per unit mass of natural uranium			X	X				
EP-2.1	Cumulative demand of natural nuclear material, i.e. (a) natural uranium and (b) thorium			X	X				
KI-3	Direct use material inventories per unit energy generated (Cumulative absolute quantities can be shown as EP-3.1)			X			X		X
Discharged Fuel[3]									
KI-4	Discharged fuel inventories per unit energy generated (Cumulative absolute quantities can be shown as EP-4.1)				X				X
Radioactive Waste and Minor Actinides									
KI-5	Radioactive waste inventories per unit energy generated[4] (Cumulative absolute quantities can be shown as EP-5.3)				X				X
EP-5.1	(a) radiotoxicity and (b) decay heat of waste, including discharged fuel destined for disposal				X				X
EP-5.2	Minor actinide inventories per unit energy generated				X				X
Fuel Cycle Services									
KI-6	(a) Uranium enrichment and (b) fuel reprocessing capacity, both normalized per unit of nuclear power production capacity						X		X
KI-7	Annual quantities of fuel and waste material transported between groups				X		X		X
EP-7.1	Category of nuclear material transported between groups						X		
System Safety									
KI-8	Annual collective risk per unit energy generation					X			
Costs and Investment									
KI-9	Levellized unit of electricity cost (LUEC)							X	
EP-9.1	Overnight cost for Nth-of-a-kind reactor unit: (a) total and (b) specific (per unit capacity)							X	
KI-10	Estimated R&D investment in Nth-of-a-kind deployment							X	X
EP-10.1	Additional functions or benefits[5]							X	

such as production of various energy products, better use of natural resources, and minimization of radioactive waste.

Four types of nuclear energy system (NES) architecture were defined and then analyzed in GAINS to evaluate the effect of implementation of innovative technologies and their influence on the considered key indicators (KIs):

– Homogeneous "business-as-usual" (BAU) scenario based on pressurized water reactors (PWRs) (94% of power generation) and heavy water reactors (HWRs) (6%) operated in a once-through fuel cycle in which the world was modelled as a single NG. A variant of this scenario included the introduction of an advanced PWR replacing conventional PWR technology (named the "BAU+" scenario).

– Homogeneous (single group) scenario for a closed cycle using thermal and fast reactors to be compared with the above mentioned scenarios. Some of these fuel-recycle scenarios included HWRs (6%) operated in a once-through mode.

– A hybrid heterogeneous-architecture scenario comprising a once-through fuel cycle strategy in NG2, a closed fuel cycle strategy in NG1 and use of thermal reactors in a once-through mode in NG3. Both synergistic and non-synergistic cases were analyzed for this scenario. In the synergistic case, NG3 receives fresh fuel from NG2 and

NG1 and returns the associated spent nuclear fuel to those groups.
- Other innovative NES scenarios in the homogeneous world model, including: (a) operation of fast-spectrum reactors or thermal-spectrum HWRs using thorium fuel cycle for the reduction of natural uranium consumption; (b) reduction of minor actinides (MAs) using accelerator driven systems (ADSs) or molten salt reactors (MSRs), and other innovative NES scenarios.

The framework measures the transition from an existing to a future sustainable nuclear energy system by the degree to which the selected targets (e.g. minimized waste, minimized amounts of direct use materials in storage, or minimized natural resource depletion, see Table 1) are approached in particular evolution scenarios. The KIs are compared to determine the more promising options for achieving the selected targets. Possible benefits and issues of different options could also be analyzed.

The framework developed in GAINS is based on the participants' experiences in implementing similar studies at national and international levels. The framework can be used for developing national nuclear energy strategies, exploring opportunities for cooperation or partnerships with other countries in nuclear fuel cycle back end, also highlighting how global trends may affect national developments. Individual countries can make use of this framework with their own national and regional data to evaluate particular approaches in a global or regional context.

3 INPRO collaborative project on synergistic nuclear energy regional group interactions evaluated for sustainability (SYNERGIES)

The ongoing collaborative project SYNERGIES [5] was started in 2012 with Algeria, Armenia, Belarus, Belgium, Bulgaria, Canada, China, Egypt, France, India, Indonesia, Israel, Italy, Japan, Republic of Korea, Malaysia, OECD-NEA, Pakistan, Poland, Romania, Russian Federation, Spain, Ukraine, USA and Vietnam as participants or observers.

The SYNERGIES project applies and amends the analytical framework developed in GAINS to examine more specifically the various forms of regional collaboration among nuclear energy suppliers and users. In particular, a database of best estimate cost data for each step of the nuclear fuel cycle and each component of the levelized unit electricity cost for nuclear reactors has been compiled and is being maintained with the project [6].

Synergies among the various existing and innovative nuclear energy technologies and options to amplify them through collaboration among countries in fuel cycle back end are being examined in SYNERGIES through case studies performed by the project participants. The project focuses on short- and medium-term collaborative actions that can help developing pathways to long-term NES sustainability.

To meet its objectives, the SYNERGIES project investigates sustainability indicators of a dynamic NES, including a variety of technologies and infrastructure-

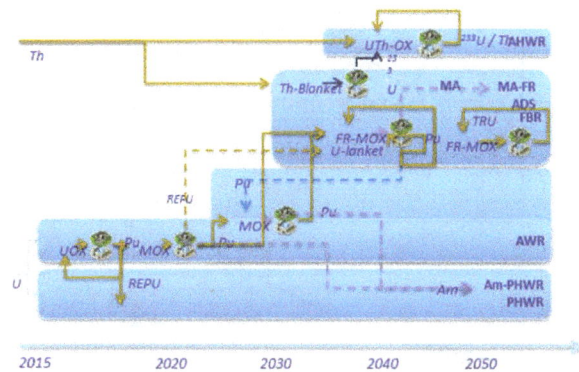

Fig. 2. Scenario families in the SYNERGIES project.

related factors, as well as the collaborative scenarios and architectures of interest to participants, involving, inter alia, fuel cycle infrastructure development with shared facilities.

Within SYNERGIES the focus is on regional studies of collaboration among countries in line with the agreed upon overall picture of the global nuclear energy system evolution in the 21st century. Summaries of 27 case studies performed by the participants are grouped in families of scenarios as follows, see Figure 2:

- Business-as-usual scenarios and scenarios with mono-recycling of U/Pu in thermal-spectrum reactors.
- Scenarios with the introduction of a number of fast reactors to support multi-recycling of Pu in light water reactors (LWRs) and fast reactors.
- Fast reactor centered scenarios — scenarios with reprocessing of thermal reactors' fuel to enable noticeable growth rate of fast reactor capacity.
- Scenarios of transition to Th/^{233}U fuel cycle and scenarios with U/Pu/Th fuel cycles.

The SYNERGIES project explores the various issues related to synergies in technology and synergistic collaborations among countries, including selection of reactor and fuel cycle options, uncertainties in the scale of nuclear energy demand growth, possible modes of collaboration among countries, the sensitivity studies of possible impacts to the market shares of countries with different nuclear fuel cycle policy and to the scale of collaboration among countries, etc.

4 Major findings and conclusions of the GAINS and SYNERGIES collaborative projects

Major findings and conclusions of the GAINS and SYNERGIES collaborative projects are as follows [1,5,7]:

- The dynamics of world's nuclear power capacity expansion indicates that in all cases low projections are more likely to meet the reality compared to the high ones.
- The sensitivity studies to the shares of country groups with different policy regarding nuclear fuel cycle back end (NG) taking into account possible synergistic

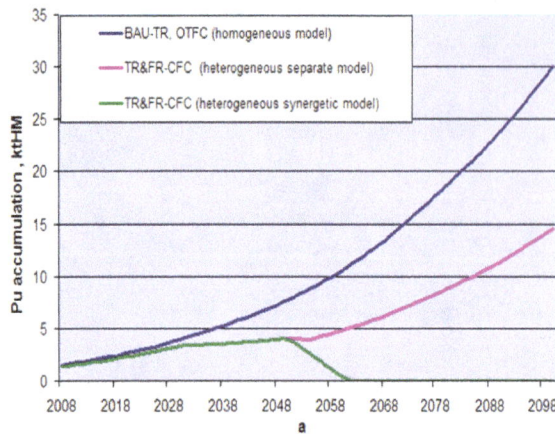

Fig. 3. Plutonium in short-term cooled spent nuclear fuel for the moderate GAINS scenario [1].

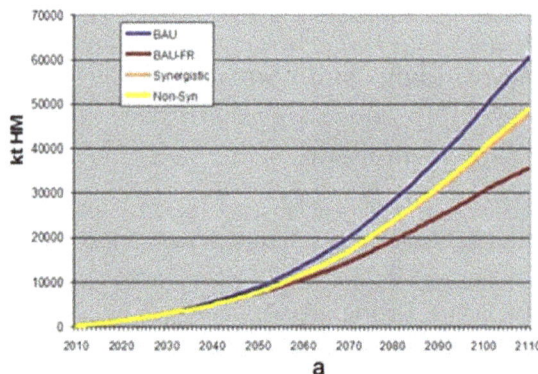

Fig. 5. Long-term spent fuel storage requirements versus time for NG1 group of countries in one of the scenarios considered in the SYNERGIES project (green colour corresponds to spent nuclear fuel imported from NG2 countries and not reprocessed because of the insufficient reprocessing capacity in NG1).

collaborations among countries indicate that LWRs will retain their position as the larger part of the overall reactor park all throughout the 21st century.

– In the present century, global nuclear energy is likely to follow a heterogeneous world model, within which most of the countries will continue to use thermal reactors in a once-through nuclear fuel cycle.

– Cooperation among countries could amplify the positive effects of technology innovation in achieving sustainable nuclear energy.

– The global fleet of fast reactors could be doubled in the synergistic case compared to the non-synergistic case; this would reduce accumulation of the discharged LWR spent fuel. This can also be of interest with respect to uranium resource savings and plutonium management options, see Figure 3.

– Natural uranium savings up to 20–40% could be achieved in heterogeneous world with synergistic collaboration among countries (NG1 countries could deploy more fast and less thermal reactors at the expense of U–Pu extracted from spent nuclear fuel of the NG3 or NG3 + NG2 countries), see Figure 4.

– The NG1 (recycling group) power demand as well as reprocessing capacity is critical for the fast reactor

introduction rate and for the capability of NG1 to reprocess all spent fuel from other NGs (once-through fuel cycle groups), see Figure 5.

– Sharing of the reprocessing facilities contributes to a reduction of the cumulative expenditures for spent nuclear fuel reprocessing; however, adequate evaluation of the resulting benefits for future generation requires an analysis performed in terms of cash flows without a discount rate, see Figure 6.

– Simulations of a transition to sustainable nuclear energy systems at national, regional, and global levels have become an essential part of the scientific work that supports the decision making process on national nuclear power programmes. To support this activity from an international perspective, the IAEA's INPRO Section provides online training sessions and workshops on nuclear energy sustainability and INPRO's activities for students at all levels, as well as faculty and research

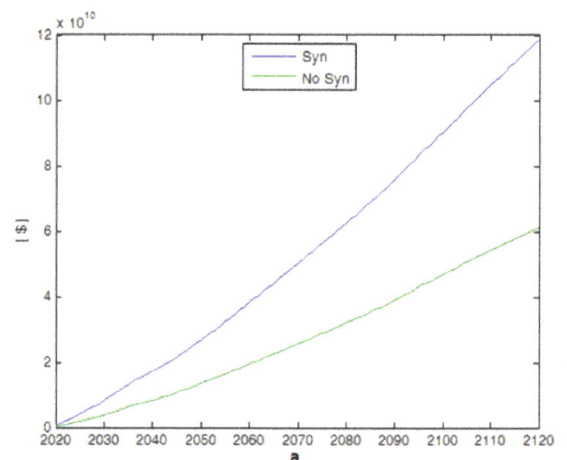

Fig. 4. Cumulative natural uranium consumption versus time in different GAINS scenarios.

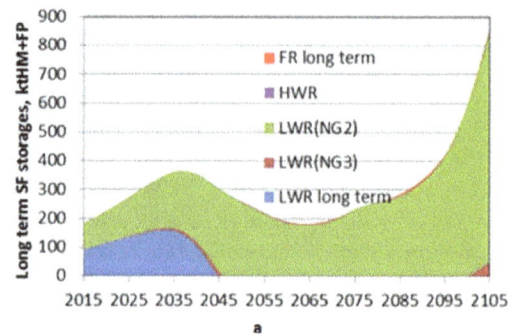

Fig. 6. Cumulative reprocessing expenditures versus time for synergistic and non-synergistic cases in one of the scenarios of the SYNERGIES project (NG1 and NG2 synergies were explored in that study).

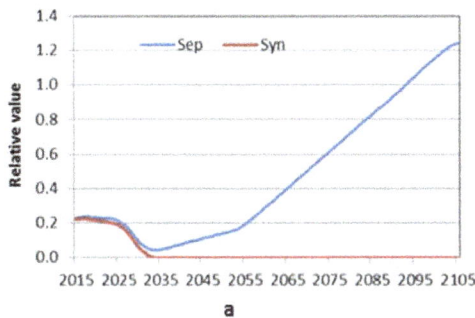

Fig. 7. Long-term spent fuel storage volume requirements versus time for synergistic (Syn) and non-synergistic (Sep) cases in one of the scenarios of the SYNERGIES project (NG1 and NG2 synergies were explored in that study).

staff of nuclear universities and research centres in interested Member States [7]. A web-based conferencing service facilitates lecturing from the IAEA to audiences in different Member States.

- Countries that do not pursue fast reactor programmes could benefit from the synergistic approach as it results in reduced requirements to long-term spent nuclear fuel storage and ultimate disposal of waste, see Figure 7. However, there are a number of important legal and institutional impediments for cooperation among countries in nuclear fuel cycle back end [8], those will be addressed in more detail in the future INPRO activity titled "Cooperative approaches to the back end of nuclear fuel cycle: drivers and legal, institutional and financial impediments".
- Achieving synergistic NFC backend architectures requires industrial, public and political consensus. For timely global answers to global challenges, building of the architecture has to be started straight away.
- Scenarios with the introduction of a limited number of fast reactors to support multi-recycling of plutonium in LWRs and in fast reactors could be a flexible and risk-balanced option under uncertainties in the scale of demand for nuclear energy and before fast reactors are proven to be reliable and competitive source of energy with a potential of broad deployment, see Figure 8; upon recommendations from participants of the SYNERGIES project such scenarios will be examined in more detail in future INPRO projects.

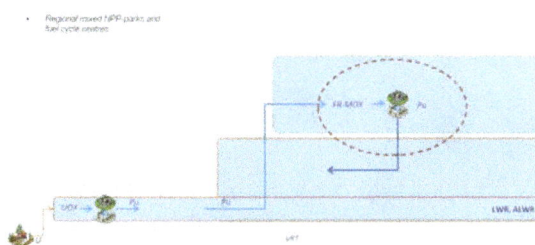

Fig. 8. Scenarios with LWRs and a limited number of fast reactors for Pu multi-recycling.

5 Conclusions

This paper summarizes the major findings and conclusions of the INPRO collaborative projects, GAINS and SYNERGIES, that performed studies related to the role of global and regional architectures of nuclear energy systems in making a transition to future sustainable nuclear energy, in terms of the assurance of sufficient nuclear material resources, minimized inventories of spent nuclear fuel and high-level radioactive waste and overcoming of the investment barriers to commercial introduction of innovative nuclear technologies.

The completed GAINS project provided IAEA Member States with the analytical framework to help explore transition scenarios to future globally sustainable nuclear energy systems that would combine the synergy of nuclear technologies with innovative institutional approaches to foster collaboration among countries to amplify the benefits of the innovation. The ongoing SYNERGIES project applies and amends this framework to explore the various issues related to synergies in technology and synergistic collaborations among countries.

The outputs of the INPRO collaborative projects on GAINS and SYNERGIES clearly indicate that the criteria for developing sustainable nuclear energy cannot be achieved without major innovations in reactor and nuclear fuel cycle technologies. Cooperation among countries could then amplify the positive effects of technology innovation in achieving sustainable nuclear energy for all interested users. Collaborative solutions in nuclear fuel cycle and, specifically, in the fuel cycle back end are a key for moving toward global sustainability of nuclear energy systems from the near (2012–2030) through the medium (2030–2050) toward the long (2050–2100) term.

Both projects indicate that, to pursue sustainability goals efficiently, national strategies would need to be harmonized with regional and global nuclear energy architectures.

The authors would like to express their gratitude to all participants of the GAINS [1] and SYNERGIES [5] collaborative projects for their contribution to the development and application of the framework for the analysis and assessment of dynamic nuclear energy systems for sustainability.

References

1. International Atomic Energy Agency, *Framework for assessing dynamic nuclear energy systems for sustainability, final report of the INPRO collaborative project on global architectures of innovative nuclear energy systems with thermal and fast reactors and a closed nuclear fuel cycle (GAINS)*, (IAEA Nuclear Energy Series NP-T-1.14, 2013)
2. International Atomic Energy Agency, *Nuclear energy development in the 21st century: global scenarios and regional trends*, (IAEA Nuclear Energy Series No. NP-T-1.8, Vienna, 2010): http://www-pub.iaea.org/MTCD/publications/PDF/Pub1476_web.pdf
3. International Atomic Energy Agency, *The role of thorium to supplement fuel cycles of future nuclear energy systems*, (IAEA Nuclear Energy Series, NF-T-2.4, IAEA, Vienna, 2011)

4. International Atomic Energy Agency, *Joint study: assessment of nuclear energy systems based on a closed nuclear fuel cycle with fast reactors (CNFC-FR)*, IAEA-TECDOC-1639 (IAEA, Vienna, 2009): http://www-pub.iaea.org/MTCD/Publications/PDF/te_1639_web.pdf

5. International Atomic Energy Agency, *Web page of the SYNERGIES collaborative project*: http://www.iaea.org/INPRO/CPs/SYNERGIES/index.html

6. Economic input data for SYNERGIES, *Web page of the SYNERGIES collaborative project*: http://www.iaea.org/INPRO/CPs/SYNERGIES/rev4_4_2_Economic_assessment_method_data.pdf

7. INPRO provides training on nuclear energy sustainability, *INPRO web-page*: https://www.iaea.org/INPRO/News/2015-05-22-inpro.html

8. Drivers and Impediments for Regional Cooperation on the Way to Sustainable Nuclear Energy Systems, *Materials of the 4th INPRO Dialogue Forum, 30 July-3 August 2012*, (IAEA, Vienna): http://www.iaea.org/INPRO/4th_Dialogue_Forum/index.html

7

Statistical model of global uranium resources and long-term availability

Antoine Monnet[1*], Sophie Gabriel[1], and Jacques Percebois[2]

[1] French Alternative Energies and Atomic Energy Commission, I-tésé, CEA/DEN, Université Paris Saclay, 91191 Gif-sur-Yvette, France

[2] Université Montpellier 1–UFR d'Économie–CREDEN (Art-Dev UMR CNRS 5281), Avenue Raymond Dugrand, CS 79606, 34960 Montpellier, France

Abstract. Most recent studies on the long-term supply of uranium make simplistic assumptions on the available resources and their production costs. Some consider the whole uranium quantities in the Earth's crust and then estimate the production costs based on the ore grade only, disregarding the size of ore bodies and the mining techniques. Other studies consider the resources reported by countries for a given cost category, disregarding undiscovered or unreported quantities. In both cases, the resource estimations are sorted following a cost merit order. In this paper, we describe a methodology based on "geological environments". It provides a more detailed resource estimation and it is more flexible regarding cost modelling. The global uranium resource estimation introduced in this paper results from the sum of independent resource estimations from different geological environments. A geological environment is defined by its own geographical boundaries, resource dispersion (average grade and size of ore bodies and their variance), and cost function. With this definition, uranium resources are considered within ore bodies. The deposit breakdown of resources is modelled using a bivariate statistical approach where size and grade are the two random variables. This makes resource estimates possible for individual projects. Adding up all geological environments provides a repartition of all Earth's crust resources in which ore bodies are sorted by size and grade. This subset-based estimation is convenient to model specific cost structures.

1 Long-term cumulative supply curves (LTCS)

The availability of natural uranium will have a direct impact on the global capability to build new nuclear reactors in the coming decades as it is forecasted that Light Water Reactors (LWRs) will remain the main nuclear technology for most of the 21st century [1,2]. The cost associated with this availability is also important. Even though its share in the electricity production cost is relatively low, it may influence the choice of fuel cycle options in the short term or the choice of reactor technologies in the long term.

1.1 Concepts and objectives

Considering natural uranium as any other mineral commodity, academics in mineral economics and decision makers in mining industries usually look at availability by the mean of two analytical tools. The first one is generally called cash-cost curve. This curve consists in plotting the cumulated production capacity (tU/year) of all known production capacities, either running mines or short-term projects, against the unit production cost ($/kgU) of those mines once they have been sorted by cost merit order. This tool essentially helps analyzing short-term to medium-term availability issues, i.e. from a couple of years to a decade or two.

Since the objective of this research is to analyze the adequacy of uranium supply to long-term demand, another tool was preferred as it suits availability problems with implications over several decades. This tool is the long-term cumulative supply curve (LTCS). It was made popular by Tilton et al. [3,4] in 1987. The curve depicts the cumulated amount (tU) of all known resources, eventually adding estimates of undiscovered resources, after they have been

* e-mail: `antoine.monnet@cea.fr`

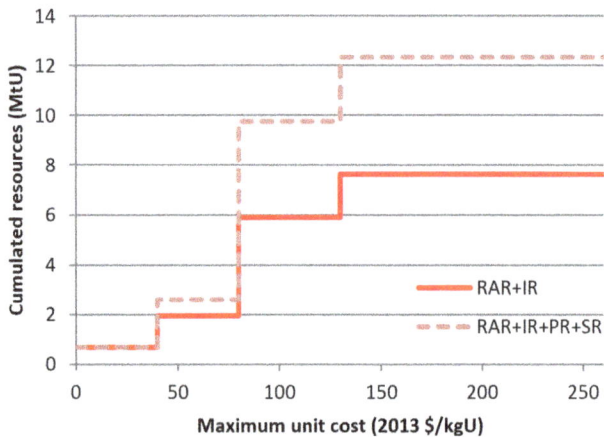

Fig. 1. Long-term supply curve built from the 2014 Red Book data [5].

sorted by rising unit production cost ($/kgU). Unlike cash-cost curve, there is no time dimension in the LTCS curve. In order to assess the adequacy of supply to demand over time, one would need to compare the LTCS curve with a time-dependent demand scenario. In this paper, the stress is put on the method used to build the LTCS curve.

1.2 Aggregated LTCS curve

The easiest way to build a LTCS curve is to aggregate existing data of cumulated resources and associated production costs published in the literature or in technical reports. Focusing on uranium, this can be achieved by gathering the resources declared by countries in the IAEA/ OECD-NEA biennial report called the Red Book [5]. The result is shown in Figure 1 for the aggregation of total known resources (Reasonably Assured Resources [RAR] and Inferred Resources [IR], red curve, and for total known and prognosticated resources [RAR + IR + Prognosticated Resources (PR) + Speculative Resources (SR)], light-red curve).

1.3 Limits of the aggregation approach

The aggregation approach to build LTCS curves is convenient provided that consistent data are available. Conversely, it can be criticized due to the aggregation of different levels of uncertainty in the example of the Red Book data. By definition, the amount and the cost of prognosticated or speculative resources are more uncertain than known resources (RAR or IR) to which they were added in the light-red curve (Fig. 1). While the analysis is usually performed by assuming that cheaper resources are extracted first, there is no guarantee that undiscovered resources between 40 and 80 $/kgU will all be discovered before RAR at below 80 $/kgU are exhausted. Conversely, if one only considers known resources (red curve, Fig. 1), it is likely that some resources at below 80 $/kgU that are not known at present will be discovered in the long term.

Finally, using aggregated data to perform analysis on LTCS curves has two limits. First, when data are incomplete, long-term resources are underestimated. Second, when data are over-aggregated, short-term resources may be overestimated while the long-term is affected by a growing uncertainty on costs. This appears on the upper part of the light-red curve for which 3 MtU of SR are missing since they have no cost estimate reported in the Red Book. These limits prompted some academics to develop alternative methods to build LTCS curve.

2 Global elastic crustal abundance models

To avoid aggregating estimates with different cost and amount uncertainties, some recent studies, mainly conducted by Schneider from University of Texas and Matthews and Driscoll from MIT [6–8], model the costs and quantities of resources of the entire Earth's crust with the same methodology. They introduce a 3-step method to build LTCS curves:

- first, they model the link between the quantity (cumulated amount) and the quality (represented by ore grade) of resources;
- second, they model the link between the unit production cost and the quality of resources;
- finally, they infer from the first two steps the general relation between cumulated amounts of resources and associated costs.

From this framework, the elastic crustal abundance model provides a LTCS curve for the entire world.

2.1 Step 1: quantity-quality relationship

The authors introduce a power relationship between the grade g and the cumulated amount of metal q according to equation (1). This results in an elastic relationship in log-scale where α is the elasticity of quantities in relation to grades and where q_0 and g_0 are calibration parameters.

$$\frac{q}{q_0} = \left(\frac{g_0}{g}\right)^{\alpha}. \tag{1}$$

As explained in the MIT study [8], an empirical relationship between cumulated uranium resources and ore grades is used to estimate α. This empirical relationship was established in 1979 by Deffeyes and Macgregor [9]. It is a well-known bell-shape relationship, as depicted in Figure 2. In the high-grade range (10^2–10^4 ppmU), the bell-shape curve is approximated by its slope denoted by α in equation (1).

2.2 Step 2: cost-quality relationship

The second relationship (Eq. (2)) introduced by the authors is also a power-relation. This time, β represents the elasticity of unit costs in relation to grades; g_0 and c_0 are calibration parameters.

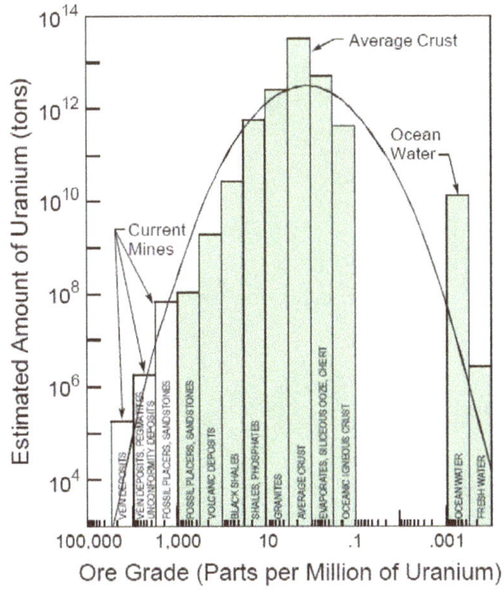

Fig. 2. Empirical bell-shape relationship between cumulated uranium resources and ore grades [9].

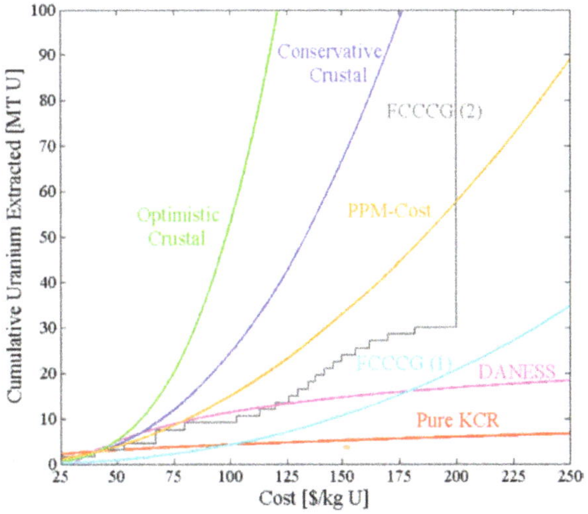

Fig. 3. Long-term cumulative supply curves for different versions of the elastic crustal abundance model [6].

$$\frac{g}{g_0} = \left(\frac{c_0}{c}\right)^{\beta}. \qquad (2)$$

Different versions of this relationship can be found in the literature. While Schneider makes the simple assumption that $\beta = 1$ before looking at sensitivity, the MIT study introduces a more complex expression of β to take account of learning effects in addition to economies of scale. Finally, different versions of the relationship can be found depending on the value of β, either imposed or fitted. A number of them are gathered in Schneider and Sailor paper [6].

2.3 Step 3: cost-quantity model

Once the previous two relations are defined, step 3 derives the cost-quantity relationship from equations (1) and (2), according to equation (3).

$$q = q_0 \left(\frac{c}{c_0}\right)^{\alpha\beta}. \qquad (3)$$

In this formula, the product denoted by $\alpha\beta$ can be interpreted as the global elasticity of supply to unit costs of production. The LTCS curve is finally obtained by plotting the relationship of equation (3), once all parameters have been fitted or calibrated. Figure 3 shows the LTCS curves presented by Schneider for different versions of the previous framework[1].

---

[1] To be more correct, FCCCG(2) and DANESS models differ from the elastic crustal abundance model. These specific characteristics are not covered by this paper.

2.4 Limits of the elastic crustal abundance models

At this stage, several shortcomings can be raised against the framework proposed by Schneider, Matthews and Driscoll. First, the results are sensitive to calibration (Sect. 2.4.1). Second, only one intrinsic parameter of the resource, i.e. its grade, is used to determine both the geological availability (Eq. (1)) (Sect. 2.4.2) and the economic value of the resource (Eq. (2)) (Sect. 2.4.3).

2.4.1 Sensitivity to calibration (Eq. (3))

The final equation (Eq. (3)) for the LTCS curve requires a calibration point denoted by (q_0, c_0). Although Schneider investigates the sensitivity of $\alpha\beta$ through different versions of his model (Fig. 3), the sensitivity to calibration is not covered. This paper conducts this sensitivity analysis according to the following methodology.

The cumulative resources (q_0) and the corresponding cost limits (c_0) were taken from various editions of the Red Book. To run the following sensitivity tests, the version of the elastic crustal abundance model that was used is Schneider's 'optimistic crustal' model ($\alpha\beta = 3.32$). Table 1 presents the different calibration points that were considered and Figure 4 shows the resulting sensitivity.

Figure 4 shows how the choice of the calibration points affects the LTCS curve.

2.4.2 Limits to quantity-grade relationship (Eq. (1))

In the late 1970s, Deffeyes and Macgregor [9] reported imperfections in the bell-shape distribution of the grades. They noted that in the case of chromium, but also uranium, certain high grades can be overrepresented compared to the theoretical model, as shown in Figure 5.

Deffeyes explained this kind of bimodal distribution by particular forms of mineralization. These would be formed by a different sequence of independent phenomena

Table 1. Calibration points (c_0, q_0).

Red Book edition	RAR (MtU)	Identified (RAR + IR) (MtU)	Identified + Undiscovered (RAR + IR + SR + PR) (MtU)
2003	3.2 < 130 \$/kgU	2.523 < 40 \$/kgU (Schneider's ref.) 4.6 < 130 \$/kgU	14.4
2007	3.3 < 130 \$/kgU	2.97 < 40 \$/kgU 4.5 < 80 \$/kgU 5.4 < 130 \$/kgU (MIT ref. for Identified)	15.9 (MIT ref. for Identified + Undiscovered)
2009	4.0 < 260 \$/kgU	0.8 < 40 \$/kgU 3.7 < 80 \$/kgU 5.4 < 130 \$/kgU 6.3 < 260 \$/kgU	16.7

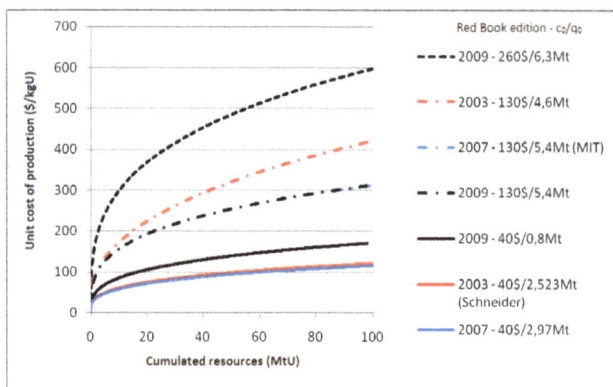

Fig. 4. Sensitivity of the elastic crustal abundance model in relation to calibration points.

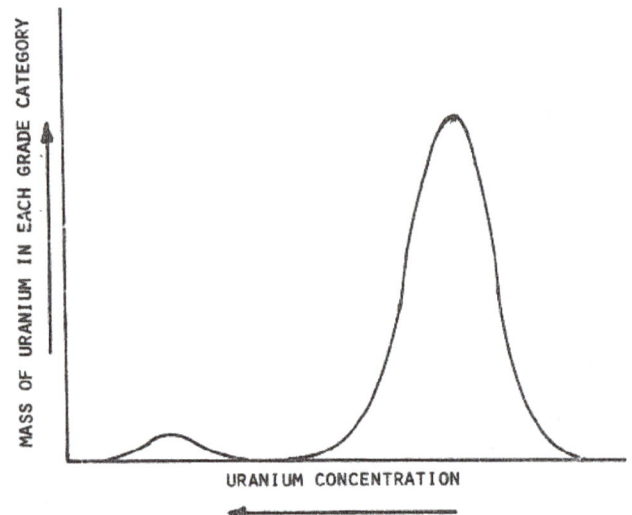

Fig. 5. Bimodal relationship between cumulated uranium resources and ore grades [9].

compared to the sequence of the main distribution and result in a separate distribution. This point is important since it was shortly after Deffeyes' publications that the main very high-grade deposits of Saskatchewan in Canada were discovered (Cigar Lake in 1981, McArthur River in 1988). The inclusion of these deposits in the diagram of Figure 2 invalidates the bell-shape model used in Schneider's and Matthews' methods.

2.4.3 Limits to the cost-grade relationship (Eq. (2))

Apart from scale effects, considering the unit cost of production as only a function of grade can be opened to criticism. Today, some running uranium mines, which must have similar total production costs to be competitive in the current market, have substantially different grades [10]:

– Cigar Lake, Canada (underground, 14.4% U, \$23/lb U_3O_8 nominal operating cost);
– South Inkai, Kazakhstan (in situ leaching, 0.01% U, \$22/lb U_3O_8 nominal operating cost).

Conversely, some projects of similar grades may have quite different production costs [10]. In the following example, the production cost for in situ leaching is mainly

operating cost, whereas for open pit, capital costs cannot be omitted:

– Carley Bore, Australia (in situ leaching, 0.03% U, \$20/lb U_3O_8 nominal operating cost);
– Letlhakane, Botswana (open pit, 0.02% U, \$58/lb U_3O_8 nominal operating cost).

As a consequence of the limits of the two previous relationships, the outputs of the model are not robust: as suggested in Figure 3, different values for the elasticity parameters $\alpha\beta$ can change the output significantly and no acceptable conclusion on available resources can be found.

While grade is certainly an important factor in the cost of a resource, there are other parameters that govern cost and it may be desirable to model them. These include the size of ore bodies and the geochemical nature of deposits. Any change in these parameters can lead to specific mining techniques and therefore specific costs. When a deposit is located in a given country with specific legislation, taxes and royalties can also be taken into account through the cost.

Thus, if the cost function keeps a limited number of parameters, it can be more realistic to calibrate each deposit category or geological environment individually (one calibration for Canadian underground mines, one calibration for Australian ISL mines, etc.).

3 A statistical approach based on geological environments

To overcome the limits of previous models, this paper proposes a statistical approach that differs on three points from the elastic crustal abundance models:

- geological availability and production costs are estimated by a bivariate model. The two variables are grade (mean grade of a deposit, denoted g) and tonnage (ore tonnage of a deposit, denoted t);
- the scope of the model is split to several regional crustal abundance estimations. These regions are called geological environments[2]. A geological environment is defined by its own geographical boundaries, resource dispersion (average grade and size of ore bodies and their variance), and cost function;
- a statistical approach is adopted. Variables g and t are treated as random variables and their probability density functions (pdfs) serve to build the corresponding relationship.

Section 3.1 briefly presents former geostatistical models, which have been applied to uranium endowment and share the same framework as the one developed in this article. Then Sections 3.2 to 3.4 describe the methodology step-by-step.

3.1 Former geostatistical models

Several bivariate or multi-variate statistical models for crustal abundance and associated costs can be found in the literature. Their objectives are the same as in Section 2 but rather than proceeding to the economic appraisal of cumulative quantities, statistical models proceed to the economic appraisal at a deposit level and then add up all the resources of deposits. The benefit of this approach is that models can be specific to each geological environment.

Among the models available in the literature, three have been applied to uranium endowment estimation. They were developed by Drew [11], Harris et al. [12–15] and Brinck [16–18]. None of them served to build a complete LTCS

[2]This terminology was first used in Drew [11]. It is convenient because the model produces an assessment of geological resources rather than reserves within the environment. Yet, the meaning of "geological" can be confusing. The boundaries do not aim to circle a single geological structure but rather groups of structures that share a maximum of common properties (types, size, grade of known deposits and also economic, political conditions) compared with other environments (e.g. US groups of deposits vs. Canada, Australia, Africa or Kazakhstan).

curve (rather they served to estimate the undiscovered resources at below a given cost, i.e. the price of U_3O_8 at the time of the studies), but some parts inspired the model developed in this paper. The general framework can be described in three parts which differ a little from the three steps described in Section 2 :

- For a specific environment, the geological abundance q can be defined using a constant q_0, the total metal endowment of the geological environment, and a probability density function $f(g,t)$ (Eq. (4)):

$$q = q_0 \iint f(g,t)dgdt. \tag{4}$$

q_0 is estimated from the mass of rock M in the geological environment and the mean grade of the crust (clarke) ($q_0 = M \times clarke$). It should be noticed that this q_0 has no embedded consideration about economics nor technical recovery, unlike the calibration values used in Section 2.3.

q is derived from the statistics of g and t among the known deposits of a given geological environment. Since these statistics are biased (high-grade and high-tonnage deposits tend to be first discovered), a specific method is required to derive the unbiased function $f(g,t)$. This method is based on economic filtering.

- The second part consists in a cost model which is similar to that of elastic crustal abundance models, except costs are estimated at a deposit level and ore tonnage is taken into account. The resulting cost-grade-tonnage relationship is of the form described by equation (5), which can be also written as in equation (6) with $x = \ln(g)$, $y = \ln(t)$ and A a constant.

$$c(g,t) = c_0\left(\frac{g}{g_0}\right)^{\beta_g}\left(\frac{t}{t_0}\right)^{\beta_t} \tag{5}$$

$$\ln(c(g,t)) - A = \beta_g x + \beta_t y. \tag{6}$$

- In part 3, Drew proposes to compute the cumulated metal resources available at below a given unit production cost C_1 by using to intermediate calculations: the numerical computation of N, the total number of deposits in the environment (the total mass of rock M divided by the mean tonnage of all deposits), and $m(C_1)$, the mean metal content of deposits that are "cheaper than C_1". Equations (7) and (8) give the analytical expressions of N and $m(C_1)$ in terms of statistical expectations.

$$N = \frac{M}{\iint_0^\infty tf(g,t)dgdt} \tag{7}$$

$$m(C_1) = \iint_{c(g,t)\leq C_1} gtf(g,t)dgdt. \tag{8}$$

Finally, the LTCS curve is built by plotting the function $C_1 \rightarrow N \times m(C_1)$.

In this paper, the numerical method used to derive the parameters of the unbiased function $f(g,t)$ (part 1) is inspired from Drew, except for the cost limit used by the economic filter (see Sect. 3.3.2). The general form of the cost-grade-tonnage relationship (part 2, Eq. (5)) is also inspired from Drew and Harris, but its calibration is a different procedure (see Sect. 3.3.1). Lastly, the numerical procedure used to compute the cumulated resources available at a given cost (part 3) is specific to this paper (see Sect. 3.4).

Apart from Harris, Brinck and Drew's models, a more advanced approach has been proposed by the United States Geological Survey (USGS): "Quantitative Mineral Resources Assessments" [19,20]. Although this methodology is often referred to as "3-part resource assessment", these parts are not exactly the same as the three parts of our general framework. Neither are the objectives: within a given geological environment (e.g. United States), tracts are delineated (e.g. a sandstone basin in New Mexico) and mapped data available on these tracts are analysed in order to find similarities with unexplored or less explored tracts. The output is not only an estimation of undiscovered resources but also the density and target location of undiscovered deposits. This localization dimension is missing in our approach since it is not in the scope of this research, without mentioning the difficulty to gather consistent and extensive mapped data for grade and tonnage over large areas such as geological environments.

3.2 Part 1: abundance model

3.2.1 Log-normal distribution of grade and tonnage

The purpose of part 1 is to characterize the density function f, i.e. the statistical distribution of grade and tonnage among the deposits of the geological environment being considered. It is common, although sometimes criticized, to assume that f follows a bivariate log-normal distribution [13]. (Since g and t follow log-normal distributions, $x = \ln(g)$ and $y = \ln(t)$ follow normal distributions.) This assumption is shared with Harris, Brinck and Drew's models. It leads to the mathematical form described by equation (9), provided that grade and tonnage are independent random variables[3].

$$f(g,t) = \frac{\exp\left(-\frac{(\ln g - \mu_x)^2}{2\sigma_x^2} - \frac{(\ln t - \mu_y)^2}{2\sigma_y^2}\right)}{2\pi g t \sigma_x \sigma_y}, \quad (9)$$

where μ_x, σ_x^2 and μ_y, σ_y^2 are the means and variances of x and y respectively.

The most technical part of part 1 is to estimate those parameters from statistical data on known deposits. In

descriptive statistics, mean and variance are computed according to equations (10) and (11).

$$\bar{x} = \frac{1}{n}\sum_{k=1}^{n} x_k \quad (10)$$

$$s_x^2 = \frac{1}{n-1}\sum_{k=1}^{n}(x_k - \bar{x})^2. \quad (11)$$

If deposits were randomly sampled and n large enough, equations (10) and (11) would be the best estimators of μ_x and σ_x^2 (μ_y and σ_y^2 respectively) ($\bar{x} \simeq \mu_x$ et $s_x^2 \simeq \sigma_x^2$). Unfortunately, deposits are not randomly sampled. Rather, the richer (high grade, high tonnage) raise economic interest first.

3.2.2 Economic filter and procedure to estimate the parameters of the unbiased distribution

The procedure used in this paper is derived from Drew [11,14]. Harris and Drew propose similar procedures to correct for the sampling bias that affects known deposits [14]. Their idea is to model an economic filter. This filter is a function that truncates the density function of deposits, i.e. f. Thus, deposits are split between observable and non-observable deposits, based on a given cost limit and their economic value.

With this filter, empirical data correspond to observable deposits. Because of truncation, grade and tonnage of observable deposits do not follow a log-normal distribution anymore. Rather, they follow a truncated log-normal distribution. The truncation limits (g_{lim} and t_{lim}) are related to a given cost limit C_{lim} through a cost-grade-tonnage relationship which characterizes the economic filter.

Drew and Harris propose to use the same kind of relationship as in part 2 (Eqs. (5) and (6)):

$$C_{lim} = c(g_{lim}, t_{lim}) = c_0\left(\frac{g_{lim}}{g_0}\right)^{\beta_g}\left(\frac{t_{lim}}{t_0}\right)^{\beta_t}$$

$$\ln(c_{lim}) - A = \beta_g x_{lim} + \beta_t y_{lim}.$$

When g_{lim} and t_{lim} are known, the probability density functions (pdfs) of truncated log-normal distributions have explicit expressions that can be related to the non-truncated pdf [14]. Indeed through mathematical manipulations, Drew showed that the statistical expectations (mean value) for grade, tonnage and metal content (respectively denoted γ_g, γ_t, γ_m) on the truncated population could be expressed in terms of the unknowns μ_x, σ_x^2 and μ_y, σ_y^2. This is shown in equations (12) to (14).

$$\gamma_g = \frac{\exp(\mu_x + \sigma_x^2/2)\int_{-\infty}^{C_{lim}}\exp\left(-\frac{1}{2}\left(\frac{c-\mu_c}{\sigma_c}\right)^2\right)dc}{\int_{-\infty}^{C_{lim}}\exp\left(-\frac{1}{2}\left(\frac{c-\mu'_c}{\sigma_c}\right)^2\right)dc}, \quad (12)$$

---

[3]The question of the independence between grade and tonnage in mineral deposits is in constant discussion. Beside, in his research [14], Harris comes to the conclusion that in the case of biased observations, if any correlation exists, it could very well be mitigated, amplified or even totally concealed by the bias filter. In this paper, assumption is made that g and t are independent.

$$\gamma_t = \frac{\exp\left(\mu_y + \sigma_y^2/2\right)\int_{-\infty}^{C_{lim}}\exp\left(-\frac{1}{2}\left(\frac{c-\mu_c'''}{\sigma_c}\right)^2\right)dc}{\int_{-\infty}^{C_{lim}}\exp\left(-\frac{1}{2}\left(\frac{c-\mu_c'}{\sigma_c}\right)^2\right)dc}, \quad (13)$$

$$\gamma_m = \exp\left(\mu_x + \sigma_x^2/2 + \mu_y + \sigma_y^2/2\right)$$
$$\times \frac{\int_{-\infty}^{C_{lim}}\exp\left(-\frac{1}{2}\left(\frac{c-\mu_c''}{\sigma_c}\right)^2\right)dc}{\int_{-\infty}^{C_{lim}}\exp\left(-\frac{1}{2}\left(\frac{c-\mu_c'}{\sigma_c}\right)^2\right)dc}, \quad (14)$$

where:

$$\mu_x = \ln(clarke) - \sigma_x^2/2$$
$$\mu_c = \beta_t\mu_y + \beta_g\left(\mu_x + \sigma_x^2\right)$$
$$\sigma_c^2 = \beta_t^2\sigma_y^2 + \beta_g^2\sigma_x^2$$
$$\mu_c' = \beta_t\mu_y + \beta_g\mu_x \qquad\qquad (15)$$
$$\mu_c'' = \beta_t\left(\mu_y + \sigma_y^2\right) + \beta_g(\mu_x + \sigma_x^2)$$
$$\mu_c''' = \beta_t\left(\mu_y + \sigma_y^2\right) + \beta_g\mu_x.$$

Since bias has been taken into account, γ_g, γ_t and γ_m are the theoretical value of the empirical estimators $\bar{g}, \bar{t}, \overline{m}$ (Eq. (10) applied to g, t, and $m = g \times t$). If γ_g, γ_t and γ_m are replaced by these empirical values in equations (12) to (14), the system consists of 3 equations and 4 unknowns. It can be solved using the additional constraint of equation (15). The solution tuple $(\mu_x, \mu_y, \sigma_x, \sigma_y)$ can be numerically found by using an optimization routine that minimizes the error Δ defined in equation (16).

$$\Delta = \left(1 - \frac{\gamma_g}{\bar{g}}\right)^2 + \left(1 - \frac{\gamma_t}{\bar{t}}\right)^2 + \left(1 - \frac{\gamma_m}{\overline{m}}\right)^2. \quad (16)$$

3.3 Part 2: cost-grade-tonnage relationship

3.3.1 Calibration of the cost-grade-tonnage relationship

The form of the cost-grade-tonnage relationship (Eq. (5)) is chosen by Harris and Drew to handle a linear form in the log-space (see Eq. (6)). This is necessary to achieve the integrations of part 1 (when the relationship is used as economic filter) and part 3 (when it is used for the economic assessment of all deposits).

To calibrate the function, Drew and Harris first compute the theoretical total cost $C_{tot}(g,t)$ of a symbolic deposit as if it was a mining project. They use the discounted cash flow (DCF) method with costs from abacus. Then parameters β_g, β_t and constants are optimized so that the unit cost $c(g,t)$ from the relationship of equation (11) best fits the unit cost $(C_{tot}(g,t)/(g \times t))$ computed for the symbolic deposit.

This paper follows the same methodology except for the computation of C_{tot}. Rather than using abacus which are not publicly available for current mines, we propose to compute C_{tot} from recent mines or recent projects whose

capital costs CC, development time DT, operating costs OP, lifetime LT, grade and tonnage are known. The corresponding formula is given by equation (17) where a is the discount rate.

$$C_{tot} = \sum_{i=-DT}^{0}\frac{CC}{(DT+1)(1+a)^i} + \sum_{i=1}^{LT}\frac{OP}{(1+a)^i}. \quad (17)$$

Once C_{tot} is computed for a set of deposits taken from the database (each having specific grade and tonnage), parameters β_g and β_t and constants were optimized so that the unit cost $c(g,t)$ from the relationship of equation (6) best fits $C_{tot}(g,t)/(g \times t)$.[4]

3.3.2 Use of the cost-grade-tonnage relationship

Once calibrated, the cost-grade-tonnage relationship is used in two different ways in part 1 and in part 3.

In part 1, it truncates the bivariate log-normal distribution in order to characterize observable deposits in today's economic conditions. To that end, unit cost is taken equal to a constant C, which can be fixed at the current long-term uranium price[5]. And from equation (6), minimal grade for any deposit of tonnage t to be observable is given by equation (18). Likewise, minimal tonnage for any deposit of tonnage g to be observable is given by equation (19).

$$g_{lim} = \exp\left((\ln(C) - A - \beta_t\ln(t))/\beta_g\right) \quad (18)$$

$$t_{lim} = \exp\left((\ln(C) - A - \beta_g\ln(g))/\beta_t\right). \quad (19)$$

In part 3, when the cost-grade-tonnage relationship is used, unit cost is the output (cf. Sect. 3.4).

3.4 Part 3: LTCS curve construction

Finally, when the distribution function f is known (Eq. (9)), any deposits from the geological environment can be simulated. In addition, once the cost-grade-tonnage relationship has been calibrated, the cost of each of these deposits can be estimated (Eq. (5)). Therefore, part 3 is the procedure that adds up the resources of deposits within a given cost range (Eqs. (7) and (8)).

---

[4]The fitting procedure is applied to the relationship of equation (6) rather than equation (5). This allows for a simple linear regression since equation (6) handles the logarithm of total costs.

[5]In their studies, Drew and Harris considered short-term prices (8 \$/lbU$_3$O$_8$ in 1977 [11] and 50 \$/lbU$_3$O$_8$ in 1988 [15]). Although no long-term index existed at that time, this choice is open to criticism, especially when spot prices fluctuated as they did in the late 1970s and more recently. Long-term price index was preferred in this study as it is more stable. The highest Red Book cost limit (260 \$/kgU) could have been considered as well but since this price has never been reached over long periods, it is expected that this cost category only contains sparse data.

The integral of equation (8) raises some difficulties as it cannot be solved analytically (essentially because the domain of integration is dependent upon g and t through $c(g,t)$). To compute a numerical approximation of the integral, Drew introduces the following variable substitution:

$$(g,t) \to (g,c) = \left(g, c_0 \left(\frac{g}{g_0}\right)^{\beta_g} \left(\frac{t}{t_0}\right)^{\beta_t}\right).$$

Using this substitution, the domain of integration of variable c is simplified (it is integrated from 0 to C_1 as defined in Eq. (8)). But Drew does not mention the new domain of integration of variable g. In fact, before the substitution, g and t were independent random variables. But g is not, in any way, independent from $c_0\left(\frac{g}{g_0}\right)^{\beta_g}\left(\frac{t}{t_0}\right)^{\beta_t}$. Therefore, the mathematical expression used to compute the statistical expectation of equation (8) cannot stand for the computation of cumulated resources since the probability distribution of c is unknown.

For those reasons, this study developed an alternative numerical method to compute the cumulated metal resources available at below a given unit production cost C_1. These quantities are estimated though a numerical approximation of the following integral derived from equation (4) with the relevant domain of integration:

$$q(C_1) = q_0 \iint\limits_{c(g,t) \leq C_1} f(g,t) dg dt. \quad (20)$$

The numerical approximation consists in applying the rectangle method and introducing the following indicator function:

$$\varepsilon(g,t,C_1) = \begin{cases} 1 \text{ if } c(g,t) \leq C_1 \\ 0 \text{ otherwise} \end{cases}. \quad (21)$$

Hence, q can be approximated by the following sum:

$$\begin{aligned} q(C_1) = q_0 \sum_i \sum_k & (g_{i+1} - g_i)(t_{k+1} - t_k) \\ & \times \varepsilon\left(\frac{g_{i+1} + g_i}{2}, \frac{t_{k+1} + t_k}{2}, C_1\right) \\ & \times f\left(\frac{g_{i+1} + g_i}{2}, \frac{t_{k+1} + t_k}{2}\right). \end{aligned} \quad (22)$$

In equation (22), (g_i) and (t_k) are used as a mesh of the domain of integration. To ensure a precise approximation, the mesh and its refinement should be carefully defined. In this paper, we used a logarithmic mesh defined as follows:

$$g_i \in [\exp(\mu_x - 10\sigma_x), \exp(\mu_x - 10\sigma_x)], i = 1 \text{ to } 400$$
$$t_k \in [\exp(\mu_y - 10\sigma_y), \exp(\mu_y - 10\sigma_y)], k = 1 \text{ to } 400.$$

The LTCS curve is finally obtained by plotting the function $C_1 \to q(C_1)$.

4 Preliminary results for the US endowment

The case of United States was chosen to validate the methodology developed in this paper. Several reasons have guided this choice. First, this country has a sustained history of uranium exploration and mining. The data required for this study are all available and generally quite extensive. Second, the United States has long experience in mineral appraisal assessment too (see the USGS "Quantitative Mineral Resources Assessments" [19,20]). Besides, Harris and Drew conducted similar economic appraisal of US resources. Although the results cannot be compared due to cost escalation since their studies (late 70 s, early 80 s), our model provides an up-date of uranium resource appraisals.

Since part 1 uses the calibrated cost relationship described in part 2, the database used for the calibration and the results of this calibration are presented first (Sects. 4.1.1 and 4.1.2). Then the database used for the deposit statistics is presented (Sect. 4.2.1). Finally, Section 4.2.2 and Section 4.3 gather the results of part 1 and part 3 applied to the US geological environment.

4.1 Calibration of the cost relationship

4.1.1 WISE Uranium cost database for US deposits

The WISE Uranium project gathers information on uranium mining activities around the world [10]. Among them are a list of mining companies, statistics of the mining industry and a list of known deposits with related recent issues. For 55 of those deposits, publicly available cost data are detailed so that for each of them capital costs CC, operating costs OP, lifetime LT, grade and resources are known. Fifteen of those deposits are located in United States. Twelve of them[6] were used to estimate the parameters of equation (6) (β_g, β_t and the constant A). Table 2 gathers the total costs of these deposits. They were computed according to equation (17) based on the following assumptions:

- tonnage t was computed as m/g where m includes all metal resources (indicated, inferred and measured, either reserves or resources) and g is the average grade of those resources;
- life time was computed as the minimum of t/K_{mill} and $m/K_{overall}$ where K_{mill} is ore processing capacity (in tonnes of ore per year) and $K_{overall}$ is the overall production capacity (in tonnes of uranium per year);
- discount rate is 10%;
- development time is 3 years.

4.1.2 Results of part 2: cost function calibration

From the data of Table 2, a linear regression gives the following results:

$$\ln(C_{tot}) = 0.501 \times \ln(t) + 11.61 \quad (R^2 = 0.85). \quad (23)$$

Unit production cost is obtained by dividing C_{tot} by $m = g \times t$ in equation (23) (where m is the metal content, g the mean grade and t the mean tonnage). Hence, we derive

---

[6]Two have incomplete data and Roca Honda mine seems to have abnormal data, perhaps because milling costs are omitted (milling occurs at White Mesa mill).

Table 2. Total cost and ore tonnage of US deposits.

Deposit name (type)	Tonnage (Mt)	C_{tot} (M\$)
Bison Basin (ISL)	3.4	229.7
Centennial (ISL)	6.4	340.7
Churchrock section 8 (ISL)	3.6	174.6
Dewey-Burdock (ISL)	2.7	360.0
Lance (ISL)	50.8	715.7
Lost Creek (ISL)	11.6	197.0
Nichols Ranch & Hank (ISL)	1.1	84.5
Reno Creek (ISL)	23.6	565.4
Sheep Mountain (OP/UG/HL)	11.8	480.2
Coles Hill (UG)	89.7	1115.3
Roca Honda (UG)	2.5	900.7
Hansen (UG)	28.1	711.8
Shirley Basin (ISL)	1.6	156.9

ISL: in situ leaching; OP: open pit; UG: underground; HL: heap leaching.

Table 3. Statistics of known US deposits (UDEPO).

Statistics	Value	Unit
$\overline{g}$	0.0015	Grade in kgU/kg of ore
s_g^2	1.69×10^{-6}	(kgU/kg of ore)^2
$\overline{t}$	4.47×10^6	Tonnage in tonnes of ore
s_t^2	9.40×10^{13}	(tonnes of ore)^2
$\overline{m}$	3506	Metal content in tU
s_m^2	3.65×10^7	(tU)^2
$\overline{x}$	-6.76	ln (kgU/kg)
s_x^2	0.59	ln (kgU/kg)^2
$\overline{y}$	14.24	ln (tonnes of ore)
s_y^2	1.94	ln (tonnes of ore)^2

equation (24) where β_g, β_t and the constant A can be identified.

$$\ln(c) - 11.61 = (0.501 - 1) \times \ln(t) - \ln(g), \quad (24)$$

where $\beta_t = -0.499$, $\beta_g = -1$ and $A = 11.61$.

4.2 Calibration of the abundance model

4.2.1 UDEPO data for US deposits

IAEA provides a large database on uranium deposits called UDEPO [21]. This database gives a resource assessment on most known deposits in the world. It is based on available information and may not always be JORC or NI 43-101[7] compliant. The database classifies the deposits based on a number of parameters including mean grade and corresponding metal content.

This paper uses the statistics of US deposits available in the UDEPO database (329 deposits[8]). Since ore tonnage is not an explicit parameter of the database it was approximated by m/g where m is the metal content and g is the mean grade. UDEPO has a lower cutoff on metal content: only deposits bigger than 300 tU are reported. Although there is a number of known deposits below this cutoff in the US, they would not influence the estimation of the log-normal parameters since only deposits above the

economic filter are taken into account during the procedure of part 1. Table 3 presents the statistics of US deposits and Figure 6 shows the tonnage and the grade of both UDEPO deposits and WISE projects; the economic filter obtained from Section 4.1.2 (Eq. (24)) is also displayed for $C_{\lim} = 125$ \$/kgU.

4.2.2 Results of part 1: estimated log-normal parameters

Using the statistics of US deposits from UDEPO database, the optimization routine described in Section 3.2.2 is run, with the additional assumptions:

– the mean grade of the crust within the geological environment, *clarke*, is taken equal to 3 ppm (eq. U_3O_8)[9] $= 2.54 \times 10^{-6}$ kgU/kg of ore;
– current[10] long-term price of uranium: 125 \$/kgU [23].

The resulting estimated parameters are:

– $\mu_x = -15.26$;
– $\sigma_x = 2.18$;
– $\mu_y = 13.63$;
– $\sigma_y = 1.08$.

The bias correction between these estimations and the original UDEPO statistics (Tab. 3) is noticeable. In particular, the mean grade is largely overestimated in UDEPO ($\mu_x = -15.26 < \overline{x} = -6.76$, see Tab. 4 for non-logarithmic comparisons) but standard deviation for grade is underestimated ($\sigma_x = 2.18 > s_x = 0.77$). Regarding deposit size (y), the bias is also significant (we tend to discover bigger deposits first) but less markedly than for grade.

Those estimations of unbiased parameters can be compared with Harris and Drew's values (Tab. 4). Since the authors use different units for grade, all results are given for variables (g,t) in ppmU and tonnes. Conversion from (x, y) parameters to (g,t) parameters is given in equations (25) to (28) according to the definition of the log-normal distribution.

---

[7] JORC and NI 41-101 are two national (Australian and Canadian respectively) sets of rules and guidelines for estimating and reporting mineral resources.
[8] Three hundred and forty-two in total but 13 deposits were discarded. Three of them have incomplete data. Seven of them correspond to regional resource assessments (e.g. Northern Great Plains, Phosphoria Formation, Central Florida). Three of them are high-tonnage and very low-grade deposits where uranium is a by-product (Bingham Canyon, Yerington, Twin Butte).

[9] This is a common value found in literature for the upper part of the Earth's crust (first 20 km below the surface) [22]. This is also the same value as Harris' [15].
[10] March 2015.

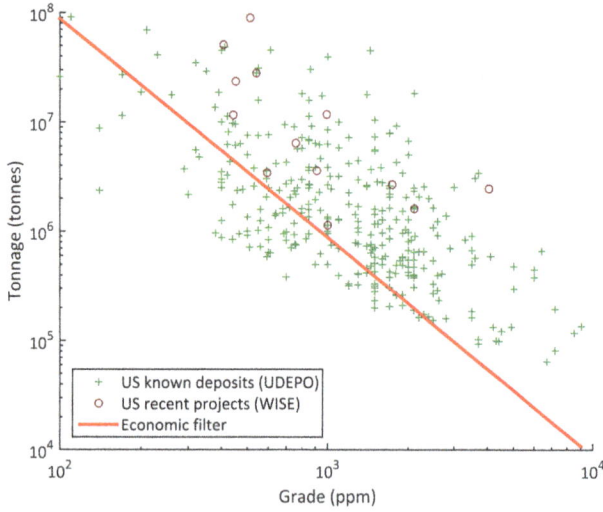

Fig. 6. Grade and tonnage of US known deposits and recent projects.

Table 4. Comparison of the biased empirical statistics with the estimated unbiased parameters for the log-normal distribution.

Parameter	$\bar{g}$ (ppm)	S_g (ppm)	$\bar{t}$ (Mt)	S_t (Mt)
UDEPO (biased)	1544	1299	4.47	9.69
Harris (biased) [15]	1560	1076	0.596	1.70
Drew (biased) [11]	2185	NA	0.993	NA
This study (unbiased)	2.54	27.42	1.48	2.20
Harris (unbiased) [15]	2.54	6.71	0.164	0.150
Drew (unbiased) [11]	1.70	12.18	0.0182	0.0422

$$\bar{g} = \exp\left(\mu_x + \sigma_x^2/2\right) \qquad (25)$$

$$s_g = \exp\left(2\mu_x + \sigma_x^2\right)\left(\exp\left(\sigma_x^2\right) - 1\right) \qquad (26)$$

$$\bar{t} = \exp\left(\mu_y + \sigma_y^2/2\right) \qquad (27)$$

$$s_t = \exp\left(2\mu_y + \sigma_y^2\right)\left(\exp\left(\sigma_y^2\right) - 1\right). \qquad (28)$$

Table 4 shows significant differences on several points. First, the average ore tonnage of deposits, $\bar{t}$, is much larger in this study than in Harris or Drew. Since this difference can already be seen in input statistics (biased statistics from known deposits), it can be explained by different definitions of deposits. Drew has certainly the most restrictive definition (probably taking only measured reserves to delineate deposits) while the UDEPO database used in this study has a less compelling definition (resources that do not

comply with JORC/NI 43-101 are considered). These differences may not impact the construction of the LTCS curve if the cost-grade-tonnage relation is calibrated using the same resource definition. In this study, the deposits from Wise Uranium that are used for calibration include all resources (including inferred and indicated resources), as specified in Section 4.1.

In addition, there are also significant differences in the standard deviations for grade, S_g. This time, the difference cannot be noticed in input statistics: $\bar{g}$ (mean grade of known deposits) is similar in this study (1544 ppmU) and Harris (1560 ppmU) and so is s_g (1299 ppmU in this study and 1076 ppmU in Harris'). This suggests that the definition of the deposit size can significantly influence the estimated standard deviation for grade during the bias correction procedure (Sect. 3.2.2).

4.3 US LTSC curve

4.3.1 Results

Finally, the calibrated cost relationship obtained in Section 4.1.2 and the parameters of the bivariate log-normal distribution obtained in Section 4.2.2 can be used to build the US LTSC curve. The procedure is described in Section 3.4. In addition to the previous assumptions, the size of the US geological environment was assumed to be the total mass of rock, M, contained in the total US area to a depth of 2 km[11]. $M = 4.24 \times 10^{16}$ tonnes. The procedure also takes account of a 75% overall recovery rate[12] (including extraction losses, ore sorting losses and processing losses).

The results are plotted in Figure 7.

Figure 7 shows the US LTCS curve obtained with the methodology developed in this paper (blue curve). It is compared with the US resource declaration available in the Red Book (2011 edition[13] [24]) (red curves). It appears that the known resources (RAR) reported in the Red Book[14] are more limited than the simulated US resource appraisal. This is expectable since past production and undiscovered resources are excluded from the Red Book RAR quantities. It is also noticeable that for costs falling below the Red Book limit of 130 \$/kgU, the simulated endowment is more conservative than the expected total resources (known and undiscovered, RAR + PR + SR) reported in the Red Book.

---

[11] A maximum depth of 2 km was preferred to Drew's value (1 km [11]) as some uranium mines are known at those depths.
[12] This choice was guided by the Red Book [5] reference values (70 to 75% for underground and ISL methods which are the most common in the United States, and 75% when no method is specified).
[13] In the 2014 edition, there is no declaration of US Prognosticated and Speculative resources (PR & SR). The 2011 edition was preferred for comparison purposes.
[14] The US does not report inferred resources (IR).

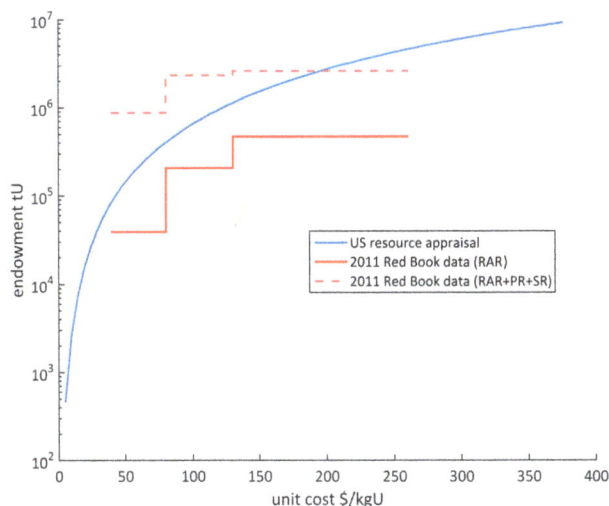

Fig. 7. US LTCS curve (logarithmic scale).

4.3.2 Discussion

These results are preliminary outputs. Before further exploitation and analysis, some sensitivity tests are still necessary. Among the sensitivity parameters which require further investigations are:

- the geological environment under study;
- parameters related to this geological environment (maximum depth that define M, mean crust grade *clarke*);
- parameters related to the cost-grade-tonnage relationship (β_g, β_t and A);
- parameters specifically related to the economic filter (cost limit C used to define observable deposits).

5 Conclusion

For the purpose of analyzing the long-term availability of uranium resources, this paper develops a methodology to build long-term cumulative supply curves. After covering existing models and stressing their limits, a methodology based on geological environments is proposed. Its statistical approach provides a more detailed resource estimation and is more flexible regarding cost modelling. In particular, both grade and tonnage are considered in the economics of deposits and an economic filter is introduced to correct the observation bias that limits our knowledge to the richest deposits.

Preliminary results for the US endowment are presented. Although the model still requires some additional sensitivity tests, these results are promising. They showed a slightly more conservative endowment than the estimated undiscovered resources reported in the Red Book. The preliminary results validate the general methodology and could maybe allow for future comparison with alternative methodologies such as the USGS "3-part resource assessment".

References

1. A. Baschwitz, C. Loaec et al., Long-term prospective on the electronuclear fleet: from GEN II to GEN IV, in *Global 2009: The nuclear fuel cycle: sustainable options & industrial perspectives, Paris, France* (2009)
2. S. Gabriel, A. Baschwitz et al., Building future nuclear power fleets: the available uranium resources constraint, Res. Pol. **38**, 458 (2013)
3. J.E. Tilton, B.J. Skinner, The meaning of resources, in *Resources and world development*, edited by D.J. McLaren, B. J. Skinner (John Wiley & Sons, New York, 1987), p. 13
4. J.E. Tilton, A. Yaksic, Using the cumulative availability curve to assess the threat of mineral depletion: the case of lithium, Res. Pol. **34**, 185 (2009)
5. OECD NEA, IAEA, in *Uranium 2014: resources, production and demand* (OECD Nuclear Energy Agency, Paris, France, 2014), p. 508
6. E.A. Schneider, W.C. Sailor, Long-term uranium supply estimates, Nucl. Tech. **162**, 379 (2008)
7. I.A. Matthews, M.J. Driscoll, in *A probabilistic projection of long-term uranium resource costs* (Massachusetts Institute of Technology, Cambridge, Massachusetts, 2010), p. 141
8. Massachusetts Institute of Technology, in *The future of nuclear fuel cycle an interdisciplinary MIT study* (Massachusetts Institute of Technology, Cambridge, Massachusetts, 2011), p. 258
9. K. Deffeyes, I. Macgregor, in *Uranium distribution in mined deposits and in the earth's crust* (Princeton University, Princeton, New Jersey, 1979), p. 509
10. World Information Service on Energy, WISE Uranium Project (website), WISE Uranium, http://www.wise-uranium.org/index.html (accessed: 03/2015)
11. M.W. Drew, US uranium deposits: a geostatistical model, Res. Pol. **3**, 60 (1977)
12. M.L. Chavez-Martinez, in *A potential supply system for uranium based upon a crustal abundance model* (University of Arizona, Tucson, Arizona, 1982), p. 491
13. D.P. Harris, in *Quantitative methods for the appraisal of mineral resources* (University of Arizona, Tucson, Arizona, 1977), p. 862
14. D.P. Harris, *Mineral resources appraisal: mineral endowment, resources, and potential supply: concepts, methods and cases* (Oxford University Press, Oxford, UK, 1984)
15. D.P. Harris, Geostatistical crustal abundance resource models, in *Quantitative analysis of mineral and energy resources*, edited by C.F. Chung, A.G. Fabbri, R. Sinding-Larsen (Springer, Netherlands, 1988), p. 459
16. J.W. Brinck, MIMIC - The prediction of mineral resources and long-term price trends in the non-ferrous metal mining industry is no longer utopian, Eurospectra **10**, 46 (1971)
17. J.W. Brinck, Calcul des ressources mondiales d'uranium, Bull. Communaute Eur. Energie At. **6**, 109 (1967)
18. H.I. de Wolde, J.W. Brinck, *The estimation of mineral resources by the computer program 'IRIS'* (Commission of the European Communities, Luxembourg, 1971)
19. D.A. Singer, *Short course introduction to quantitative mineral resource assessments* (US Geological Survey, Menlo Park, California, 2007)
20. D.A. Singer, W.D. Menzie, in *Quantitative mineral resource assessments — An integrated approach* (Oxford University Press, New York, 2010), p. 219

21. IAEA, World Distribution of Uranium Deposits (UDEPO) (website), https://infcis.iaea.org (extensive copy of the database accessed: 11/2013)

22. K. Hans Wedepohl, The composition of the continental crust, Geochim. Cosmochim. Acta **59**, 1217 (1995)

23. UX Consulting, "UxC" (website), http://www.uxc.com (accessed: 03/2015)

24. OECD NEA, IAEA, in *Uranium 2011: resources, production and demand* (OECD Nuclear Energy Agency, Paris, France, 2012), p. 489

8

Lock-in thermography for characterization of nuclear materials

Alexandre Semerok[*], Sang Pham Tu Quoc, Guy Cheymol, Catherine Gallou, Hicham Maskrot, and Gilles Moutiers

Den-Service d'Études Analytiques et de Réactivité des Surfaces (SEARS), CEA, Université Paris-Saclay, 91191 Gif-sur-Yvette, France

Abstract. A simplified procedure of lock-in thermography was developed and applied for characterization of nuclear materials. The possibility of thickness and thermal diffusivity measurements with the accuracy better than 90% was demonstrated with different metals and Zircaloy-4 claddings.

1 Introduction

Lock-in thermography is a non-destructive method which may be applied to test and to ensure remote control over materials in severe environment (e.g. nuclear installations) in a wide temperature range. The method is based on the laser heating of a sample with a modulated laser power at a given frequency f(Hz) followed by measurements of a thermal radiation emitted by the sample. The phase shifts $\Delta\varphi$ between the laser power and the thermal radiation measured at different modulated frequencies are then compared with those obtained with an analytical (3D + t) model developed at the LISL (DEN/DANS/DPC/SEARS) in case of the heating of a sample covered by a deposited layer [1,2]. Thus, it is possible to provide a tool to characterize some sample properties (thickness, thermal diffusivity, deposited layer/surface thermal contact resistance, characterization of under-surface defects and their evolution with time). The phase shift of heating temperature is presented in Figure 1.

2 Model for the heating of a plate

In a thermal model for homogeneous and isotropic plate with infinite dimensions, we supposed that:

- variations of the optical and thermal properties for a surface covered by a deposited layer during its heating are negligible;
- the surface roughness effect and heat exchange due to the sample surface/air convection are also negligible.

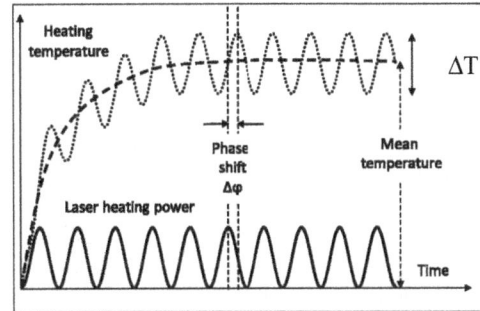

Fig. 1. Phase shift of heating temperature.

In the heating models [1,2] for a surface with a deposited layer, we supposed that the layer/surface thermal resistance ($\Re$) and the layer optical thickness (αL) on laser wavelength are very high ($\Re \to \infty$ and $\alpha L >> 1$). By applying the Fourier series analysis to the intensity of the laser beam and the temperature in the stationary regime of the laser heating [2,3], the complex temperature amplitude of the front face of a plate can be written as:

$$\Delta T(z,r) = \frac{\alpha(1-R)\tilde{I}}{k} \int_0^{+\infty} \frac{\Theta(\xi)}{\xi^2 - \alpha^2 - \frac{2\pi i f C_v}{k}} \left(\left(\frac{\alpha e^{-\Psi L}}{\Psi(e^{\Psi L} - e^{-\Psi L})} + \frac{\alpha}{\Psi} \right) e^{-\Psi z} \right.$$
$$\left. + e^{-\Psi z} + \frac{\alpha e^{-\Psi L}}{\Psi(e^{\Psi L} - e^{-\Psi L})} e^{\Psi z} - e^{-\alpha z} \right) J_0(\xi r) d\xi,$$

$$(1)$$

with $\quad \Theta(\xi) = \frac{\xi r_0^2}{2} \exp\left(-\frac{\xi^2 r_0^2}{4} \right) \quad$ for the Gaussian beam;

$\Psi = \sqrt{\xi^2 - 2\pi i f C_v / k}$; $\qquad \tilde{I} = i\sqrt{a_1^2 + b_1^2} e^{-i\Phi_{LP}}$;

$\Phi_{LP} = \mathrm{atan}\left(\frac{b_1}{a_1} \right)$; $\qquad J_0(\xi r) = \frac{1}{\pi} \int_0^\pi \cos(-\xi r \sin\tau) d\tau$;

$a_1 = 2f \int_0^{1/f} I(t) \cos(2\pi f t) dt$;

$b_1 = 2f \int_0^{1/f} I(t) \sin(2\pi f t) dt$; $I(t) = I_0(1 - \cos(2\pi f t))$,

[*] e-mail: alexandre.semerok@cea.fr

where: z and r, respectively, are the propagation direction of the laser beam and the radial distance from the center of the heated zone at the sample surface; C_v, k and L: the volumetric specific heat, the thermal conductivity and the thickness of the sample; α and R: the laser absorption coefficient and the reflectivity of the sample surface; r_0: the laser beam radius at $1/e$ intensity, $I(t)$ and I_0: the intensity of the laser beam and its amplitude; f, t and i: the repetitive rate frequency of the laser, the time and the complex unity; ξ and τ: the variables of the integration; Φ_{LP}: the phase of the laser power.

The phase shift between the laser and the thermal power can be found by the expression:

$$\Delta\varphi = \mathrm{atan}(Re(\Delta T)/\mathrm{Im}(\Delta T)). \qquad (2)$$

2.1 Environmental effect

The heat exchange with the environment by convection mechanism can be introduced by the conditions of limits at $z=0$ and $z=L$ [4]:

$$\left.\frac{\partial \Delta T}{\partial z}\right|_{z=0} = \mu(T_{z=0} - T_a); \quad \left.\frac{\partial \Delta T}{\partial z}\right|_{z=L} = \mu(T_{z=L} - T_a), \quad (3)$$

where: $\mu = h/k$, $h(\mathrm{Wm}^{-2}\,\mathrm{K}^{-1})$ is a coefficient of thermal exchange with environment by convection. The environment temperature T_a is supposed to be equal to the initial temperature of the plate, thus: $T_{z=0} - T_a = \Delta T(t, z = 0, r)$; $T_{z=L} - T_a = \Delta T(t, z = L, r)$. The losses by thermal emission are supposed to be negligible.

The solution of the heat equation:

$$\rho c \frac{\partial \Delta T}{\partial t} = k\left(\frac{\partial^2 \Delta T}{\partial z^2} + \frac{\partial^2 \Delta T}{\partial r^2} + \frac{1}{r}\frac{\partial \Delta T}{\partial r}\right) \\ + \alpha(1 - R_c)I(t,r)e^{-\alpha z}, \qquad (4)$$

on $0 \le z \le L$ with the initial condition $\Delta T(t = 0, z, r) = 0$ was obtained for the nth harmonic of the laser repetition frequency:

$$\Delta T_n(z=0, r) = \frac{\alpha(1-R)\tilde{I}_n}{k} \int_0^\infty \frac{\Theta(\xi)}{\xi^2 - \alpha^2 - 2\pi \times i \times n \times f \times \frac{\rho c}{k}} \times J_0(\xi r)$$

$$\times \left(\frac{e^{-\Psi L} - e^{-\alpha L}}{e^{\Psi L} - e^{-\Psi L}}\left(\frac{\mu + \alpha}{\mu - \Psi} - \frac{\mu + \alpha}{\mu + \Psi}\right) - \frac{\mu + \alpha}{\mu + \Psi} + 1\right)d\xi, \qquad (5)$$

where: $\Psi = \sqrt{\xi^2 - 2\pi \times i \times n \times f \times \rho c/k}$; $\mu = h/k$;

$$\Theta(\xi) = \begin{cases} \dfrac{\xi r_0^2}{2}\exp\left(-\dfrac{\xi^2 r_0^2}{4}\right) - \text{the Gaussian beam;} \\ r_0 J_1(\xi r_0) - \text{top-hat beam;} \end{cases}$$

$J_n(\xi r) = \frac{1}{\pi}\int_0^\pi \cos(n\tau - \xi r \sin\tau)d\tau$; $\tilde{I}_n = i\sqrt{a_n^2 + b_n^2}\,e^{-i\varphi_L^n}$;

$\varphi_L^n = a\tan(b_n/a_n)$; $a_n = 2f\int_0^{1/f} I(t)\sin(2\pi n f t)dt$;

$b_n = 2f\int_0^{1/f} I(t)\cos(2\pi n f t)dt$.

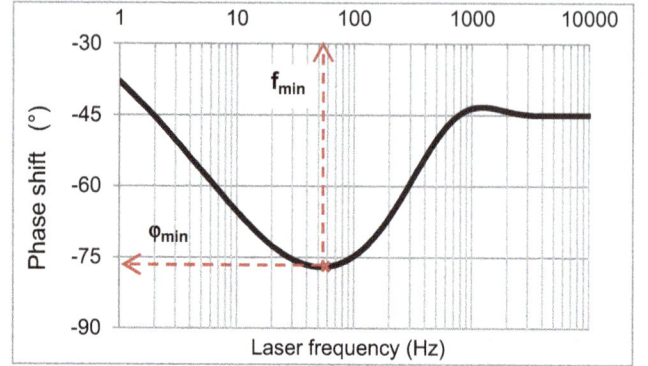

Fig. 2. Phase shift for SS 304L plate of $400\,\mu$m thickness.

The phase shift can be written as:

$$\Delta\varphi = \mathrm{atan}(Re(\Delta T_n)/\mathrm{Im}(\Delta T_n)). \qquad (6)$$

It depends on the parameters k, ρc, α, L, r_0, h. For thin metal plates, the environmental effect on the phase shift may be considered as negligible. For example, for SS 304L of $400\,\mu$m thickness, the phase shifts are not affected by the environment even for a material with $h = 100\,\mathrm{W/m^2\,K}$ (water) [4].

2.2 Simple analytical expressions

The numerical simulation of laser heating is used to fit the calculated phase shifts with the experimental ones by adjusting the material properties. A typical dependence of a phase shift on a laser modulation frequency is presented in Figure 2. For the laser beam with a diameter satisfying $r_0/100 \le L \le r_0/2$, one may observe a minimum on the phase shift curve with the corresponding values $\varphi_{\min}$ and $f_{\min}$.

Multiparameter simulation of laser heating enables one to determine the effect of the interaction parameters and the material properties on the phase shifts [3–7]. Two analytical expressions were derived to relate laser parameters, sample properties, $\varphi_{\min}$ and $f_{\min}$. This inter-relationship may provide rapid measurements of thickness L and diffusivity D of a sample with 99% accuracy:

$$L[\mu\mathrm{m}] = (r_0[\mu\mathrm{m}]/\zeta_\varphi) \times \ln(90°/\varphi_{min}), \qquad (7)$$

$$D[\mu\mathrm{m}^2/s] = (1/\zeta_f) \times r_0[\mu\mathrm{m}] \times L[\mu\mathrm{m}] \times f_{min}[\mathrm{Hz}], \qquad (8)$$

with ζ_φ and ζ_f calculated values (Fig. 3).

3 Experimental

The experimental setup with a compact fiber laser (low divergent near the Gaussian beam with $M^2 = 1.1$, beam radius $r_0 \cong 1\,\mathrm{mm}$, 1060 nm wavelength) is simple in its

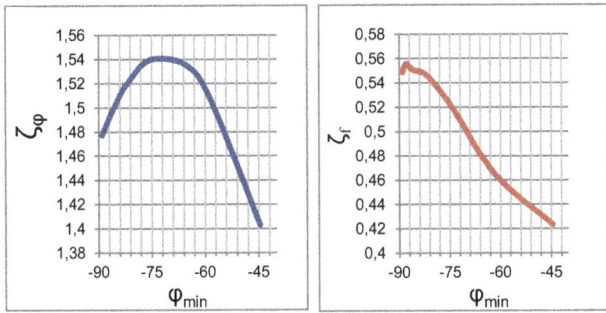

Fig. 3. Calculated values of ζ_φ and ζ_f parameters.

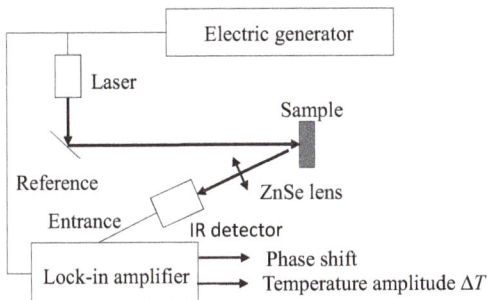

Fig. 4. Scheme of the experimental setup.

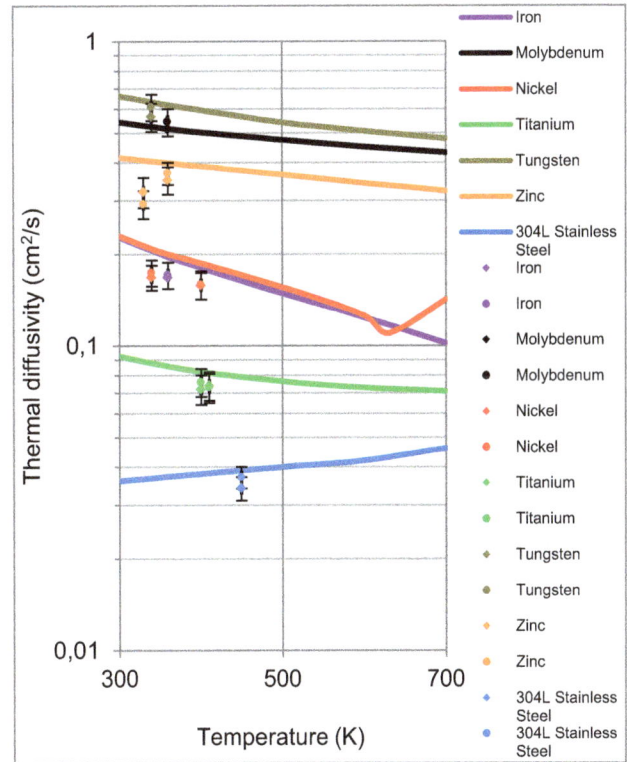

Fig. 5. Comparison of the measured thermal diffusivities along with the referenced values from literature (presented as solid lines) [8]. The square points – D_{m2} and the circle points – D_{m1} (see Tab. 1 for D_m definition).

arrangement (Fig. 4) and enables one to make remote measurements with a millimeter lateral resolution in a wide distance range (from some centimeters up to several meters).

3.1 Metal plates characterization

The lock-in thermography for a plate characterization was validated with a set of etalon samples. The obtained results on plate thickness and diffusivity measurements are presented in reference [3]. These results have demonstrated ≈90% accuracy of thickness and thermal diffusivity measurements. The measured thermal diffusivities along with the referenced values from literature are summarized in Figure 5.

3.2 Zircaloy-4 cladding characterization

After the characterization of the metal plates by the above procedure, the method was used for studying Zircaloy-4 claddings (Fig. 6). The schematic comparison of the cladding diameter and the one of the heated zone (with the tested zone on it) is presented in Figure 6. As the heated zone diameter is smaller than the one of Zy4 cladding, the heated zone may be considered as a plane surface, and thus, the same procedure as the one for metal plates may be followed.

Table 1. Thickness and thermal diffusivity measurements for Zy4 claddings with oxide layers.

		Zircaloy-4 claddings with oxide layers		
Oxide thickness	μm	5	10	15
Laser mean power	W	2.7	2.7	2.7
Temperature	°C	100 ± 5	100 ± 5	100 ± 5
D_{ref} reference	cm^2/s	0.073 ± 0.003	0.073 ± 0.003	0.073 ± 0.003
L_{ref} reference	μm	570 ± 2.5	570 ± 2.5	570 ± 2.5
r_0 measured	μm	1740 ± 30	1740 ± 30	1740 ± 30
φ_{min} measured	°	-53.3 ± 0.5	-53.1 ± 0.5	-54.7 ± 0.5
f_{min} measured	Hz	3.1 ± 0.3	3.1 ± 0.3	3.4 ± 0.3
L_m measured	μm	594 ± 15	598 ± 15	564 ± 15
D_{m1} measured with L_m	cm^2/s	0.068 ± 0.006	0.069 ± 0.006	0.071 ± 0.006
D_{m2} measured with L_{ref}	cm^2/s	0.066 ± 0.006	0.066 ± 0.006	0.071 ± 0.006

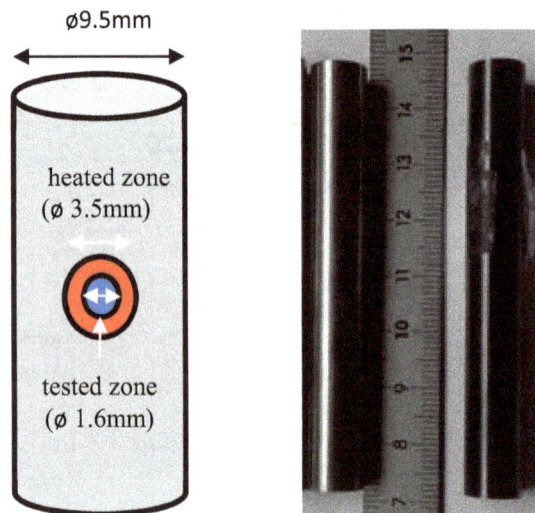

Fig. 6. On the left: schematic comparison of the Zy4 cladding diameter and those of the heated zone (in red) and of the tested zone on it (in blue). On the right: the picture of Zy4 claddings. Some Zy4 claddings were artificially oxidized (on the right in the picture).

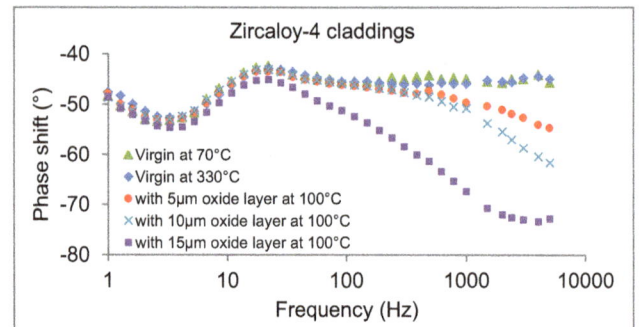

Fig. 7. Phase shifts for Zy4 claddings as a function of modulated frequency.

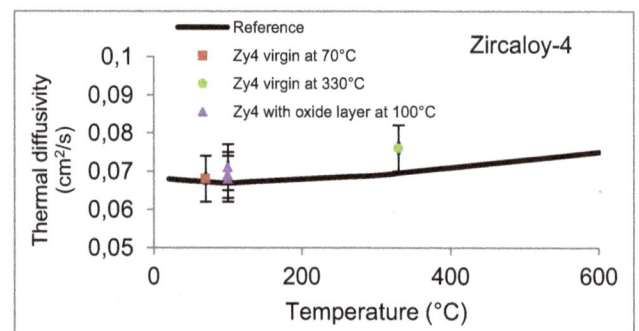

Fig. 8. Measured and reference values of thermal diffusivity as a function of temperature (°C) for Zy4 claddings.

To study the effect of the oxide layer on the measured thermal diffusivity, some Zy4 claddings were artificially oxidized in a furnace at different regimes (temperature, environment, time) to obtain oxide layers of different thickness (5–15 μm). For 5 μm, 10 μm, and 15 μm oxide layer thickness, the regimes, respectively, were as follows: (500 °C, in air, for 37 hours), (550 °C, in air, for 23 hours), and (550 °C, in water vapor, for 51 hours). The surface of Zy4 with the oxide layer of 10 μm thickness has suffered nitriding effect, while the samples with 5 μm and 15 μm oxides were of a good quality.

The results on Zy4 claddings characterization are presented in Figures 7 and 8 and in Table 1. At low modulation frequency ($f < 20$ Hz), the phase shift is poorly affected by the presence of the oxide layer (Fig. 7). Thus, for Zircaloy-4 claddings, φ_{min} and f_{min} method may be used to determine thickness and thermal diffusivity (see Tab. 1). The measurement relative deviations were less than 10%.

At higher modulation frequency ($f > 20$ Hz), a clear effect of the oxide layer on the phase shifts is observed (Fig. 7). Due to the fact that ZrO_2 layers are semitransparent, the theoretical models for phase shift calculation [1,2] are not applicable in this case. However, there are all the reasons to suppose that further development of the thermal model of heating a semi-transparent layer on a metal plate will provide an adequate on-line in situ characterization of oxide formation.

4 Conclusions

The homemade thermal model of the local heating of a homogeneous and isotropic plate with infinite dimensions was developed and verified by characterizing different metal plates and Zircaloy-4 claddings. Two analytical expressions (7) and (8) for sample thickness L and thermal

diffusivity D were derived. These expressions and the lock-in thermography measurements (the minimal phase shift φ_{min} and the corresponding minimal modulated frequency f_{min} [6,7]) are used to measure the thickness and the thermal diffusivity of the samples. The obtained results are in agreement with experimental data within an accuracy of 90%. The developed method may be applied for any material with a high absorption coefficient α, that is, for any plate with $\alpha L \gg 1$. Based on the results obtained, we may conclude that a rapid remote in situ control over components in nuclear installations may be ensured with a good spatial resolution (of the order of a laser beam diameter $2r_0$).

Authors acknowledge DEN/DANS/DMN/SRMA/LC2M team for Zircaloy-4 claddings supply.

References

1. A. Semerok, F. Jaubert, S.V. Fomichev et al., Laser lock-in thermography for thermal contact characterisation of surface layer, Nucl. Instrum. Methods Phys. Res. A **693**, 98 (2012)

2. A. Semerok, S.V. Fomichev, F. Jaubert et al., Active laser pyrometry and lock-in thermography for characterisation of deposited layer on TEXTOR graphite tile, Nucl. Instrum. Methods Phys. Res. A **738**, 25 (2014)

3. S. Pham Tu Quoc, G. Cheymol, A. Semerok, New contactless method for thermal diffusivity measurements using modulated photothermal radiometry, Rev. Sci. Instrum. **85**, 054903 (2014)

4. S. Pham Tu Quoc, Caractérisation des propriétés d'un matériau par radiométrie photothermique modulée, CEA PhD Thesis, France, 2014

5. D. Melyukov, Étude et développement d'une méthode de caractérisation in-situ et à distance de dépôts en couches minces par pyrométrie active laser, CEA PhD Thesis, France, 2011

6. D. Melyukov, P.-Y. Thro, Patent CEA FR2980846-A1, Procédé de détermination sans contact de l'épaisseur d'un échantillon, 2013

7. S. Pham Tu Quoc, G. Cheymol, A. Semerok, Patent CEA FR1355905, Procédé de détermination de la diffusivité thermique et système pour la mise en œuvre, 2013

8. Y.S. Touloukian, R.W. Powell, C.Y. Ho, M.C. Nicolaou, *Thermal diffusivity* (IFI/Plenum, New York, 1973)

9

Evaluations of Mo-alloy for light water reactor fuel cladding to enhance accident tolerance

Bo Cheng[1*], Peter Chou[1], and Young-Jin Kim[2]

[1] Electric Power Research Institute (EPRI), Palo Alto, CA 94304, USA
[2] GE Global Research Center, Schenectady, NY 12309, USA

Abstract. Molybdenum based alloy is selected as a candidate to enhance tolerance of fuel to severe loss of coolant accidents due to its high melting temperature of ~2600 °C and ability to maintain sufficient mechanical strength at temperatures exceeding 1200 °C. An outer layer of either a Zr-alloy or Al-containing stainless steel is designed to provide corrosion resistance under normal operation and oxidation resistance in steam exceeding 1000 °C for 24 hours under severe loss of coolant accidents. Due to its higher neutron absorption cross-sections, the Mo-alloy cladding is designed to be less than half the thickness of the current Zr-alloy cladding. A feasibility study has been undertaken to demonstrate (1) fabricability of long, thin wall Mo-alloy tubes, (2) formability of a protective outer coating, (3) weldability of Mo tube to endcaps, (4) corrosion resistance in autoclaves with simulated LWR coolant, (5) oxidation resistance to steam at 1000–1500 °C, and (6) sufficient axial and diametral strength and ductility. High purity Mo as well as Mo + La$_2$O$_3$ ODS alloy have been successfully fabricated into ~2-meter long tubes for the feasibility study. Preliminary results are encouraging, and hence rodlets with Mo-alloy cladding containing fuel pellets have been under preparation for irradiation at the Advanced Test Reactor (ATR) in Idaho National Laboratory. Additional efforts are underway to enhance the Mo cladding mechanical properties via process optimization. Oxidation tests to temperatures up to 1500 °C, and burst and creep tests up to 1000 °C are also underway. In addition, some Mo disks in close contact with UO$_2$ from a previous irradiation program (to >100 GWd/MTU) at the Halden Reactor have been subjected to post-irradiation examination to evaluate the chemical compatibility of Mo with irradiated UO$_2$ and fission products. This paper will provide an update on results from the feasibility study and discuss the attributes of the coated Mo cladding design to meet the challenging requirements for improving fuel tolerance to severe loss of coolant accidents.

1 Introduction

Zr-based alloys have served as the fuel cladding for light water reactors due to unique combination of low neutron cross-section, adequate corrosion resistance and mechanical properties. The reliability of Zr-alloy cladding has been steadily improved over the last five decades and has reached an excellent status in recent years. The Fukushima Daiichi accident triggered by the tsunami following an earthquake has illustrated the vulnerability of Zr-alloys to rapid steam oxidation during a severe accident when the flow of coolant into the reactor core is interrupted. Without availability of coolant flow to remove the nuclear decay heat, the fuel cladding temperature will rise rapidly. At temperatures exceeding 700–1000 °C, depending on the steam pressure, exothermic reaction of Zr with steam will release hydrogen and enthalpy or heat when ZrO$_2$ is formed. Due to the high

packing density of fuel rods in reactor cores, where a typical size core may have ~50,000 fuel rods, the total heat from zirconium oxidation may exceed that of the nuclear decay heat, if oxidation heat is all released within a short duration of an hour or so. The excessive oxidation heat may contribute to earlier melting of some reactor core components and subsequently the fuel pellets. The amount of hydrogen generated by zirconium oxidation can be in the order of exceeding 1,000 kg, and, hence, can complicate efforts by plant operators to recover the cooling system to stabilize the plant [1].

For current LWRs, it is most essential to maintain availability of coolant flow into the core under any accident conditions, and the FLEX program initiated by the US NRC is targeted to achieving that objective. Another potential defense is to replace the zirconium alloy fuel cladding with another material having substantially higher resistance to steam oxidation and capability of maintaining fuel rod integrity at elevated temperatures, which may provide meaningful coping time for plant operators to

* e-mail: bcheng@epri.com

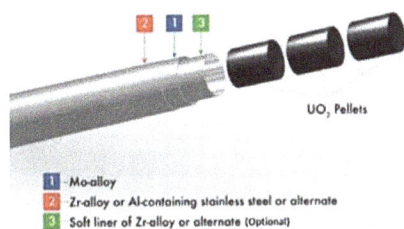

Fig. 1. Schematic of coated Mo-alloy cladding.

restore the core cooling systems [2,3]. The US Department of Energy (DOE) launched a multi-year R&D program with funding to the national laboratories and nuclear fuel vendors to develop enhanced accident tolerant fuel (ATF) in 2012 [2]. Various international programs have also been launched over the last 3 years [4]. Candidate new cladding materials have included: coated Zr-alloy, SiC-SiC$_f$ composite, Al-containing stainless steel and refractory metal (primarily molybdenum alloy).

EPRI initiated conceptual designs of coated molybdenum alloy cladding by utilizing the high temperature strength (1500 °C and beyond) of molybdenum to maintain core geometry for coolability during a design based and beyond the design based loss of coolant accidents, as illustrated in Figure 1 [1,3]. Metallurgically bonded surface coating with Al-containing stainless steel or Zr-alloy is to provide corrosion resistance during normal operation and steam oxidation resistance during loss of coolant accident with a target of surviving in steam at 1200–1500 °C for at least 24 hours. The inner surface may have a soft liner layer as an option, but its need will be determined following completion of the feasibility study.

This paper outlines the scope of the feasibility study for coated Mo-alloy cladding and shares the results obtained to date, as well as discusses the challenges ahead for completion of the feasibility study.

2 Scope of feasibility study and results

The feasibility study focuses on the following topics:

- fabricate Mo-alloy tubes with 0.2–0.25 mm wall thickness and characterize mechanical properties;
- form metallurgically bonded coating with Al-containing stainless steel or Zr-alloy of ~0.05 mm;
- weld Mo-alloy tube to endcaps;
- characterize corrosion resistance of coated and uncoated Mo-alloy tubes in autoclaves with simulated LWR coolants;
- characterize steam oxidation of coated and uncoated Mo-alloy tubes in steam at 1000–1500 °C.

Rodlets with coated Mo-alloy cladding containing enriched fuel pellets are also being fabricated for irradiation at the ATR reactor under funding by the US DOE and the Halden Reactor under its base program funding. Post-irradiation characterization of high purity Mo disks and cladding rings which were previously irradiated to high burnup at the Halden Reactor is also included in this feasibility study.

The effects of higher neutronic absorption cross-sections of Mo on fuel economics and fuel cycle designs were assessed and reported previously [1].

2.1 Fabrication of thin wall Mo-alloy tubes and mechanical properties

Mo-based alloys have been used in reducing environments as containers or thermocouples at elevated high temperatures as high as ~2000 °C. An extensive test program was undertaken by Bettis and Oak Ridge National Laboratories to evaluate the irradiation properties of various Mo-alloys in coupon forms for potential space reactor fuel material applications [5,6]. Thin wall Mo-alloy cladding suitable for LWR applications was not previously available. Under this ATF program, thin wall Mo-alloy tubes with the outer diameter of 9.4 or 10 mm (0.37 or 0.40 inches) and wall thickness of 0.2–0.25 mm have been fabricated in length of 1.5 meters (5 ft). Tubes have been made of pure Mo, including from low carbon arc cast (LCAC) and powder metallurgy (PM) billets, as well as oxide dispersion strengthened Mo-alloy (Mo-DOS) in which Mo is doped with dispersed La_2O_3. Typical La_2O_3 concentration in weight is ~0.3%, but some tubes with 1% were fabricated. Both LCAC Mo and Mo-ODS were found to retain small residual ductility following irradiation performed by the Bettis/Oak Ridge Program. Excellent tube straightness and uniform wall thickness have been achieved (Fig. 2).

Because of its high melting temperature and single phase structure (i.e., absence of phase transition) until melting at ~2600 °C, the mechanical strength will be maintained to much higher temperatures than that of common structural alloys including Zr, Fe and Ni based alloys. Pure Mo can maintain tensile strength of ~69 MPa (~10 ksi) at 1500 °C, while most other alloys would have lost their strength below 1000 °C.

The thin wall Mo and Mo-ODS tubes can achieve a tensile strength of 500–600 MPa in partially recrystallized form with good axial fracture elongation of 20–25% when measured with 3.8 cm (1.5") gauge length tube at room temperature as shown in Table 1. The strength can be controlled at 250–500 MPa depending on the final annealing condition.

One common issue associated with high strength, thin wall tubes is their propensity to axial split. A test rig with internally pressurized argon gas has been developed to test the diametral properties, including the diametral failure strength and diametral strain and creep rate at temperatures

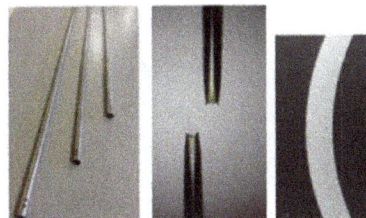

Fig. 2. Mo-alloy tubes with wall thickness of 0.2 mm and OD of 1.0 cm.

Table 1. Mechanical properties of partially recrystallized Mo-alloy tubes tested at room temperature.

	Tensile property (1.5" gauge length), 320 °C		
	PM Mo partial recrystallization		
	1	2	3
0.2% yield (ksi, MPa)	44.5 (307)	50.6 (349)	53.7 (370)
UTS (ksi, MPa)	60.1 (415)	57.7 (398)	61 (421)
Uniform elongation, %	17	18	15
Total elongation, %	20	22	24

up to 900 °C. Figure 3 illustrates the internally pressurized test results of samples of a partially recrystallized LCAC Mo tube. The measured diametral failure strength at 350 °C is ~380 MPa (55 ksi) and the diametral strain was measured as 1.4–16%.

Optimizations of the mechanical properties, particularly the diametral properties, are in progress via two approaches: firstly, modification of the tube thermo-mechanical reduction process and, secondly, controlling the microstructure of the finished tubes. For microstructure control, an induction heat treatment chamber has been set up to heat treat tube samples to various temperatures in the 1000–1700 °C range for 5–30 seconds, in order to determine the optimum heat treatment temperature for achieving the desired diametral and axial mechanical properties. Preliminary data show that diametral strain of 6–18% can be achieved at lower diametral strength.

Attempts have also been made to control the microstructure with very fine grain size, which has been reported to increase both the strength and ductility of molybdenum [7]. Data to date has indicated that rapid heat treatment and cooling via induction heat treatment cannot accomplish grain size control for high purity Mo or Mo-ODS. Further evaluation with alloying of the Mo matrix has been planned.

2.2 Formation of surface protective coating and interface stability

Molybdenum is susceptible to accelerated corrosion and oxidation in oxidizing environments at $>\sim 300$ °C forming soluble and volatile MoO_3. Due to the lack of technical basis for alloying Mo to improve its corrosion and oxidation resistance at present, surface protection via a corrosion resistant outer layer, as depicted in Figure 1, has been pursued for this ATF design. A limited effort of evaluating corrosion resistant Mo-alloys, such as Mo-Nb binary alloys, has been explored in parallel.

FeCrAl alloys with ~20%Cr and ~6%Al has been known to possess excellent corrosion resistance in simulated LWR coolants relying on the formation of a protective Cr_2O_3, as well as excellent oxidation resistance in steam to ~1450 °C owing to the formation of a thin Al_2O_3 protective oxide. Zr-alloys can be optimized to possess excellent corrosion resistance in LWRs, and will convert to ZrO_2 rapidly at >1000 °C, which is expected to be stable in steam to protect the underlying Mo cladding.

This ATF project was designed to first prepare Mo-alloy tube samples with an outer protective layer utilizing suitable surface deposition techniques for various testing to

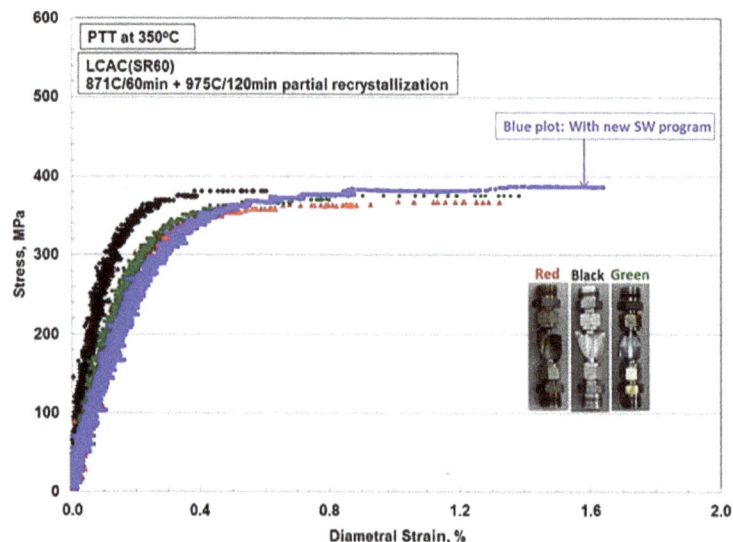

Fig. 3. Internally pressurized test of samples from a partially recrystallized LCAC Mo tube at 350 °C.

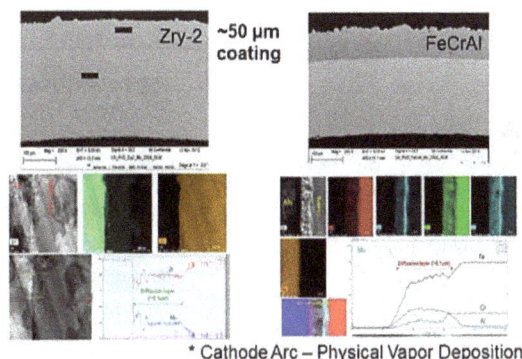

Fig. 4. Cross-section views of Mo tube with 0.2 mm wall thickness coated with a Zircaloy-2 or FeCrAl layer of ~0.05 mm. The pictures at the bottom show the presence of an inter-diffusion layer of 0.3 and 0.1 μm at the interface of the Zircaloy-2 and FeCrAl coating, respectively [10].

demonstrate feasibility of the ATF design. Tasks have been implemented to evaluate the feasibility of mechanical co-reduction. The outer coating is required to form a metallurgical bonding with the Mo cladding in order to achieve integrity in all conditions.

Formation of a corrosion and oxidation resistant outer layer of Al-containing stainless steel or Zr-alloy have been successfully developed using various deposition techniques. Cathode Arc Physical Vapor Deposition, or CA-PVD, has been found to achieve the deposited layer with (1) excellent thickness uniformity, (2) excellent adhesion of the coating to the Mo tube, (3) excellent metal density with no visible porosity of the coating. Examples of CA-PVD coated Mo tubes are illustrated in Figure 4. It can be seen an inter-diffusion layer of 0.3 and 0.1 μm forms at the interface of the Zircaloy-2 and FeCrAl coating, respectively. The inter-diffusion layer provides excellent metallurgical bonding of the coated layer to the Mo tube. Figure 5 shows a uniform coating of FeCrAl formed on a welded Mo tube.

2.3 Tube to endcap welding

Mo tube to endcap welding has been successfully demonstrated via plasma and tungsten inert gas welding as well as electron beam (EB) welding. EB welding has been found to produce smaller weld and heat affected zones, as illustrated in Figure 6a, and has been used to fabricate rodlets for irradiation at the Advanced Test Reactor (ATR) in Idaho National Laboratory (INL).

Resistance projection welding has been demonstrated to fuse Mo tube to endcap at the interface without forming a

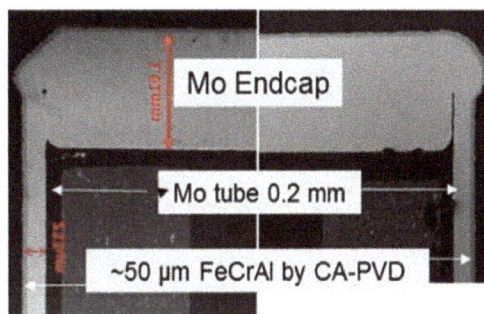

Fig. 5. Uniform FeCrAl coating thickness of 50 μm on a welded Mo tube (similar results for Zircaloy coating).

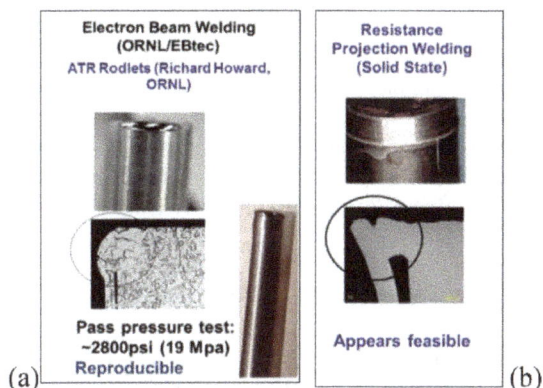

Fig. 6. Welded samples fabricated by (a) electron beam welding and (b) resistance projection welding.

fusion zone, as shown in Figure 6b. This technique is preferred for commercial deployment and will require further development.

2.4 Corrosion resistance in simulated LWR coolants

The corrosion resistance of pure Mo and Mo-ODS has been characterized in simulated BWR and PWR coolants at 288 and 330 °C, respectively, in long-term autoclave tests [8,9]. Table 2 summarizes the corrosion data in LWR coolants. It is noted that the corrosion resistance of bare Mo-alloys in simulated BWR and PWR environments is excessively high and hence will require protection with a coating of Zr-alloy or (Cr, Al)-containing stainless steel.

Preliminary results on Mo-alloys containing Nb have been found to significantly reduce the corrosion rate, as shown in Figure 7. Mo-alloy C containing 10% Nb has an

Table 2. Summary of the corrosion rate (in μm per month) of bare and coated Mo-alloy tubes in simulated LWR coolants.

Test condition	Mo/ML	Zr-coated	FeCrAl-coated
PWR – 330 °C, 3.6 ppm H_2	~5	Very low	Very low
BWR-HWC – 288 °C, 0.3 ppm H_2	~9	Very low	Very low
BWR-HWC – 288 °C, 1 ppm O_2	~40	Very low	Very low

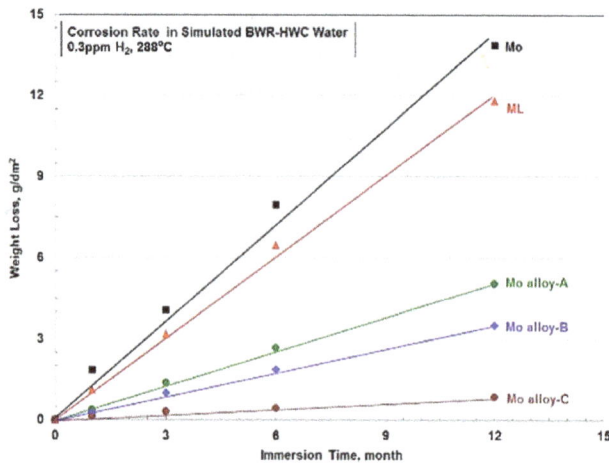

Fig. 7. Corrosion resistance of pure and ODS Mo and Nb-containing Mo samples in simulated a BWR-HWC coolant.

order of magnitude lower corrosion rate ($\sim$0.5 μm/mo) than that of the bare, pure Mo tube in a simulated BWR water containing 0.3 ppm dissolved hydrogen.

2.5 Oxidation resistance in 1000–1500 °C steam

The oxidation resistance of uncoated Mo in oxidizing steam, i.e. steam containing free oxygen, has been known to be poor due to the formation of volatile MoO_3 at temperature $>\sim$600 °C. In a severe loss of coolant accident, it is anticipated that the reactor core will have excess hydrogen due to corrosion and oxidation of various metallic components in the core. The Zr-alloy and FeCrAl coatings are to provide protection from steam oxidation at elevated temperatures.

In pure steam and steam plus 10% hydrogen at 1000 °C, uncoated Mo-alloy cladding has been found to have reasonable oxidation rate with a thicken loss of $\sim$20–25 μm per day, which is two order of magnitude lower than that of Zr-alloys [9,10].

Coated Mo tubes with both-end welded received full protection from FeCrAl in 1000 °C for 7 days, shown in Figure 8. The Zircaloy-4 coated Mo tube showed delamination of the ZrO_2 and suffered some loss in the Mo wall thickness, but the Mo tube remains intact after 7 days.

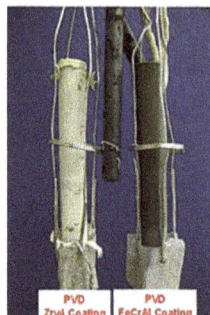

Fig. 8. Mo tubes with both-ends welded to endcaps and coated with either FeCrAl or Zircaloy-4 after testing in 1000 °C steam for 7 days.

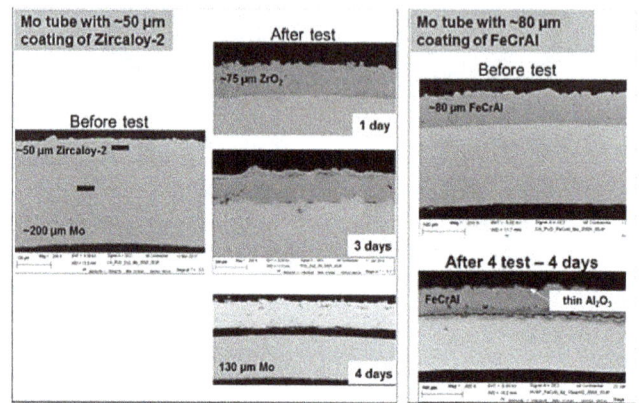

Fig. 9. Open-ended Mo-alloy tube samples with coating after oxidation test in 1000 °C for up to 4 days.

Figure 8 shows cross-sections of open-ended Mo tube samples coated with either Zircaloy-2 or FeCrAl after being tested in 1000 °C steam for up to 4 days. The Zircaloy-2 coated Mo tube shows the conversion of Zircaloy-2 to a dense ZrO_2 which protects the Mo tube from steam oxidation for up to 4 days. The good integrity of the oxide formed from Zircaloy-2 differs from the flaky oxide formed from Zircaloy-4 as shown in Figure 9.

Investigation of the integrity of the ZrO_2 oxide formed from the Zircaloy-4 coating in Figure 8 and Zircaloy-2 coating in Figure 9 has found that slightly over half of the alloying elements, Fe, Cr, Sn and Ni (Zircaloy-2 only), added to zirconium for corrosion and oxidation resistance were substantially lost during the CA-PVD coating process. This likely explains the flaky oxide observed in Figure 8. New target materials with substantially higher alloying concentrations have been obtained to improve the alloy chemistry of the coating. For the longer term, mechanical co-reduction should avoid the difficulty of controlling the alloy chemistry.

An interesting observation of the 1000 °C steam tested samples is the formation of an inter-diffusion zone in the FeCrAl-coated samples. The thickness was measured to be 2, 4, and 5 μm after 1, 3, and 7 days, respectively. Clearly, the inter-diffusion was not significant enough to impact the cladding integrity at 1000 °C for 7 days or much longer.

Tests in steam at 1200–1500 °C has been in process.

2.6 Irradiation properties

Previous irradiation tests have shown Mo and its alloys, like other metals and alloys, will suffer from irradiation embrittlement and the mechanism has been attributed to grain boundary weakness. It is anticipated that improving the diametral mechanical properties through process and micro-structure optimization, as discussed above, may lead to better properties after irradiation. Furthermore, additions of minor concentrations of alloying elements, such as B, Al, Ti, and Zr have been suggested to strengthen the grain boundaries. New dilute Mo-alloys are being fabricated for testing.

Short rodlets with Mo and Mo-ODS cladding and 4% enriched fuel pellets have been under preparation for irradiation in the ATR beginning in 2016. Additional irradiation in the Halden Reactor has been under preparation to begin in 2016.

3 Summary on results to date and challenges ahead

This feasibility study of coated Mo-alloy cladding for accident tolerant fuel has demonstrated excellent corrosion in simulated LWR coolants and great oxidation properties in 1000 °C. The thin wall Mo tube has adequate mechanical strength and ductility, and further improvement in the diametral ductility is continuing to provide better resistance to pellet-to-cladding mechanical interaction. Fabrication of rodlets for irradiation is underway.

For commercial deployment, it may be necessary to fabricate the coated tubes via mechanical co-reduction, rather than relying on the PVD process, as the latter can be prohibitively expensive and challenging for quality control. A feasibility study on forming metallurgical bonding between the outer coating and the Mo-alloy via hot hydrostatic pressuring (Hipping), and subsequent mechanical co-reduction to produce coated Mo tubes has been underway. It is also necessary to incorporate fabrication process control to achieve optimized mechanical properties.

While the corrosion resistance of the coated cladding may be adequate, it is desirable to improve the corrosion resistance of the Mo tube to ensure full reliability of the cladding in the event of loss of the outer coating. To achieve this feature, Mo-alloys, primarily Nb-containing ones, are being studied.

The authors gratefully acknowledge the important contributions of Todd Leonhardt (Rhenium Alloys Inc.) for fabrication of thin wall Mo-alloy tubes; of Stu Malloy and Andy Nelson (LANL) for steam test studies; of Sam Armijo and Peter Ring for evaluation of induction heat treatment and hipping and metal bonding; of Kristine Barret of INL and Richard Howard of ORNL for supporting ATR irradiation of Mo cladded rodlets. The authors thank Jeff Deshon (EPRI) for management support. EPRI Fuel Reliability Program and EPRI Technology Innovation Program provide funding support. Areva has entered collaboration with EPRI on future Mo-alloy cladding development beginning in 2015. Generous collaborations from the US Department of Energy and the national laboratories are also appreciated.

References

1. B. Cheng, P. Chou, Y.-J. Kim, Enhancing fuel resistance to severe loss of coolant accidents with molybdenum-alloy fuel cladding, Paper 100075, in *WRFPM2014, Sendai, Japan, September, 2014* (2014)

2. S. Bragg Sitten, Application of MELCOR to ATF concepts for severe accident analysis, in *EPRI/INL/DOE workshop on Accident Tolerant Fuel, San Antonio, February, 2014* (2014)

3. B. Cheng, Fuel behavior in severe accidents and Mo-alloy based cladding to improve accident tolerance, Paper A0034, in *TopFuel 2012, Birmingham, UK* (2012)

4. *IAEA Symposium on "Accident Tolerant Fuel Concepts For Light Water Reactors", Oak Ridge National Laboratory, Tennessee, October 13–16, 2014* (Proceedings to be published by IAEA)

5. B.V. Cockeram, R.W. Smith, K.J. Leonard, T.S. Byun, L.L. Snead, J. Nucl. Mater. **382**, 1 (2008)

6. T.S. Byun, M. Li, B.V. Cockeram, L. Snead, J. Nucl. Mater. **376**, 240 (2008)

7. G. Liu et al., Nanostructured high-strength molybdenum alloys with unprecedented tensile ductility, Nat. Mater. **12**, 344 (2013)

8. Y.-J. Kim, B. Cheng, P. Chou, Molybdenum alloys for accident tolerant fuel cladding: high temperature corrosion and oxidation behavior, Paper 100144, in *TopFuel 2014, September 14–17, 2014, Sendai, Japan* (2014)

9. A.T. Nelson, E.S. Sooby, Y.-J. Kim, B. Cheng, S.A. Maloy, High temperature oxidation of molybdenum in water vapor environments, J. Nucl. Mater. **448**, 441 (2013)

10. Y.-J. Kim, B. Cheng, P. Chou, Steam oxidation behavior of protective coatings on LWR Mo cladding for enhancing accident tolerance at high temperatures, Paper A0172, in *TopFuel 2015, September 13–17, Zurich, Switzerland* (2015)

10

Eddy current testing system for bottom mounted instrumentation welds

Noriyasu Kobayashi[1*], Souichi Ueno[1], Naotaka Suganuma[1], Tatsuya Oodake[2], Takeshi Maehara[3], Takashi Kasuya[3], and Hiroya Ichikawa[4]

[1] Power and Industrial Systems Research and Development Center, Toshiba Corporation, 8, Shinsugita-cho, Isogo-ku, Yokohama 235-8523, Japan
[2] Power and Industrial Systems Research and Development Center, Toshiba Corporation, 1, Komukaitoshiba-cho, Saiwai-ku, Kawasaki 212-8581, Japan
[3] Keihin Product Operations, Toshiba Corporation, 2-4, Suehiro-cho, Tsurumi-ku, Yokohama 230-0045, Japan
[4] Isogo Nuclear Engineering Center, Toshiba Corporation, 8, Shinsugita-cho, Isogo-ku, Yokohama 235-8523, Japan

Abstract. The capability of eddy current testing (ECT) for the bottom mounted instrumentation (BMI) weld area of reactor vessel in a pressurized water reactor was demonstrated by the developed ECT system and procedure. It is difficult to position and move the probe on the BMI weld area because the area has complexly curved surfaces. The space coordinates and the normal vectors at the scanning points were calculated as the scanning trajectory of probe based on the measured results of surface shape on the BMI mock-up. The multi-axis robot was used to move the probe on the mock-up. Each motion-axis position of the robot corresponding to each scanning point was calculated by the inverse kinematic algorithm. In the mock-up test, the probe was properly contacted with most of the weld surfaces. The artificial stress corrosion cracking of approximately 6 mm in length and the electrical-discharge machining slit of 0.5 mm in length, 1 mm in depth and 0.2 mm in width given on the weld surface were detected. From the probe output voltage, it was estimated that the average probe tilt angle on the surface under scanning was 2.6°.

1 Introduction

Eddy current testing (ECT) techniques to detect a defect, especially a stress corrosion cracking (SCC), on a reactor vessel (RV) and reactor internals have been developed as one of the surface inspection methods for nuclear power plants [1–7]. As a part of maintenance methods for the RV and reactor internals, laser peening and underwater laser beam welding techniques to prevent and repair from the SCC have been developed [8–11]. These inspection and maintenance techniques can contribute to shorten their work period, including the initial set-up because it is possible to work underwater without draining the reactor coolant. In order to provide faster services, the defect detection capability of the ECT probe using the cross coil has been estimated for the inspection before and/or after the underwater laser beam welding for the dissimilar metal welding area at the RV nozzle in pressurized water reactors (PWRs) [12–14]. The ECT system, including the small ECT probe and the probe moving equipment based on the portable laser peening system, has been developed for the bottom mounted instrumentation (BMI) weld area in PWRs [15]. In this development, the SCC detection capability of the system was demonstrated by moving the probe on the area of 10 mm × 6 mm of the BMI mock-up [15].

More precise probe action control is required to move the probe on the whole BMI weld area because the area has complexly curved surface and the narrow spaces. We measured the surface shape of weld area using the laser displacement meter and made the scanning trajectory of the probe based on the shape measurement data of the complex surface. As a BMI mock-up test, the ECT probe was automatically moved on the whole BMI weld area by the multi-axis robot. From the test results, we evaluated the defect detection capability of the ECT system and the probe tilt angle on the weld surface under scanning. In this paper, we describe the procedure of BMI mock-up test; the results of measuring weld surface shape and defect detection tests.

*e-mail: noriyasu.kobayashi@toshiba.co.jp

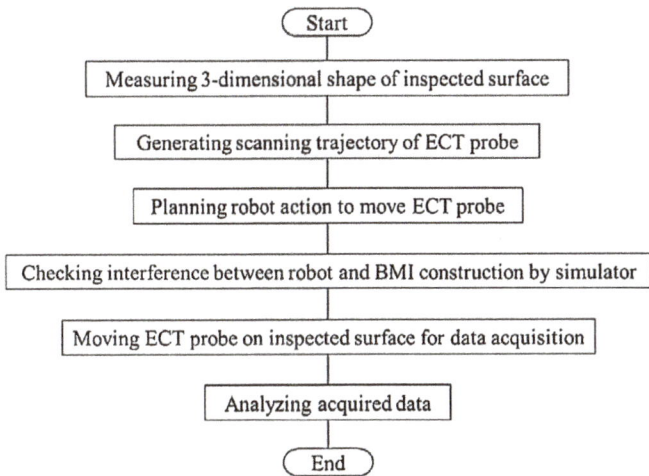

Fig. 1. Process flow diagram of ECT for BMI welds.

2 Weld surface shape measurement

2.1 Procedure of BMI mock-up test

A process flow diagram of an ECT for BMI welds is shown in Figure 1. The three-dimensional shape of inspected weld surface was measured in order to generate the precise scanning trajectory of ECT probe. After generating the trajectory, the action of multi-axis robot, which moves the probe along the scanning trajectory on the weld surface, was planned and checked for the interference between the robot and the BMI mock-up. The probe was moved on the whole BMI weld area by the robot to acquire the ECT defect detection data. Finally, we analyzed the acquired data, including the signal processing for noise rejection and signal identification.

2.2 Measurement of surface shape

We measured the surface shape of the weld area on the BMI mock-up for generating the scanning trajectory of ECT probe. A half of weld area was the target for scanning by the ECT probe because the mock-up is axisymmetric. The measurement range of surface shape is the half side of weld area and within approximately 60 mm in radius centering on the BMI nozzle as shown in Figure 2. The sensor head of laser displacement meter (KEYENCE, LJ-G200) mounted on the multi-axis robot measured the three-dimensional surface shape in an approximately 0.5 mm interval at the points of approximately 60 mm from the center of nozzle within the measurement range. The sensor head rotated round trip half side around the nozzle. The laser spot size is $180\ \mu m \times 70$ mm. The base work distance is 200 ± 48 mm.

2.3 Scanning trajectory of probe

The space coordinates and the normal vectors at the scanning points as the scanning trajectory of ECT probe on welds were generated based on the measured results of weld surface shape on the BMI mock-up. The calculated results of the trajectory were shown in Figure 3. The blue range in Figure 3 is the measured surface shape. The red arrows indicate the calculated results of normal vector for determining the probe angles at the scanning points. The probe is set on the inspected surface, as the probe central axis is adjusted to coincide with the normal vector at each scanning point.

2.4 Multi-axis robot

The multi-axis robot, which moves the ECT probe, is shown in Figure 4. The robot has three translation axes and

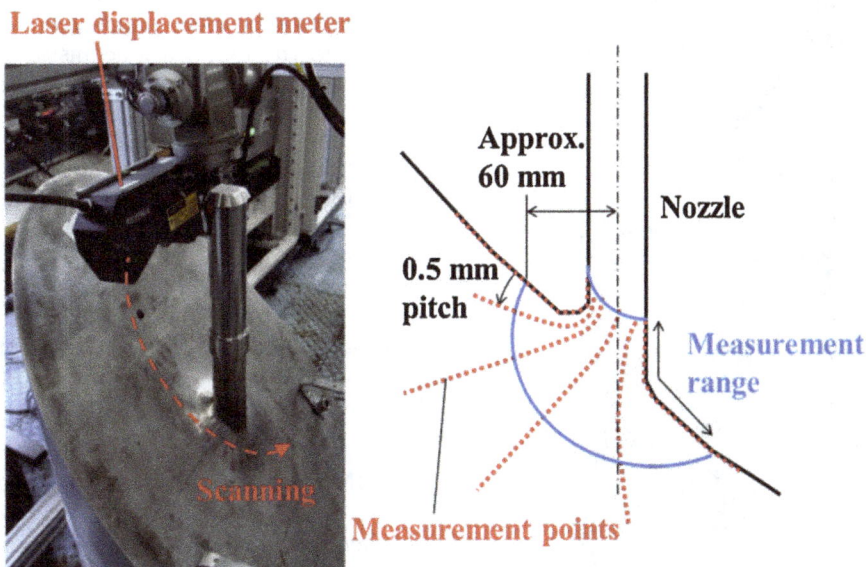

Fig. 2. Measurement of weld surface shape on BMI mock-up.

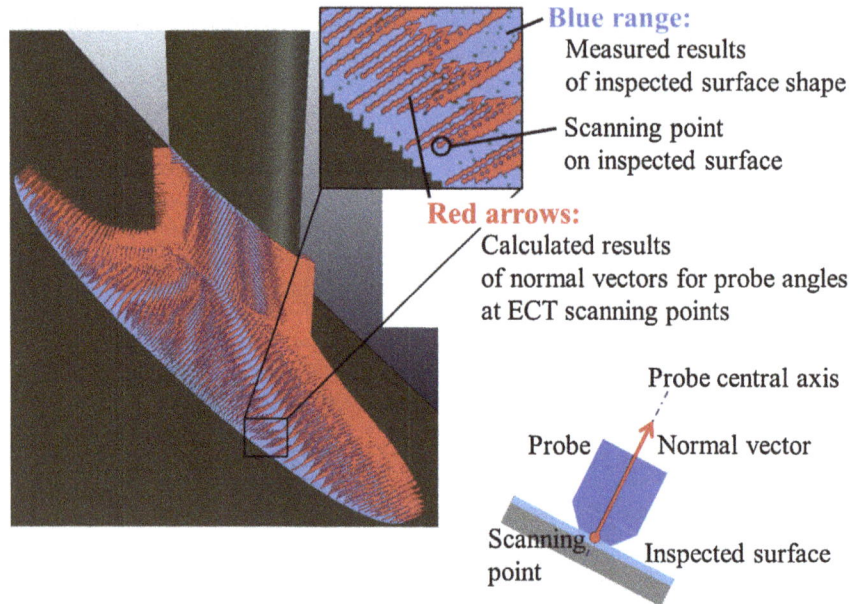

Fig. 3. Scanning trajectory of ECT probe.

four rotation axes. The probe was mounted on the end of the robot arm. Each motion axis position as a robot action corresponding to each scanning point was calculated by the inverse kinematic algorithm. After the two rotating motion axis positions were provided as the constant values, the other motion axis positions were led in the calculation. The probe-scanning trajectory shown in Figure 3 was divided into the three ranges (nozzle, J-welds and build-up welds)

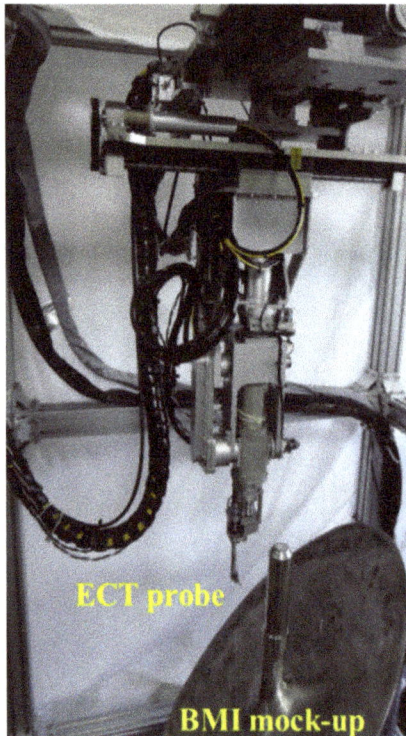

Fig. 4. Multi-axis robot.

as shown in Figure 5. The three different algorithms for the three ranges were used to prevent from the interference between the robot and the BMI mock-up. It was confirmed not to interfere between the robot and the mock-up using the three-dimensional simulator before the mock-up test.

3 Experimental apparatus and methods

3.1 ECT system

A block diagram of the ECT system is shown in Figure 6. This system consists of the ECT probe, the multi-axis robot, the robot controller, the ECT data acquisition system and the ECT data analysis system. The probe was moved to a start point of scanning manually. As soon as a scanning was started under the order from the robot controller, the ECT data acquisition system received the coordinate data of the start point from the controller and voltage signals from the probe. After the acquisition system paired the coordinate data with the voltage signals and saved them into a memory, the acquisition system sent an acquisition end signal at the start point to the controller. The controller automatically moved the probe to the next scanning point using the multi-axis robot based on the probe-scanning trajectory. These movements were repeated until the entire scanning is completed. The ECT data analysis system read the scanning coordinate data and the probe output signals, and conducted the signal processing and displayed the inspected results.

3.2 BMI mock-up and scanning range

A schematic of the BMI mock-up simulating the outermost nozzle at the bottom of RV is shown in Figure 7 [15]. The nozzle was fixed to the bottom of RV by a tungsten inert gas

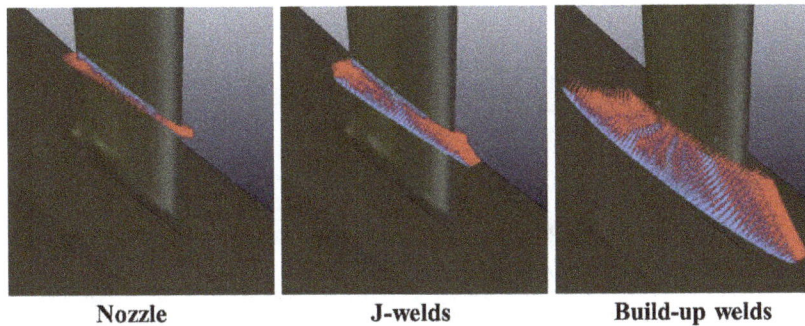

Fig. 5. Divided scanning trajectories.

(TIG) welding. The surface of weld area was machined smoothly. Both the nozzle and the weld metal are made of alloy 600. Artificial and circumferential defects were given on the weld surface at the points of 10 mm from the outer surface of nozzle. The type and size of defects are described in Table 1. It was defined that the top of the mock-up is at 0° in circumferential angle as shown in Figure 7. The length of SCC shown in Table 1 is the value of indication on penetrant testing (PT).

The scanning range by the ECT probe is shown in Figure 8. The start point of scanning is on the outer surface of nozzle at 0° in circumferential angle and approximately 3 mm above the J-weld. The probe was moved in less than 0.5 mm interval within the scanning range in a circumferential direction and made several round trips half side around the nozzle. An end point of scanning is on the build-up weld surface at 0° in circumferential angle and approximately 40 mm from the center of nozzle. This scanning range includes the nozzle, the J-welds, the build-up welds and the artificial defects.

3.3 Experimental and calibrating conditions

The experimental and calibrating conditions are shown in Table 2. We used the developed ECT probe [15] that has small-sized cross coil and the higher directional

characteristics of magnetic field in the mock-up test. The diameter of the probe tip that has contact with an inspected surface is 3.4 mm. The probe operated with the differential mode at the frequency of 250 kHz, 500 kHz and 1 MHz. The calibration block made of alloy 600 has an EDM slit of 80 mm in length, 1 mm in depth and 0.3 mm in width. The thickness of calibration block is 20 mm. We calibrated the output voltage and the phase angle to 2 V and 90° using this block in air, respectively.

4 Experimental results of mock-up test

The C scope images as seen through the signal processing for the absolute values of imaginary part of ECT output voltages at a frequency of 250 kHz, 500 kHz and 1 MHz are shown in Figure 9. At a frequency of 250 kHz, the clear signals from the defect A (SCC), the defect B (EDM) and the defect C (EDM) were confirmed. It was considered that the signals from the defect D (EDM) and the defect E (SCC) were not detected because the volumes of the defect D and the defect E are smaller than those of the other

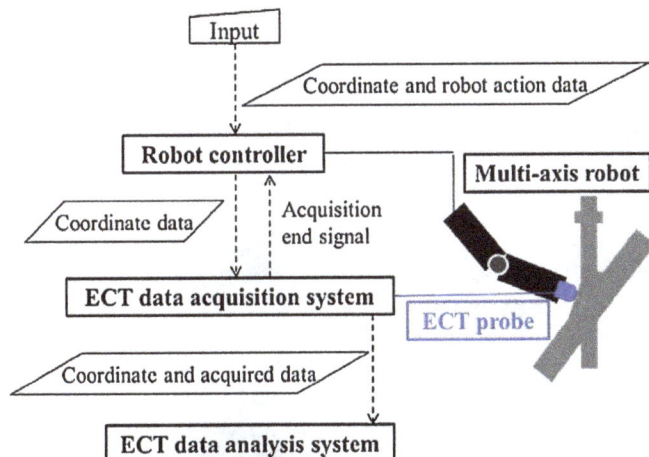

Fig. 6. Block diagram of ECT system.

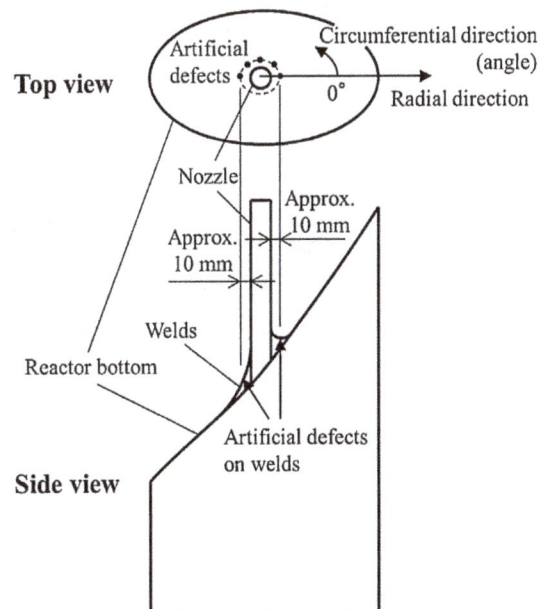

Fig. 7. Schematic of BMI mock-up [15].

Table 1. Artificial defects.

Defect	Type	Length (mm)	Depth (mm)	Width (mm)
A	SCC	Approx. 6	No data	No data
B	EDM slit	0.5	1.5	0.2
C	EDM slit	0.5	1.0	0.2
D	EDM slit	0.3	1.0	0.2
E	SCC	Approx. 3	No data	No data

EDM: electrical-discharge machining; Approx.: approximately.

defects. The maximum output voltages of the detected defects were 0.93 V in the defect A, 0.33 V in the defect B and 0.24 V in the defect C. The ratio of maximum output voltages between the defects B and C was 1.4. This value was roughly equal to the ratio of the volumes between the defects B and C (1.5). On the other hand, the maximum output voltage of the noises was 0.25 V. Under the following three assumptions:

a. the maximum output voltage from the defect is proportional to the defect volume;
b. the criterion for defect detection is that the signal to noise ratio is more than 2;
c. the ECT can detect the defect of 0.5 mm and more in depth,

it is estimated that the minimum EDM slit size that this ECT system can detect is approximately 2.3 mm in length, 0.5 mm in depth and 0.2 mm in width. The output voltage of the defect E was less than 0.125 V. It was difficult to recognize the figure of the defect E visually. If the width of the defect E was 0.05 mm, it is evaluated using the above assumption (a) that the depth of the defect E is less than 0.38 mm. Although the length of the defect E is longer than the lengths of the other EDM slits, it is considered that the signal from the defect E was not detected because the width and depth are smaller than those of the other EDM slits.

Table 2. Experimental and calibrating conditions.

ECT probe	Cross coil
Operation mode	Differential
Frequency (kHz)	250, 500, 1000
Atmosphere	In air
Calibration block	Alloy 600 (20 mm in thickness)
EDM slit	
Length (mm)	80
Depth (mm)	1
Width (mm)	0.3
Calibrated	
Output voltage (V)	2
Phase angle (°)	90

The noises increased at higher frequencies. It was considered that the sensitivity of ECT probe for the change of surface shape was increased by the dense eddy current on the mock-up surface layer because of shallower skin depths at higher frequencies. The skin depth of alloy 600 at each frequency is shown in Table 3. The skin depth at each frequency is the same or less than the depth of the EDM slit given on the calibration block, 1.0 mm. When a defect depth is the same or more than 1.0 mm, a phase angle of a signal from a defect indicates the near-calibrated value, approximately 90° or −90°. Positive and negative values mean that directions of defects are mutually orthogonal. A phase angle of an eddy current lags to the direction of material depth [16]. Therefore, a phase angle of a signal from a defect may lag behind the calibrated value if a defect depth is less than 1.0 mm. The measured phase angles of the signals from the defects (A, B and C) and the noises (F, G and H) in Figure 9 are shown in Table 4. It was reasonable that the phase angles of the signals from the defects A, B and C were approximately 90° or −90°. It was considered that the noises F and G were caused by the change of

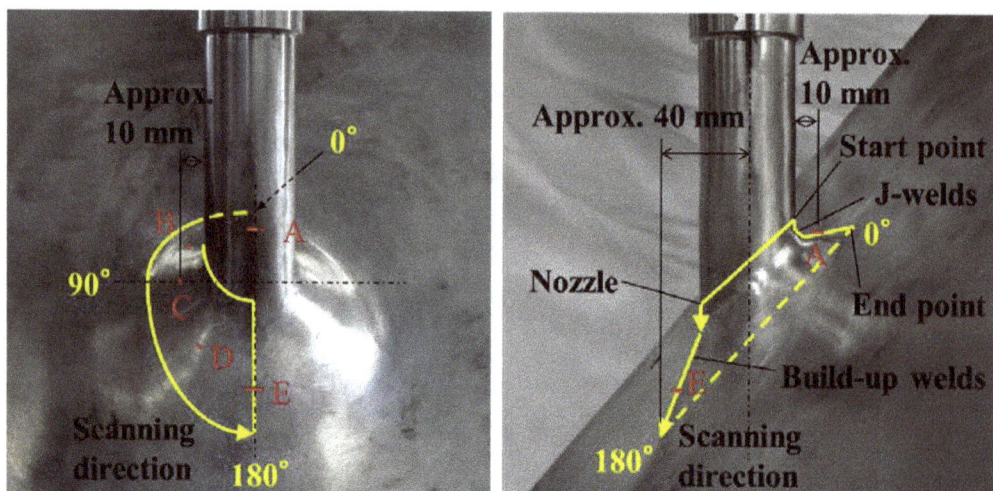

Fig. 8. ECT scanning range.

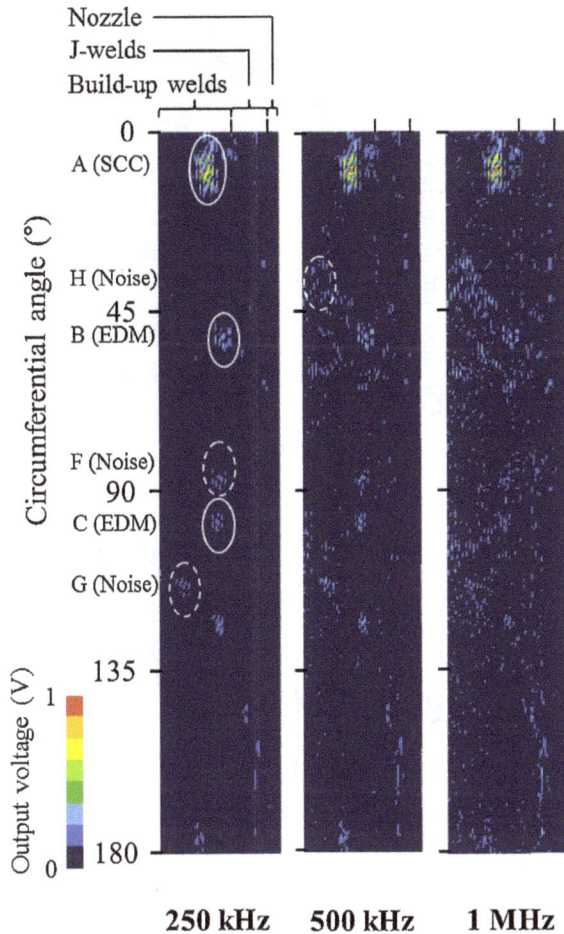

Fig. 9. C scope images of ECT output voltages.

Table 4. Measured phase angles.

Frequency (kHz)	250	500	1000
Phase angle (°)			
Defect A	99	91	80
Defect B	103	100	88
Defect C	90	87	83
Noise F	98	90	83
Noise G	−95	−94	−112
Noise H	Out of measure	−160	−105

scanning in this mock-up test could be roughly estimated. First of all, we investigated the relationship between the probe tilt angle on the flat surface specimen and the output voltage of single coil. The single coil means one of two coils that compose the cross coil and is more sensitive for the probe tilt angle than the differential mode of cross coil using the two coils. The area without the EDM slit of the calibration block shown in Table 2 was used as a flat surface specimen in this measurement. We defined the meaning of the probe tilt direction and angle as shown in Figure 10, respectively.

The measured result of the relationship between the probe tilt direction and the output voltage of single coil at a frequency of 500 kHz is shown in Figure 11. The measured result at a frequency of 500 kHz provided the smallest output voltage variation in the prior confirmation. The tilt angle is 9° at constant angle. The output voltage was normalized by the value at the tilt direction of 0° because the sensitivity of the ECT data acquisition system in this measurement was different from that in the mock-up test. The output voltage was constant within the variation of 15% by a change in the tilt direction. We measured the

surface shape more than 1.0 mm in depth because their phase angles were approximately 90° or −90° as in the case of the defects A, B and C. The phase angle of noise H at the frequency of 500 kHz was largely lagging behind −90°. It was considered that the phase lag was observed at the frequency of 500 kHz having the deeper skin depth because the depth of the surface shape change was much less than 1.0 mm. The maximum output voltages from the defects are roughly equal at each frequency. It was estimated that the best frequency for the defect detection by the used ECT probe in this BMI mock-up test is 250 kHz.

5 Discussions

5.1 Relationship between probe tilt angle and output voltage of single coil

Because the probe tilt angle on the inspected surface influences a defect detection capability, the tilt angle under

Table 3. Skin depth of alloy 600 at each frequency.

Frequency (kHz)	250	500	1000
Skin depth (mm)	1.0	0.71	0.50

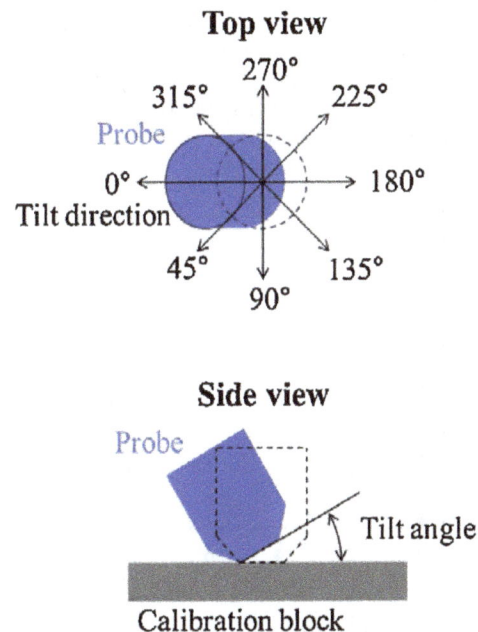

Fig. 10. Probe tilt direction and angle.

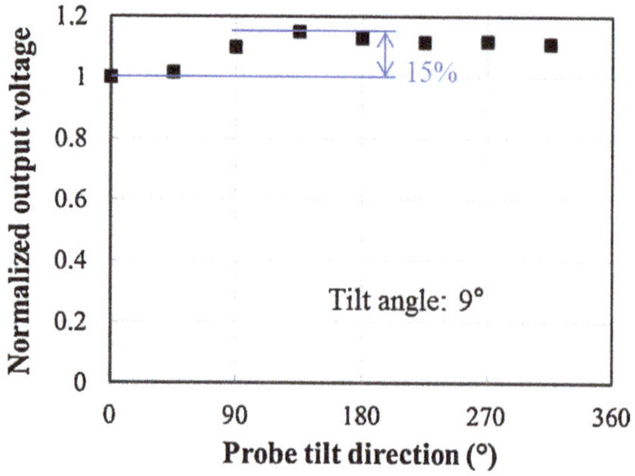

Fig. 11. Relationship between probe tilt direction and output voltage of single coil.

relationship between the probe tilt angle and the output voltage of single coil under the condition of 0° in probe tilt direction. The sensitivity of the ECT data acquisition system in this measurement was the same as that in the mock-up test. The measured result of the relationship between the probe tilt angle and the output voltage of single coil at a frequency of 500 kHz is shown in Figure 12. The tilt direction is 0° at constant direction. The output voltage monotonically increased by the increase of tilt angle. We assumed that the output voltage is proportional to the tilt angle and used the function of linear approximation as an evaluation formula to estimate the tilt angle from the output voltage in the mock-up test.

5.2 Estimation of probe tilt angle under scanning in mock-up test

The output voltage of single coil was measured while moving the ECT probe on the J-welds of the BMI mock-up at a frequency of 500 kHz. The probe tilt angle under scanning from the measured output voltage could be roughly estimated using the evaluation formula as described previously. The estimated probe tilt angle is shown in Figure 13. The average and the maximum angles were 2.6° and 8.5°, respectively. As the photographs show in

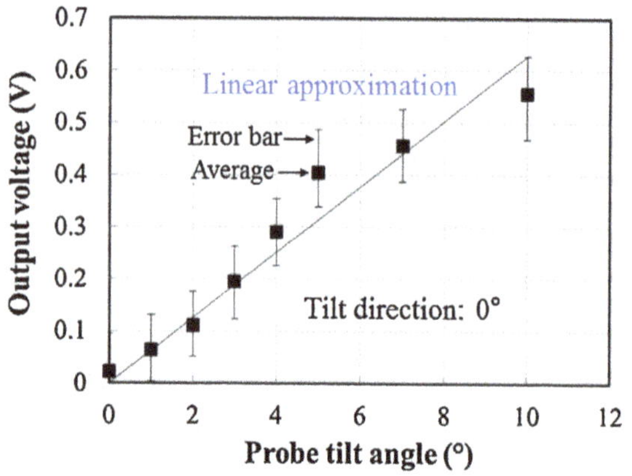

Fig. 12. Relationship between probe tilt angle and output voltage of single coil.

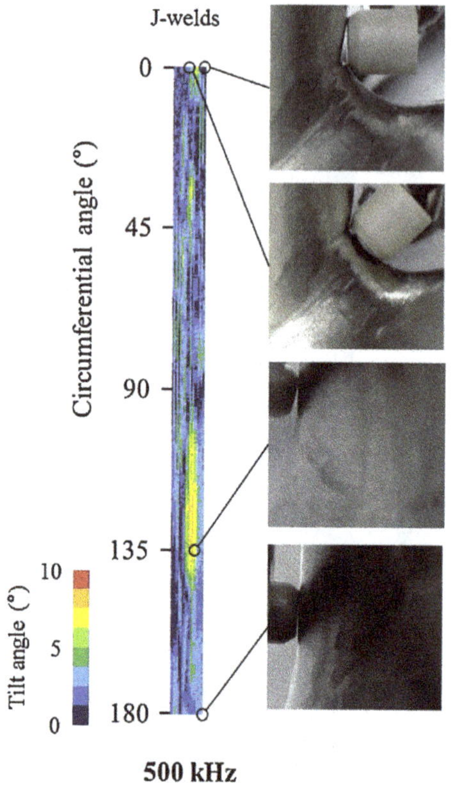

Fig. 13. Estimated probe tilt angles under scanning on J-welds of BMI mock-up.

Figure 13, the probe came into contact with the surfaces of J-welds on most of the scanning range. Our goal for the probe tilt angle is within 3° on all the scanning range because it was previously confirmed that the sensitivity of this developed ECT probe for the machined slit of 0.5 mm in depth and 0.4 mm in width decreased nearly 1 dB when the probe tilt angle increased from 0° to 3° [15]. It was considered that the cause of tilt angle of more than 3° on the partial scanning range is the accuracy of installation position between the multi-axis robot and the BMI mock-up. The highly accurate measurement and the correction of installation position are the action assignments for the inspection of actual plant.

6 Conclusions

The ECT for the whole weld area on the BMI mock-up was demonstrated using the developed ECT system and procedure in order to verify the defect detection capability for the BMI welds. The surface shape of weld area on the BMI mock-up was measured for generating the scanning trajectory of ECT probe. The space coordinates and the normal vectors at the scanning points as the scanning trajectory were calculated based on the measured results of weld surface shape. Each motion-axis position of the multi-axis robot corresponding to each scanning point was calculated by the inverse kinematic algorithm. The BMI mock-up test was performed using the developed ECT probe with the cross coil in the differential mode. The artificial SCC and EDM slits given on the build-up weld area were detected in the mock-up test. From the result of detecting defects, it is shown that this ECT system can detect a defect of approximately 2.3 mm in length, 0.5 mm in depth and 0.2 mm in width as the defect detection capability for the BMI welds. It was estimated that the average and the maximum probe tilt angles were 2.6° and 8.5°, respectively. The highly accurate measurement and the correction of installation position between the multi-axis robot and the inspected BMI for controlling the probe tilt angle are the action assignments for the actual use.

References

1. T. Kasuya, T. Uchimoto, T. Takagi, H. Huang, Simulation of Shroud Inspection based on Eddy Current Testing, Maintenology **3**, 51 (2004)

2. A.L. Hiser Jr. Cracking in Alloy 600 Penetration Nozzles - A Regulatory Perspective, in *Proceedings of the 12th International Conference on Nuclear Engineering (ICONE12), ICONE12-49226, Arlington, USA, 2004* (2004)

3. W. Bamford, J. Hall, A Review of Alloy 600 Cracking in Operating Nuclear Plants Including Alloy 82 and 182 Weld Behavior, in *Proceedings of the 12th International Conference on Nuclear Engineering (ICONE12), ICONE12-49520, Arlington, USA, 2004* (2004)

4. L. Chatellier, S. Dubost, F. Peisey, B. Richard, L. Fournier, Taking Advantage of Signal Processing Techniques for the Life Management of NPP Components, in *Proceedings of the ASME 2006 Pressure Vessels and Piping Conference (PVP2006), PVP2006-ICPVT-11-93313, Vancouver, Canada, 2006* (2006)

5. Z. Chen, L. Janousek, N. Yusa, K. Miya, A Nondestructive Strategy for the Distinction of Natural Fatigue and Stress Corrosion Cracks Based on Signals from Eddy Current Testing, J. Press. Vessel Technol. **129**, 719 (2007)

6. P. Anderle, L. Skoglund, R.S. Devlin, J.P. Lareau, H. Lenz, D. E. Seeger Jr., F.G. Whytsell, Reactor Vessel Head Penetration Inspection–Past, Present and Future, in *Proceedings of the 8th International Conference on NDE in Relation to Structural Integrity for Nuclear and Pressurized Components, We.2.C.3, Berlin, Germany, 2010* (2010)

7. Z. Kuljis, B. Lisowyj, Characterizing Austenitic Materials and Nickel Alloys with Electro-Magnetic Imaging, in *Proceedings of the ASME 2012 Pressure Vessels and Piping Conference (PVP2012), PVP2012-78502, Toronto, Canada, 2012* (2012)

8. Y. Kanazawa, M. Tamura, Underwater YAG Laser Welding Technique, Toshiba Review **60**, 36 (2005)

9. M. Yoda, M. Tamura, Underwater Laser Beam Welding Technology for Reactor Vessel Nozzles of PWRs, Toshiba Review **65**, 36 (2010)

10. I. Chida, K. Shiihara, T. Fukuda, W. Kono, M. Obata, Y. Morishima, Study on Laser Beam Welding Technology for Nuclear Power Plants, Transactions of the Japan Society of Mechanical Engineers Series B **78**, 445 (2012)

11. I. Chida, T. Uehara, M. Yoda, H. Miyasaka, H. Kato, Development of Portable Laser Peening Systems for Nuclear Power Reactors, in *Proceedings of the 2009 International Congress on Advances in Nuclear Power Plants (ICAPP'09), ICAPP09-9029, Tokyo, Japan, 2009* (2009)

12. N. Kobayashi, T. Kasuya, S. Ueno, M. Ochiai, Y. Yuguchi, C. S. Wyffels, Z. Kuljis, D. Kurek, T. Nenno, Utility Evaluation of Eddy Current Testing for Underwater Laser Beam Temperbead Welding, in *Proceedings of the 8th International Conference on NDE in Relation to Structural Integrity for Nuclear and Pressurized Components, We.2.B.2, Berlin, Germany, 2010* (2010)

13. S. Ueno, N. Kobayashi, T. Kasuya, M. Ochiai, Y. Yuguchi, Defect Detectability of Eddy Current Testing for Underwater Laser Beam Welding, in *Proceedings of the 19th International Conference on Nuclear Engineering (ICONE19), ICONE19-43658, Osaka, Japan, 2011* (2011)

14. N. Kobayashi, T. Kasuya, S. Ueno, M. Ochiai, H. Ichikawa, Feasibility Assessment of Eddy Current Testing in Underwater Laser Beam Welding, Journal of the Japanese Society for Non-Destructive Inspection **61**, 475 (2012)

15. N. Kobayashi, S. Ueno, I. Chida, M. Ochiai, T. Fujita, H. Ichikawa, Development of Eddy Current Testing System for Bottom-Mounted Instrumentation Nozzle in Reactor Pressure Vessel, Maintenology **13**, 106 (2014)

16. H.B. Libby, *Introduction to Electromagnetic Non-destructive Test Methods* (Wiley-Interscience, 1971)

11

Preliminary accident analysis of Flexblue® underwater reactor

Geoffrey Haratyk and Vincent Gourmel*

DCNS, 143 bis, avenue de Verdun, 92442 Issy-les-Moulineaux, France

Abstract. Flexblue® is a subsea-based, transportable, small modular reactor delivering 160 MWe. Immersion provides the reactor with an infinite heat sink – the ocean – around the metallic hull. The reference design includes a loop-type PWR with two horizontal steam generators. The safety systems are designed to operate passively; safety functions are fulfilled without operator action and external electrical input. Residual heat is removed through four natural circulation loops: two primary heat exchangers immersed in safety tanks cooled by seawater and two emergency condensers immersed in seawater. In case of a primary piping break, a two-train safety injection system is actuated. Each train includes a core makeup tank, an accumulator and a safety tank at low pressure. To assess the capability of these features to remove residual heat, the reactor and its safety systems have been modelled using thermal-hydraulics code ATHLET with conservative assumptions. The results of simulated transients for three typical PWR accidents are presented: a turbine trip with station blackout, a large break loss of coolant accident and a small break loss of coolant accident. The analyses show that the safety criteria are respected and that the reactor quickly reaches a safe shutdown state without operator action and external power.

1 Introduction

Flexblue® is a small modular reactor delivering 160 We to the grid. The power plant is subsea-based (up to 100 m depth and a few kilometres away from the shore) and transportable. It is entirely manufactured in shipyard (no large outdoor activities) and requires neither levelling nor civil engineering work, making the final cost of the output energy competitive. Thanks to these characteristics and its small electrical output, Flexblue® makes the nuclear energy more accessible for countries where regular large land-based nuclear plants are not adapted, and where fossil-fuelled units currently prevail on low-carbon solutions. Immersion provides the reactor with an infinite heat sink – the ocean – around the containment boundary, which is a cylindrical metallic hull hosting the nuclear steam supply systems (Tab. 1).

Several modules can be gathered into a single seabed production farm and operate simultaneously (Fig. 1). The reactor is meant to operate only when moored on the seabed. Every three years, production stops and the module is emerged and transported back to a coastal refuelling facility, which hosts the fuel pool. This facility can be shared between several Flexblue® modules and farms. During operation, each module is monitored and possibly controlled from an onshore control centre. Redundant

submarine cables convey both information and electricity output to the shore. A complete description of the Flexblue® concept, including market analysis, regulation and public acceptance, security and environmental aspects, is found in Haratyk et al. [1]. The purpose of this paper is to present the first accident analysis of Flexblue® and to discuss the performance of its innovative passive safety systems.

2 The reactor and its safety features

2.1 The reactor

The reactor and all the nuclear systems carrying primary coolant are hosted in one of the four watertight compartments of the module (other compartments host the turbo generator, an onboard control room, I&C control panels, a living area and process auxiliaries) see Figure 2. The reactor compartment boundary forms the third barrier of confinement. The reference design of Flexblue® includes a loop-type pressurized water reactor (PWR), with two horizontal steam generators (SGs) and two motor coolant pumps. This technology enjoys a long experience, both in civil power production and in naval propulsion. Primary loops are designed to ease natural circulation when coolant pumps are turned off: pumps are plugged directly on steam generators outlet in order to eliminate the usual U-shape

*e-mail: vincent.gourmel@dcnsgroup.com

Table 1. Flexblue® module main characteristics.

Parameter	Value
Unit power rating (MWe)	160
Length (m)	150
Diameter (m)	14
Immersion depth (m)	100
Fuel cycle length (months)	40
Lifetime (years)	60

Table 2. Flexblue® reactor characteristics.

Parameter	Value
Thermal power	530 MW$_{th}$
Reactor core	77 fuel assemblies
Fuel assembly	17 × 17 rods, 2.15 m high
Enrichment	<5%
Average power density	70 kW/L
Hot rod peaking factor	2.26
Reactor coolant pressure	155 bar
ΔT core	30 °C
Steam generators	2 recirculation SGs
SGs pressure	62 bar (saturated)

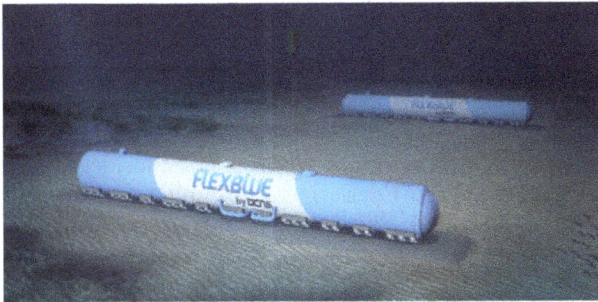

Fig. 1. Artist view of a Flexblue® farm.

pipe between SGs and pumps. The reactor core uses classical fuel assembly technology: 17 × 17 fuel bundles with an enrichment below 5%. Active length of the core is 2.15 m. Reactivity is controlled without soluble boron and only with burnable poison and control rods. This feature is very important because it allows major space savings (no boron tank). The core design is deeply described in [2] (Tab. 2).

2.2 The safety systems

The safety systems of Flexblue® are designed in order to operate passively according to the IAEA passivity definition [3]. All safety functions are fulfilled without any operator action and external electrical input. The little amount of energy needed for actuation and monitoring is supplied by onboard, redundant, rechargeable emergency batteries featuring two weeks of autonomy.

Chain reaction can be stopped by two diversified devices: the control rods and an emergency boron injection system, which is actuated only in case of anticipated transient without scram (ATWS). Both these devices can independently shut down the reactor and keep it subcritical up to cold shutdown state [2].

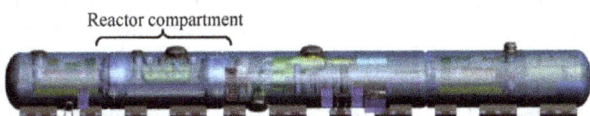

Fig. 2. Profile view of a Flexblue® module.

Residual heat removal is performed by four cooling loops, each one able to remove 50% of decay heat:

- two primary chains are connected to the primary circuit: each one includes an inlet pipe connected to a hot leg, a heat exchanger (PPHX) immersed in a large safety water tank, and an outlet pipe connected to a cold leg. The intermediate heat sinks formed by the two safety tanks are cooled by the ocean through the metallic hull;
- two secondary chains are connected to the secondary circuit: each one includes an inlet pipe connected to a main steam line, an emergency condenser directly immersed in seawater and an outlet pipe connected to a feedwater line.

Thanks to the infinite heat sink – seawater – and to the elevation difference of the heat sink with respect to the heat sources, the four chains operate passively by natural circulation. In normal conditions operation, they are closed by pneumatic valves and open to their fail-safe position when electrical load is lost. The targeted long-term safe state of the reactor is a shutdown state where continuous cooling of the reactor core is achieved by natural circulation (Fig. 3).

Protection against loss of coolant accidents is ensured by two passive safety injection trains. Each one includes a direct vessel injection (DVI) line fed by three injection sources: a core makeup tank (CMT) pressurized by the primary circuit, a classical accumulator pressurized at 50 bar by nitrogen and a large safety tank, which feeds the primary circuit by gravity when primary pressure has decreased to near containment pressure. In addition,

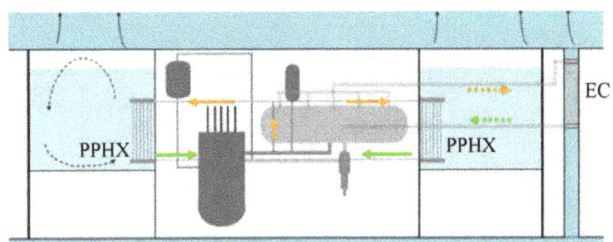

Fig. 3. Targeted safe state when primary circuit is intact.

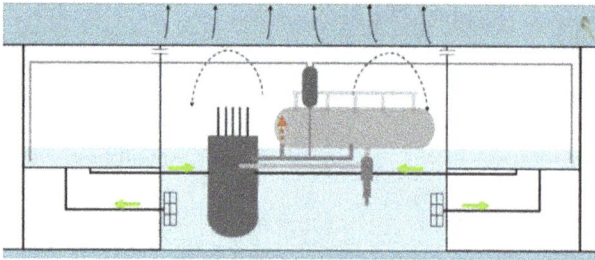

Fig. 4. Targeted safe state when primary circuit has failed.

a two-train automatic depressurization system (ADS) is connected to the pressurizer (PZR) and to the hot legs to generate a controlled depressurization of the primary circuit, which enables faster injection. Once these systems have actuated, the long-term equilibrium state is reached when the safety tanks are empty and the reactor compartment is flooded (Fig. 4). At that point, a passive recirculation path is in place: water boils off the core, is released in the containment, condensates on the containment walls, collects in the sump and is injected back into the reactor pressure vessel through sump screens and DVI lines by gravity. Decay heat is transported and removed through the metallic hull. Thanks to the unlimited heat sink (the ocean), grace period is theoretically infinite for both targeted states, which is a breakthrough in nuclear safety.

The two large safety tanks not only play the two roles of intermediate heat sinks and injection sources, but also a third role of suppression pools – when a leak leads to a quick containment pressurization. They also act as radiation shield to protect workers and systems located in the adjacent compartments. Confinement of the radioactive isotopes is guaranteed by three hermetic barriers: fuel cladding, primary circuit and containment boundary formed by the hull and the compartment walls (Fig. 5). The capability of the containment to reject decay heat to seawater has been investigated by Santinello et al. [4]. Results show that the process is satisfactory and enables all decay heat removal.

3 Analysis tool and reactor model

3.1 ATHLET

ATHLET (Analysis of Thermal-Hydraulics of LEaks and Transients) is a thermal-hydraulic system code developed by the German technical safety organization GRS. It is applicable to the analysis of PWR and BWR, and has already been used for the analysis of transients involving

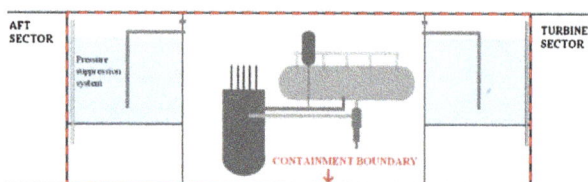

Fig. 5. Limit of the containment boundary.

horizontal SGs, similar to the ones of Flexblue®. It is composed of four main calculation modules: thermo-fluid dynamics, heat transfer and heat conduction, neutron kinetics, and control & balance of plant. ATHLET validation work (including for passive systems) is presented in [5].

3.2 Modelization

Flexblue® reactor is modelled (see Fig. 6) with ATHLET in accordance with GRS guidelines [5,6]. The nodalization of the circuits is performed in order to get both a sufficient accuracy and an acceptable calculation time. Two core channels are modelled: an outer ring and an inner channel where power density is higher. In this latter one, the hot fuel pin is modelled to calculate peak clad and fuel temperatures. The two loops are modelled, as well as all the safety systems with the exception of the emergency boron injection system (failure of scram is not considered in the studied transients). Pressurizer and piping are considered perfectly insulated. The injection sources (tanks and accumulators) are not borated. The active auxiliary systems and the regulations are not modelled. There are three fluid dynamics systems in the model: the primary one (primary circuit and connected systems), the secondary one (secondary circuit and connected systems) and seawater.

The model considers a 2.5-second delay between the scram signal and the full insertion of control rods. Decay heat calculation is based on formulas from Todreas and Kazimi [7], extracted from standards of American Nuclear Society [8], and then conservatively increased by 20% to respect NRC guidelines [9]. Figure 7 presents the considered decay heat for the accident analyses.

4 Main hypotheses

Reactor core is at 100% of its nominal power (530 MW$_{th}$) at the beginning of each transient. The initiating event always leads to a turbine trip (or is the turbine trip itself), which is followed 3 s later by the loss of electrical load. The only electrical sources available are the emergency batteries, which are able to monitor and control the safety systems, and to open or close some valves. The action of other active components and systems is not considered. It is a conservative assumption because the active systems would only have a favourable effect in the performed transients. In a future work, active systems will be modelled to study more transients (for example, active injection should be considered after a steam generator tube rupture).

The opening time of the valves is 2 s with the exception of the ADS valves, which have a longer, preset opening time. Pressurizer and steam generators safety valves setpoints are respectively 171 bar and 83 bar, with a one-second opening time. Even if it is planned to install flow restrictors in the pipes, their effects are not taken into account in the accident analysis, which is a conservative measure. To provide a sufficient core flow when a pump coast down happens, coolant pumps models include a rotating inertia represented in Figure 8: the driving

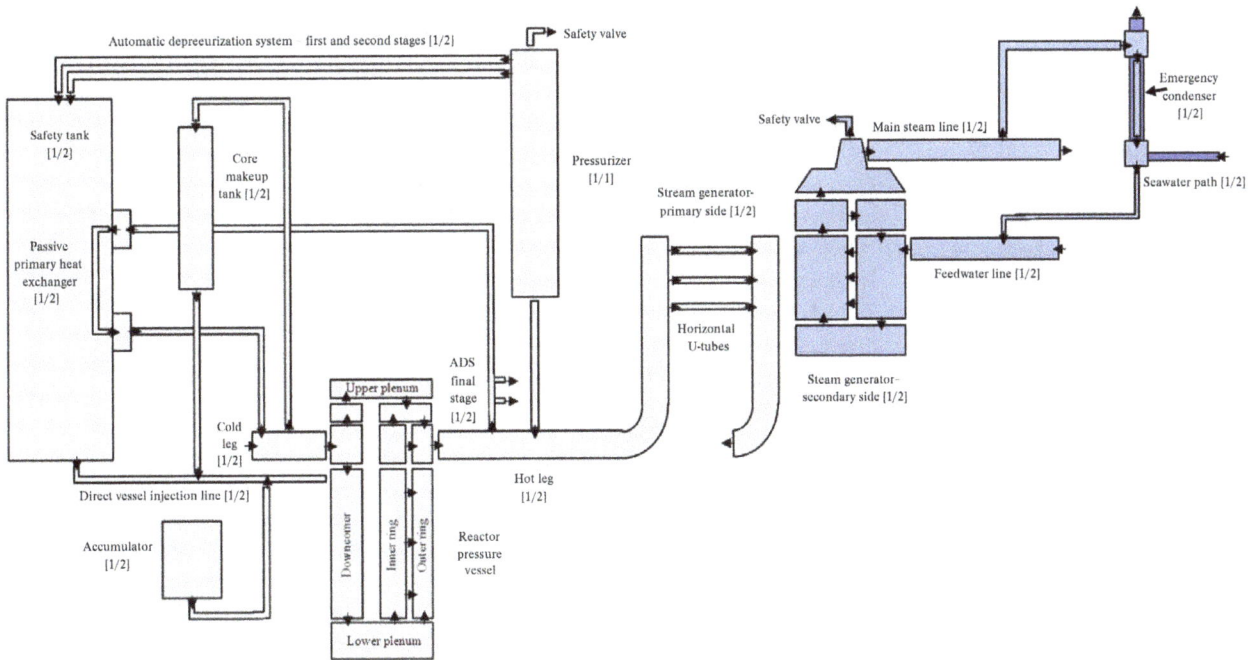

Fig. 6. ATHLET model. Dimensions are not representative. The model includes about 200 objects composed of about 1000 control volumes.

pressure reaches 50% of the nominal value after 5 s and 0% after 30 s.

The containment pressure is set constant at 1 bar during the transients, so the leak flow is maximized when a break occurs. Heat sink temperature (seawater) is conservatively

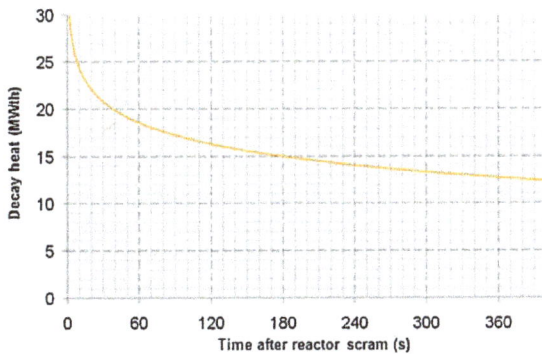

Fig. 7. Decay heat of Flexblue® core.

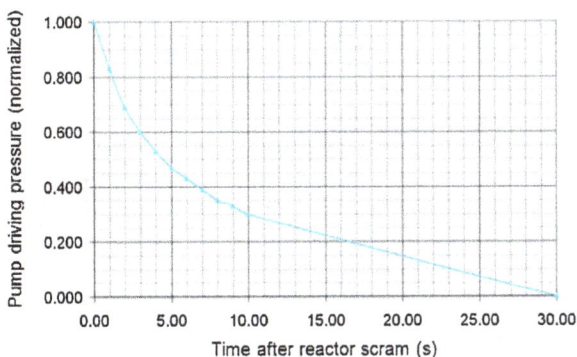

Fig. 8. Coolant pumps driving pressure after reactor scram.

set at 35 °C. Heat transfer between safety tanks and seawater through the metallic hull is not modelled, which is conservative. None of the steam generators tubes is considered clogged. The detailed design of the Flexblue® core was not yet available when these analyses have been conducted. As a consequence, the neutronic data of a typical German Konvoi have been used. The conservative nature of these input data is not established. As mentioned in Section 8, core behaviour is to be watched closely with accurate neutronic data when available. Average burn-up is 8.1 GWD/t and maximal burn-up is 45 GWD/t. The actuation logic of emergency signals and passive systems with the treatment delays considered are presented in Table 3.

5 Turbine trip

The simulated transient starts with a turbine trip that causes a loss of offsite power.

5.1 Results

The results are described in Table 4 and Figures 9–13.

5.2 Discussion

When turbine trip is triggered, steam and feedwater lines are immediately closed (0.15 s). Reactor scram happens more than 4 s later. During this time interval, primary and secondary pressures strongly increase (Figs. 9 and 10) because core is at full nominal power and heat is not removed to any heat sink. After reactor scram, core power

Table 3. Safety signals (conservative delays for actuation).

Signal	Trigger(s)	Delay (s)
Reactor protection	High containment pressure *or* low pressurizer pressure	0.9
Reactor scram	Reactor protection *or* low pump speed *or* high pressurizer pressure	1
Coolant pump stop	Reactor protection *or* reactor scram *or* ADS first stage opening *or* low pressurizer level	3
Feed and steam lines isolation	Reactor protection *or* turbine trip	0.15
Core makeup tank injection	Reactor protection *or* low pressurizer level	2
Emergency condensers actuation	SG high pressure *or* passive primary cooling actuation	0.5
Passive primary cooling actuation	CMT injection *or* high pressurizer level	4
ADS first stage opening	CMT injection *and* low level in both CMTs	20
ADS second stage opening	ADS first stage opening	70
ADS final stage opening	ADS second stage opening *and* very low level in both CMTs	250

Table 4. Sequence of turbine trip accident.

Time	Event
0 s	Turbine trip
0.15 s	Steam line and feedwater line isolation
3 s	Station blackout. Coolant pumps coast down with their inertia. Minimum DNBR is reached (3.87)
4.6 s	Reactor scram actuated by pumps low speed
6 s	Emergency condensers are connected to SGs
7.3 s	Maximum primary pressure and temperature are reached (167 bar, 322 °C)
14 s	Maximum secondary pressure and temperature are reached (83 bar, 298 °C)
8 min	Heat removed by ECs becomes greater than heat removed by SGs which is greater than decay heat
90 min	Low pressurizer level leads to CMTs injection and passive primary cooling actuation
100 min	Primary temperature falls below 215 °C
150 min	CMTs natural circulation stops
167 min	End of simulation

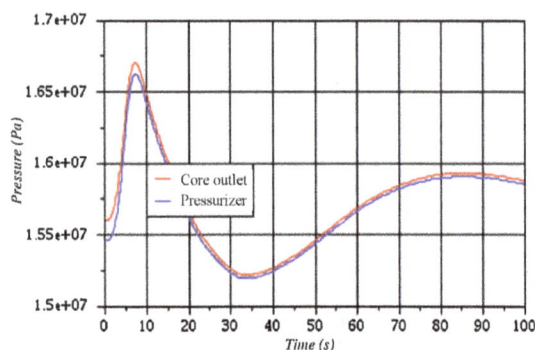

Fig. 10. Secondary pressure (Pa) with focus on first 40 s.

quickly decreases (Fig. 7) and high pressure in SGs leads to the connection of both emergency condensers (ECs) that transfer almost 16 MW$_{th}$ to seawater in the first minutes of the transient (Fig. 11). Maximum primary conditions are reached at $t = 7.3$ (167 bar, 322 °C) and maximum secondary conditions are reached 7 s later (82.7 bar, 298 °C). Both pressurizer pressure and SGs pressure remain lower than their safety valves opening setpoints.

Concerning the boiling crisis risk in this transient, the results provide a minimum departure of nucleate boiling ratio (DNBR) of 3.87 at $t = 3$ s. Clad surface temperature

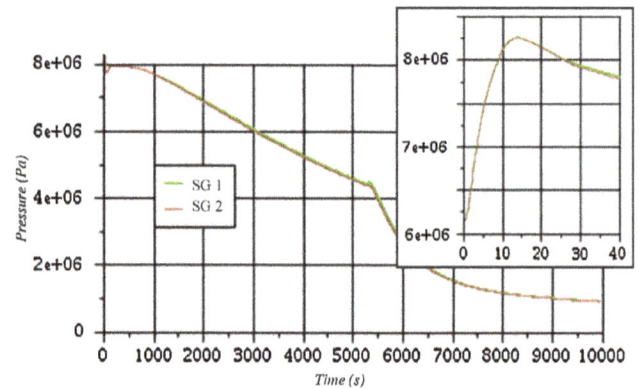

Fig. 9. Primary pressure (Pa) during first 100 s.

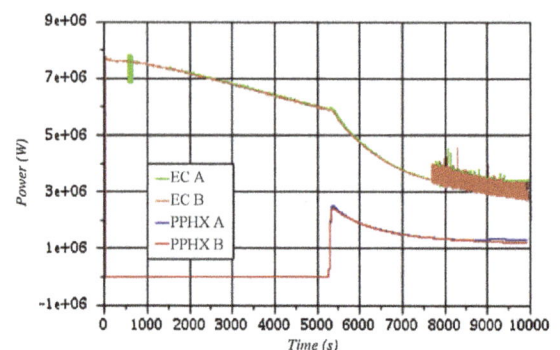

Fig. 11. Emergency heat removal by ECs and PPHXs (W).

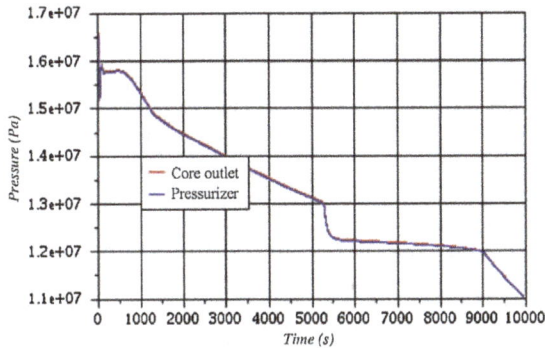

Fig. 12. Primary pressure (Pa).

does not exceed 400 °C and fuel centreline temperature does not exceed 1350 °C (melting temperature is 2700 °C). Thus, first barrier safety criteria are comfortably respected. However, system code ATHLET is not a very refined code to investigate core thermal-hydraulics. Deeper investigation of core behaviour is needed with a core analysis code (e.g. COBRA – COolant Boiling in Rod Arrays).

Eight minutes after turbine trip, the emergency heat removal by the condensers becomes greater than the heat removed by the steam generators, which is already greater than core decay heat. This situation will not change later: starting from this point, the thermal-hydraulic conditions in primary and secondary systems continuously decrease. The critical phase of the transient has passed. Natural circulation is now well established and core is passively cooled. Primary flow is around 200 kg/s. As primary fluid temperature decreases, water density lowers and pressurizer water level falls. At $t = 90$ min (5400 s), this level reaches the CMTs injection setpoint.

Cold water (50 °C) contained by CMTs flows into the vessel through direct vessel injection lines while hot water from primary circuit fills back the CMTs. This circulation causes a sudden drop of primary pressure and temperature (Figs. 12 and 13). At $t = 150$ min (9000 s), natural circulation in the CMTs stops and core is once again only cooled by passive exchangers. Cooling is very efficient because CMTs injection signal also leads to passive primary heat exchangers (PPHXs) actuation. PPHXs and ECs remove together 8.5 MW$_{th}$ (Fig. 10), while decay power is around 6.5 MW$_{th}$ at that point.

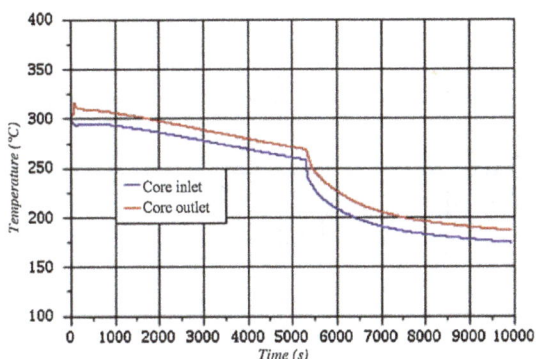

Fig. 13. Primary temperature (°C).

During the entire simulated transient, void fraction at core outlet is zero. Primary fluid remains monophasic in all the primary circuit and in the CMTs – with the exception of the pressurizer where primary fluid is at saturation conditions. Its means that low CMT level signal– which would open the ADS – is not close to be actuated. Saturation margin is always greater than 30 °C, and the liquid water in the vessel upper head does not flash. Primary temperature reaches the EPRI criterion for safe shutdown (215 °C [10]) after 100 min, far earlier than the EPRI objective of 24 h. At the end of the simulation (2 h 47 min), primary temperature at core outlet has decreased down to 175 °C. Following the PPHXs actuation, safety tanks have heated up by only 11 °C, demonstrating an important thermal inertia.

Analysis of ATHLET results shows that safety systems of Flexblue® reactor can handle a turbine trip followed by a station blackout without any operator action. After a tense sequence during the first minutes due to a high core power, the passive cooling systems quickly remove a power greater than decay heat. Safety criteria of the first barrier are fully respected and safety valves of primary and secondary circuits are not challenged. A safe shutdown state is reached in less than 2 h where primary circuit is pressurized and core is durably cooled. This quick cooling raises a concern about the thermo-mechanical stresses in the pressurizer surge line. The adiabatically modelled pressurizer stays quite hot (above 300 °C), so the temperature difference between bottom and top of the surge line is important. A refined model of the pressurizer and a thermo-mechanical study of the surge line are needed to address this concern.

6 Large break loss of coolant accident

Even if such a break could be excluded by application of the "break preclusion" concept, a double-ended guillotine break of a cold leg (400 mm diameter primary pipe between pump and pressure vessel) is postulated here.

6.1 Results

The results are presented in Table 5 and Figures 14–19.

6.2 Discussion

The double-ended guillotine break of a reactor primary leg is a very brutal transient. Leak flow reaches immediately 18,500 kg/s and core is entirely uncovered after 4.5 s (Fig. 14), which stops the chain reaction and brings down the fuel centreline temperature (Fig. 15) because of the loss of moderator. Primary pressure drops from 155 bar to 1 bar in 20 s, which triggers the reactor protection signal. It leads to reactor scram, secondary lines isolation and connection of CMTs on DVI lines following the sequence presented in Table 5. But the first injection sources that feed the vessel are the accumulators. Indeed, CMTs pressure (which is equal to primary pressure) is quickly lower than accumulators

Table 5. Sequence of large break loss of coolant accident (LOCA).

Time	Event
0 s	Double-ended guillotine break of cold leg A
0.1 s	Chain reaction is stopped by lack of moderation
3 s	Low PRZ pressure causes a reactor protection signal
3.15 s	Steam line and feedwater line isolation
4 s	Reactor scram
4.5 s	Reactor vessel has lost all coolant
5 s	Opening of CMT injection valves
5.5 s	Accumulators injection starts
6 s	Coolant pumps stop on their inertia
9 s	Passive primary cooling actuation
9.5 s	Emergency condensers (ECs) are connected to steam generators (SGs)
45 s	Maximum clad surface temperature reached (725 °C)
77 s	End of accumulators injection
98 s	CMTs injection starts
13 min	CMTs low level causes opening of the ADS 1st stage
15 min	Opening of the ADS 2nd stage
16 min	Core is flooded again by primary coolant
43 min	Opening of the ADS final stage
90 min	End of simulation

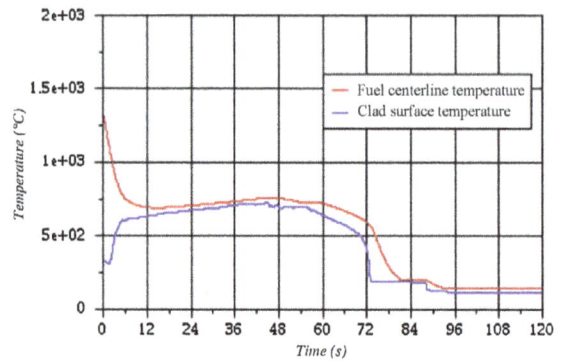

Fig. 15. Fuel centreline and clad surface temperatures at hot spot.

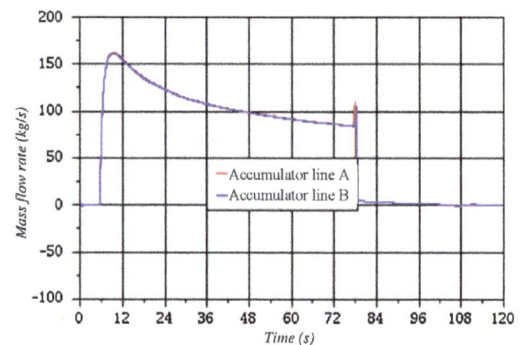

Fig. 16. Accumulators injection mass flow (kg/s).

pressure (50 bar). At $t = 5.5$ s, injection starts with a high mass flow (160 kg/s per line, Fig. 16). It stops the vessel water level fall (Fig. 14) and slows down the heat up of the fuel cladding (Fig. 15). Water level is stabilized for 15 s at rod bundles bottom (Fig. 14) so that vapour formation can initiate cooling of the fuel by single-phase gas heat transfer. In the following 50 s, accumulators injection enables the partial replenishment of the core water inventory (Fig. 14). Clad surface temperature reaches a maximum (705 °C) and starts decreasing. Maximum clad oxidation is 0.4% (far below the authorized 17%) and hydrogen generation does not

exceed 0.03% of authorized value. First barrier safety criteria are fully respected. During this time interval, primary temperature varies from 100 °C to 300 °C (superheated steam), but eventually falls down to saturation conditions at 100 °C (Fig. 17).

At $t = 77$ s, accumulators injection stops but CMTs and safety tanks quickly resume injection 20 s later. Liquid level in the core slightly decreases again without any consequence on clad surface temperature. At $t = 98$ s, CMTs injection starts. Total flow rate is limited (around 20 kg/s) but sufficient to resume vessel reflood. Until $t = 13$ min, direct vessel injection is dominated by CMTs flow and exhibits wide oscillations between 0 and 30 kg/s per line (Fig. 18). CMTs injection is driven by primary pressure, which is impacted by steam generation and consequently primary flow. This feedback could be the

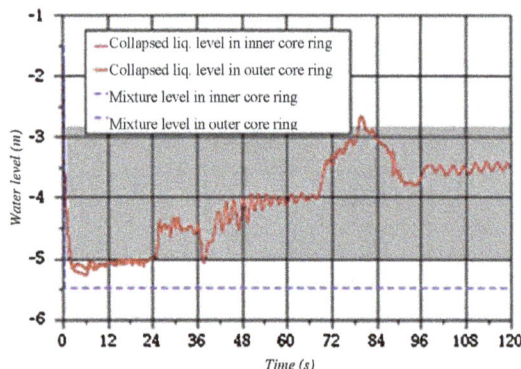

Fig. 14. Water level in the vessel. Grey area represents core zone.

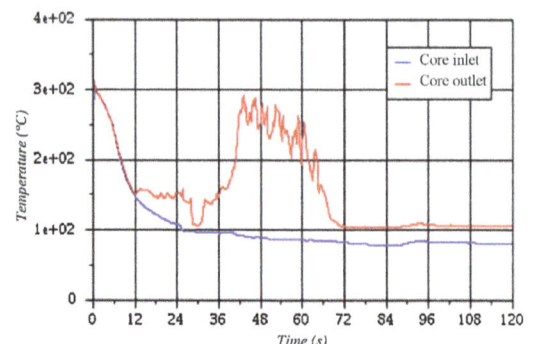

Fig. 17. Primary temperature (°C).

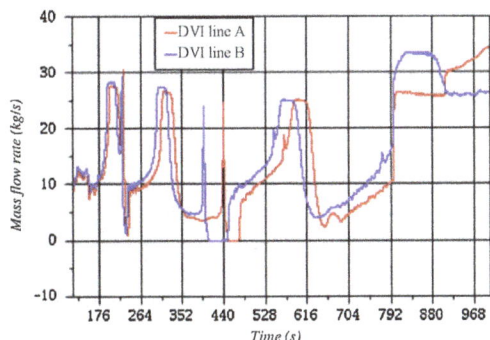

Fig. 18. Total injection mass flow from 120 s to 1000 s (kg/s).

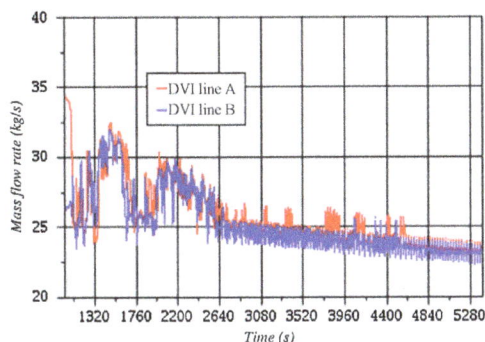

Fig. 19. Total injection mass flow from 1000 s to 5400 s (kg/s).

source of these density-wave oscillations. Low CMTs levels are reached at $t = 13$ min (780 s) and trigger the automatic depressurization system. The actuation of the ADS three stages does not impact the primary pressure – which is already very low – but eases significantly the injection mechanism by opening the hot legs of the loops. The single failure criterion is applied on one of the ADS final stage valves, without significant effect. Safety tanks now dominate the injection and total DVI flow is quite steady during the following hour, from 30 to 23 kg/s per line (Fig. 19). At $t = 16$ min, core coolant inventory is definitely recovered.

Concerning the passive heat exchangers, their actuation is very quick (9 s). Thanks to ECs heat removal, secondary pressure is kept below SG safety valves setpoint and falls to 1 bar in around 30 min. However, ECs and PPHXs play a minor role in this transient; most of decay heat is released to the containment through the break.

At the end of the simulation, liquid level in the vessel is 1 m above the top core, primary pressure is close to 1 bar and primary temperature at core outlet is 88 °C. Void fraction at core outlet oscillates between 5% and 35%. Injection flow is 46 kg/s. The remaining water inventory in the safety tanks enables an 8-h injection at this magnitude. Despite the severity of the transient, these results prove that Flexblue® passive safety systems can handle a loop double-ended break accident during the first hour and a half, without any operator action and with only emergency batteries as electrical input. Regarding the long-term mitigation of the accident, the expected safe shutdown state is represented in Figure 26. The feed and bleed process

slowly floods the containment up to a level where sump natural circulation actuates passively. Steam exiting the core through the ADS is condensed on the hull internal side and comes back down into the sump. Heat is eventually removed by seawater. AP1000 design has already been licensed with a comparable but time-limited strategy [11]. A specific study is to be conducted to prove its effectiveness with the Flexblue® design.

7 Small break loss of coolant accident

The postulated break is a 10 mm diameter break on one of the two direct vessel injection lines.

7.1 Results

The results are described in Table 6 and Figures 20–25.

Table 6. Sequence of small break loss of coolant accident.

Time	Event
0 s	Small break (10 mm diameter) on DVI line B
188 s	Low PRZ level leads to CMTs injection, PPHX actuation and coolant pumps stop
190 s	Reactor scram triggered by pumps low speed (90%)
193 s	Low PRZ pressure actuates reactor protection signal which triggers secondary lines isolation
194 s	Emergency condensers are connected to SGs
5 min	Apparition of a vapour phase in vessel upper head. Bulk boiling occurs in the reactor vessel
25 min	Flashing of water into steam in the CMTs
47 min	Low CMT levels actuate ADS opening
58 min	ADS final stage opening
59 min	Break flow turns into vapour phase. Beginning of accumulators injection (finished 100 s later)
1 h 4 min	Beginning of safety tanks injection
2 h 47 min	End of simulation

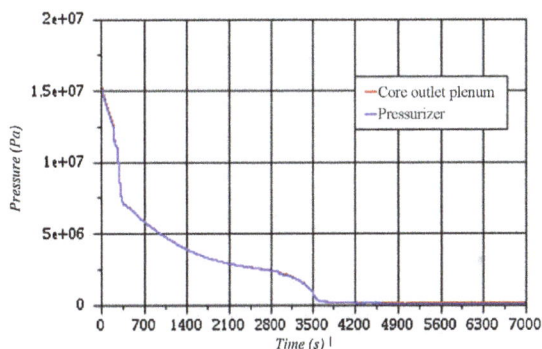

Fig. 20. Primary pressure (Pa).

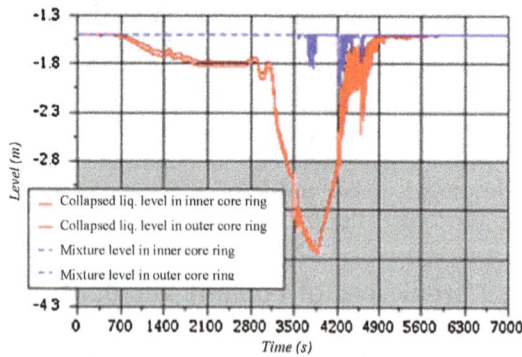

Fig. 21. Water level in the vessel. Grey area represents core zone.

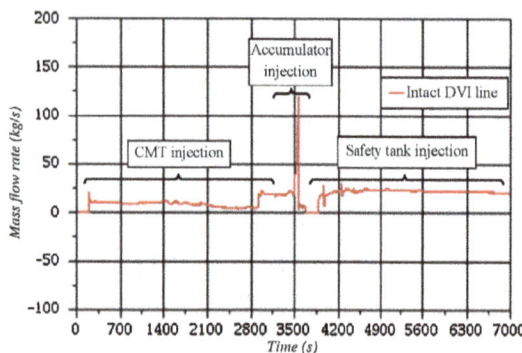

Fig. 22. Injection mass flow through intact DVI line (kg/s).

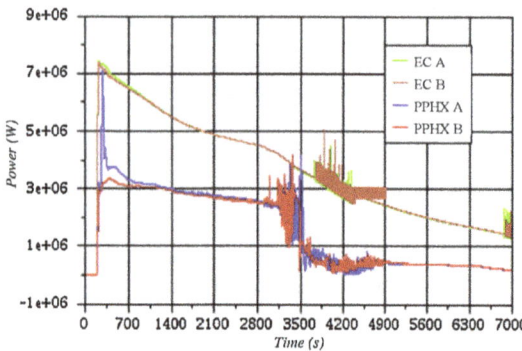

Fig. 23. Emergency heat removal by ECs and PPHXs (W).

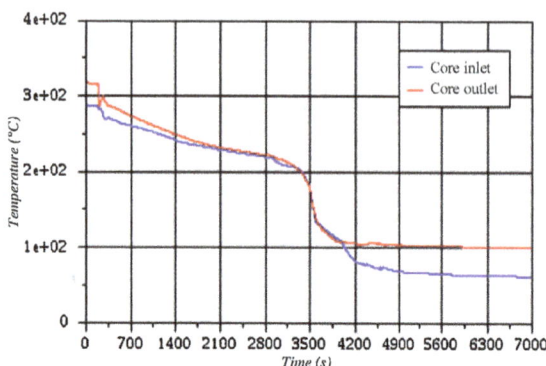

Fig. 24. Primary temperature (°C).

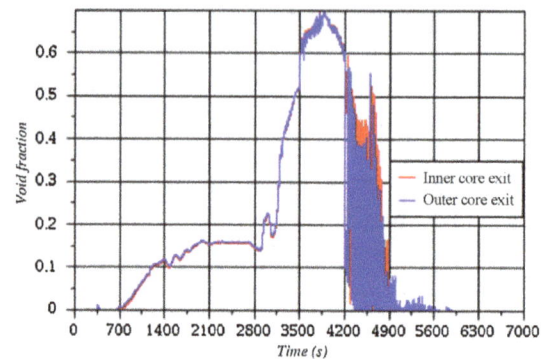

Fig. 25. Core coolant void fraction.

7.2 Discussion

The beginning of the small break transient is quite smooth. Leak mass flow rate does not exceed 18 kg/s. Primary pressure (Fig. 20) goes down to 130 bar within 3 min. Meanwhile, water inventory in the pressurizer counterbalances the leak discharge, so water level in the vessel does not fall (Fig. 21). At $t = 188$ s and 193 s, low PRZ level and low PRZ pressure safety signals are actuated. They successively trigger reactor scram, coolant pumps stop, CMTs injection, secondary lines isolation and passive heat exchangers actuation (Tab. 6). At that time, the loss of coolant is less than 6% of the primary inventory.

Injection of cold water from the CMTs starts (Fig. 22), heat removal by passive exchangers is very efficient (close to 20 MW$_{th}$ at $t = 700$ s, Fig. 23) and the break removes between 4 and 5 MW$_{th}$. The combination of these three actions causes a slow decrease of primary temperature (Fig. 24) and quickly brings down primary pressure (Fig. 20). At $t = 5$ min, primary fluid in the vessel reaches saturation conditions and boiling starts in the core. Void fraction at core outlet remains lower than 20% (Fig. 25), but a vapour bubble appears in the vessel upper head. Circulation in the CMTs is then monophasic: cold water flows to the vessel while hot primary water fills back the tanks. At $t = 25$ min (1500 s), a flashing occurs in both CMTs upper heads. Vapour phase replaces the liquid phase and CMTs start draining out. Low CMTs levels signal actuates ADS opening at $t = 47$ min (2820 s).

The opening of the first two stages of ADS (located at PZR top) causes a sudden jump of liquid level in the pressurizer but no liquid water fills up the ADS lines. Meanwhile, core void fraction strongly increases and collapsed liquid level goes down to 1 m below fuel rods top (Fig. 21). This does not significantly affect core cooling because liquid water is still wetting the fuel rods. Primary temperature is decreased by 100 °C within 15 min. Shortly after, the break flow turns into vapour phase. Accumulators injection is very brief (100 s) and is followed by a 3-min pause of DVI flow (Fig. 22). This pause is counter-balanced by the pressurizer draining into the vessel.

At $t = 1$ h 4 min (3860 s), shortly after the opening of the ADS final stage, primary pressure has finally decreased enough to enable safety tank gravity-driven injection. Reactor vessel is quickly refilled (Fig. 21), and injection flow

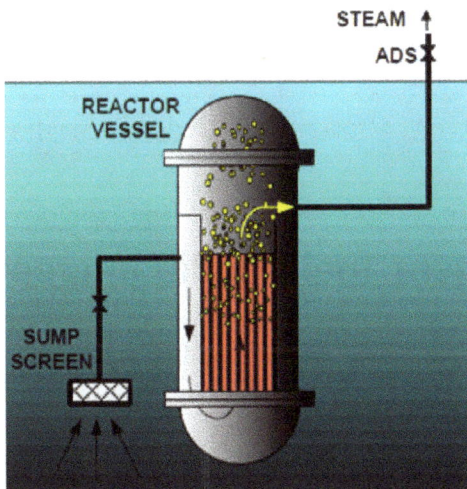

Fig. 26. Core cooling by sump natural circulation [3].

is very steady around 23 kg/s through the intact DVI line (Fig. 22). At the end of the simulation, 2 h 47 min after the break, vessel liquid level is 1.3 m above fuel rods top, core void fraction is zero, core outlet temperature is 100 °C and primary pressure is close to 1 bar. These final conditions are very similar to large break LOCA final conditions. The targeted safe state is the same one: a flooded containment with a sump natural circulation passing through the core (see end of Sect. 6.2 and Fig. 26).

8 Conclusion

The purpose of this study was to investigate the capability of Flexblue® reactor and its passive safety systems to respect safety criteria when typical PWR design-basis accidents occur. The thermal-hydraulics system code ATHLET was used to model the reactor and its safety systems with conservative assumptions. The results of the three chosen transients (turbine trip, large break LOCA and small break LOCA) prove that safety systems are appropriately designed to handle such accidents. The safety criteria are respected with significant margin and the three simulations end on a safe and stable shutdown state. It is worth noting that in the analyses, no credit was taken for operator action or external electrical input. Safe shutdown states are not limited to a given mission time because heat sink around the containment is infinite. The passive safety systems performances and their resilience to extended loss of offsite power constitute a very promising path to enhance nuclear safety.

The analysis also raised some vigilance points that deserve deeper investigations. Firstly, core behaviour is to be watched closely with accurate neutronic data and an appropriate computer code, particularly when coolant flow is suddenly lost at high core power like in the turbine trip transient. This will be possible, thanks to the progress made concerning the Flexblue® core design [2]. Secondly, ATHLET results sometime exhibit oscillations during

passive injection and natural circulation. It is crucial to check that instabilities do not jeopardize the fulfilment of the safety functions. Lastly, it will be necessary to study thermo-mechanical stresses during transients, especially in the natural circulation loops and the pressurizer surge line.

In future works, it will be interesting to study the capability of safety systems to handle a steam generator tube rupture, a main steam line break and a feedwater line break. It is also necessary to study containment and reactor coupling during break transients to confirm the pressure suppression system sizing.

The authors would like to thank GRS for their technical support in the use of ATHLET. The authors are also grateful for the comments and the review provided by other members of Flexblue® development team.

Nomenclature

ADS	automatic depressurization system
ATWS	anticipated transient without scram
CMT	core makeup tank
BWR	boiling water reactor
DNBR	departure from nucleate boiling ratio
DVI	direct vessel injection
EC	emergency condenser
LOCA	loss of coolant accident
PPHX	passive primary heat exchanger
PWR	pressurized water reactor
PZR	pressurizer
SG	steam generator

References

1. G. Haratyk, C. Lecomte, F.X. Briffod, Flexblue®: a subsea and transportable small modular power plant, in *Proceedings of ICAPP Charlotte, USA* (2014)
2. J.J. Ingremeau, M. Cordiez, Flexblue core design: optimisation of fuel poisoning for a soluble boron free core with full or half core refuelling, in *Proceedings of ICAPP Nice, France* (2015)
3. IAEA, Passive safety systems and natural circulation in water cooled nuclear power plants, IAEA-TECDOC-1624, 2009
4. M. Santinello, et al., CFD investigation of Flexblue® hull, in *Proceedings of NUTHOS-10 Okinawa, Japan* (2014)
5. GRS, *ATHLET 3.0 cycle A – user's manual*, GRS-P-1 (2012), Vol. 1, Rev. 6
6. GRS, *ATHLET 3.0 cycle A – models and methods*, GRS-P-1 (2012), Vol. 3, Rev. 3
7. N. Todreas, M. Kazimi, *Nuclear Systems I*, (CRC Press 1990)
8. American Nuclear Society, American National Standard, Decay heat power in light water reactors, ANSI/ANS-5.1-2005, 2005
9. US NRC, *ECCS Evaluation model* (2010), 10 CFR Appendix K to Part 50
10. Electric Power Research Institute, Utility Requirements Document, 1999
11. Westinghouse Electric Company, Accident analyses, *AP1000 European Design Control Document* (2009), Chap. 15.

12

Thermalhydraulics of advanced 37-element fuel bundle in crept pressure tubes

Joo Hwan Park[*] and Yong Mann Song

Korea Atomic Energy Research Institute, 989-111 Daedukdaero, Yuseong-gu, Taejon, 305-353, Korea

Abstract. A CANDU-6 reactor, which has 380 fuel channels of a pressure tube type, is suffering from aging or creep of the pressure tubes. Most of the aging effects for the CANDU primary heat transport system were originated from the horizontal crept pressure tubes. As the operating years of a CANDU reactor proceed, a pressure tube experiences high neutron irradiation damage under high temperature and pressure. The crept pressure tube can deteriorate the Critical Heat Flux (CHF) of a fuel channel and finally worsen the reactor operating performance and thermal margin. Recently, the modification of the central subchannel area with increasing inner pitch length of a standard 37-element fuel bundle was proposed and studied in terms of the dryout power enhancement for the uncrept pressure tube since a standard 37-element fuel bundle has a relatively small flow area and high flow resistance at the central region. This study introduced a subchannel analysis for the crept pressure tubes loaded with the inner pitch length modification of a standard 37-element fuel bundle. In addition, the subchannel characteristics were investigated according to the flow area change of the center subchannels for the crept pressure tubes. Also, it was discussed how much the crept pressure tubes affected the thermalhydraulic characteristics of the fuel channel as well as the dryout power for the modification of a standard 37-element fuel bundle.

1 Introduction

A CANDU-6 fuel bundle is composed of the 37 fuel elements. Spacers and bearing pads are used to prevent direct contact of the fuel elements and/or the pressure tube during the operation. In addition, the end plates are welded on both sides of the fuel bundle to configure a bundle geometry, as shown in Figure 1. For a CANDU-6 reactor such as Wolsung nuclear power plant in Korea, twelve fuel bundles are loaded into a horizontal pressure tube. Because the fuel bundles sit on the bottom inside of the horizontal pressure tube, an open gap on the top section of the fuel channel exists even at the beginning of the reactor operation. Hence, the coolant tends to flow into the open gap rather than the fuel bundle section because of the low flow resistance in the open gap.

One of the most important aging parameters of a CANDU reactor is originated from the horizontal crept pressure tubes. When the reactor becomes older, an open gap becomes wider because it is expanding radially as well as axially during its life time, as a result of the creep of the pressure tube, which has experienced with high neutron

irradiation damage under high temperature and pressure exposure conditions. It allows a by-pass flow on the top section inside the pressure tube. Hence, the crept pressure tube deteriorates the Critical Heat Flux (CHF) of the fuel channel and finally decreases the reactor operating performance.

During the last decades, there have been several studies to overcome the CHF deterioration caused by the pressure tube creep. One of the studies to enhance the CHF was the development of a CANFLEX fuel bundle, which is composed of two pin sizes and attached CHF enhancement buttons on the surfaces of 43 element fuels [1]. It is known that the critical channel power (CCP) enhancement of the CANFLEX fuel bundle can achieve about 4%, 8%, and 13% for the 0%, 3.3% and 5.1% crept pressure tubes, respectively, compared to the standard 37-element fuel bundle (37S fuel bundle) [2]. However, it has not been commercialized yet.

On the other hand, it is known that most CHF of a 37S fuel bundle have occurred at the central area because it has a relatively small flow area and high flow resistance at the peripheral subchannels of its center element compared to the other subchannels [3]. Considering such CHF characteristics of a 37S fuel bundle, there can be two approaches to enlarge the flow areas of the peripheral subchannels of a

---

[*] e-mail: jhpark@kaeri.re.kr

center element to enhance the CHF. To increase the center subchannel areas, one approach was the reduction of the diameter of a center element [4], and the other was an increase of the inner pitch length [5]. The former can increase the total flow area of a fuel bundle and redistributes the power density of all fuel elements as well as the CHF. On the other hand, the latter can reduce the gap between the elements located in the middle and inner pitch circles owing to the increasing inner pitch circle. This can also affect the enthalpy redistribution of the fuel bundle and finally enhance the CHF or dryout power. Both studies were found to be very effective at enhancing the CHF or dryout power through moving the first CHF location occurring at the center subchannels to the other subchannels of a 37S fuel bundle [6]. CHF experiments have been performed at Stern Laboratory to introduce a 37S fuel bundle with a small center element to the commercial reactors [4]. But the detail information of its CHF characteristics has not been published yet.

Recently, a 37S fuel bundle with the inner pitch length modification was studied and its dryout power enhancement was introduced in reference [5], but the creep effects of the pressure tube on the dryout power were not discussed yet. This paper investigated the pressure tube creep effects of the 37A fuel bundle on the dryout power with increasing the inner pitch length. In addition, the thermalhydraulic characteristics of the crept fuel channel were also presented.

2 Analysis modelling

2.1 37A fuel bundle

A 37S fuel bundle is composed of 37-element fuels and four pitch circles such as the center, inner, middle, and outer pitches to configure the bundle geometry, as shown in Figure 1. Recently, a 37S fuel bundle with the inner pitch length modification (here-in-after a 37A fuel bundle) was proposed to enhance the CHF of a 37S fuel bundle [5]. The 37A fuel bundle is defined as a 37S fuel bundle with an inner pitch length modification, which is increased from 14.98 to 15.38 mm in 0.1 mm steps to enlarge the center subchannel area. Each pitch length of the 37S and 37A fuel bundles is summarized in Table 1.

2.2 Pressure tube creep

The pressure tube of a CANDU reactor is made of Zr-2.5% Nb alloy. Since it is vulnerable to the irradiation of the fast neutron flux, it will be crept during the reactor operation. When the reactor operating age increases, the pressure tube will be expanded radially as well as axially. The radial creep of the pressure tube makes its diameter increase. Because a CANDU fuel bundle sits on the inside of a horizontal pressure tube during the dwelling time in the reactor, the flow area at the upper section becomes larger than at the bottom section. It is known that the creep rates of the pressure tube for a CANDU reactor can be increased to 3.3% and 5.1% at the middle and end of its lifetime,

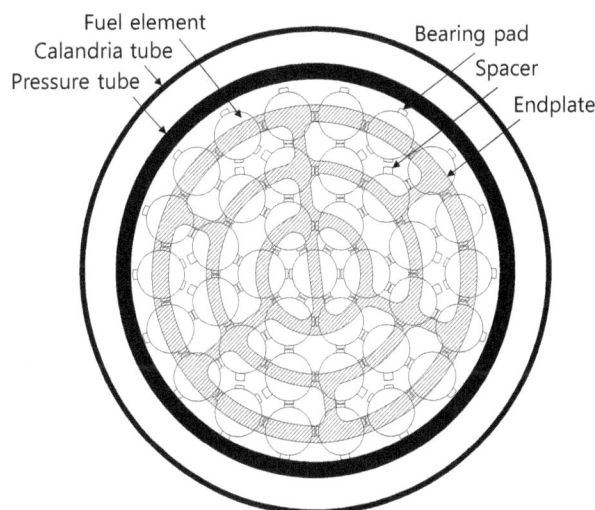

Fig. 1. The configuration of the CANDU fuel channel with a 37-element fuel bundle.

Table 1. Pitch lengths of the 37S and 37A fuel bundles.

Pitch identification	Pitch length, mm		No. of elements
	37S fuel	37A fuel	
Center	0.0	0.0	1
Inner	14.88	14.98 ~ 15.38	6
Middle	28.75	28.75	12
Outer	43.33	43.33	18

respectively. In addition, it will become more serious on the CHF deterioration as its diameter increases. As shown in Figure 2, the flow area of the outer subchannels numbered from #43 to #60 can be increased as the pressure tube is crept or its diameter is increased. But the flow areas of the upper subchannels (i.e. green colored region in Fig. 2) of the fuel bundle can be increased more than those of the lower subchannels (i.e. pink colored region in Fig. 2) because the fuel bundle sits inside of the pressure tube horizontally. These geometric characteristics can divert the coolant from the bundle section to the wider upper section due to low flow resistance. Also, such a flow distortion from bundle to upper sections can become more serious for the higher creep rate of the pressure tube. This study considered such a radial creep rather than an axial creep, which mainly affects the thermalhydraulic performance of the fuel channel.

Figure 3 shows the typical diameter profile of the pressure tube along the axial location of the fuel channel for the creep rates such as 0%, 3.3%, and 5.1%. It has a skewed cosine-shaped profile along the fuel channel. Thus, the subchannel analyses were conducted for the 3.3% and 5.1% crept pressure tubes as well as the 0% crept pressure tube as a reference. The maximum diameters for the 3.3% crept and 5.1% crept tubes were located at an axial distance of 4.3 m and 4.8 m from the entrance of the fuel channel, respectively, as shown in Figure 3. These profiles

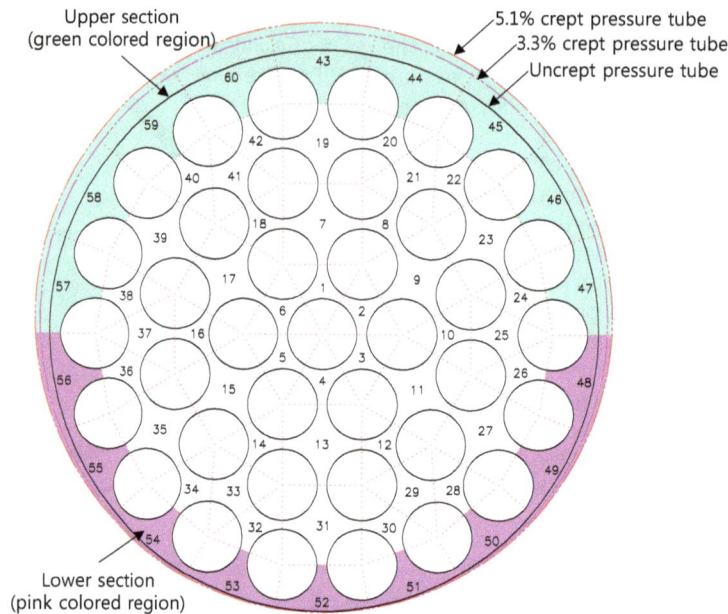

Fig. 2. Subchannel configuration of a 37-element fuel bundle in the crept pressure tubes.

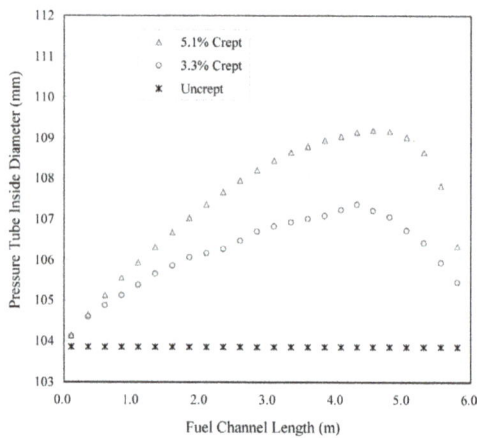

Fig. 3. Axial profile of the pressure tube diameter creep [1].

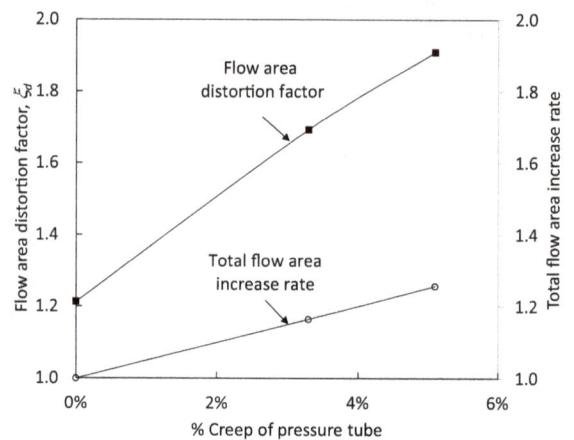

Fig. 4. Flow area distortion according to the pressure tube creep.

representatively simulate the prototypical fuel channels with plant ageing and were used for the water CHF tests as well [1].

To investigate the flow area changes in the top section due to the pressure tube creep and bundle eccentricity horizontally, the flow area distortion factor, ξ_d, is defined as follows:

$$\xi_d = \frac{A_{\text{upper outer subchanel area}}}{A_{\text{lower outer subchanel area}}},$$

where the upper and lower outer subchannel areas of a 37 fuel bundle are shown by the shaded green and pink color areas in Figure 2, respectively. The ξ_d for the 0%, 3.3% and 5.1% crept pressure tubes were shown to be 1.21, 1.69, and 1.91, respectively, while the total flow area was increased by 16% and 26% for the 3.3% and 5.1% crept pressure tubes at the axial peak creep location, as shown in Figure 4.

2.3 Subchannel analysis

A subchannel analysis was performed for a 37S fuel bundle with/without the inner pitch length modification using the ASSERT PV code [7]. The ASSERT code is originated from the COBRA-IV computer program [8,9]. It has been developed to meet the specific requirements for the thermalhydraulic analysis of two-phase flow in horizontally oriented CANDU fuel bundles. Especially, it is distinguished from COBRA-IV in terms of following features [7]:

- the lateral momentum equation is also considered with the gravity term in order to allow gravity driven lateral recirculation;
- the five-equation model was applied to the two-phase flow model in consideration of the thermal non-equilibrium and the relative velocity of the liquid and vapour phases.

Thermal non-equilibrium is calculated from the two-fluid energy equations for the liquid and vapour. Relative velocity is obtained from semi-empirical models;

– the relative velocity model accounts for the different velocities of the liquid and vapour phases in both axial and lateral directions. As well, the lateral direction modelling contains features that consider:

- gravity driven phase separation or buoyancy drift in horizontal flow,
- void diffusion turbulent mixing,
- void drift (void diffusion to a preferred distribution).

To find the subchannel and axial locations of the first CHF occurrence in a fuel channel, the calculation will continue until the convergence tolerance is reached at the specified criteria, 'ODVTOL' in the ASSERT code. Once the first CHF for the given mass flow and inlet temperature has occurred at any subchannel and axial location during iteration, the calculation is stopped and all flow parameters are printed out. Onset-of-dryout iteration for the first CHF occurrence can be found as follows:

$$\text{MCHFLO} \leq \text{MCHFR} \leq \text{MCHFUP}, \qquad (1)$$

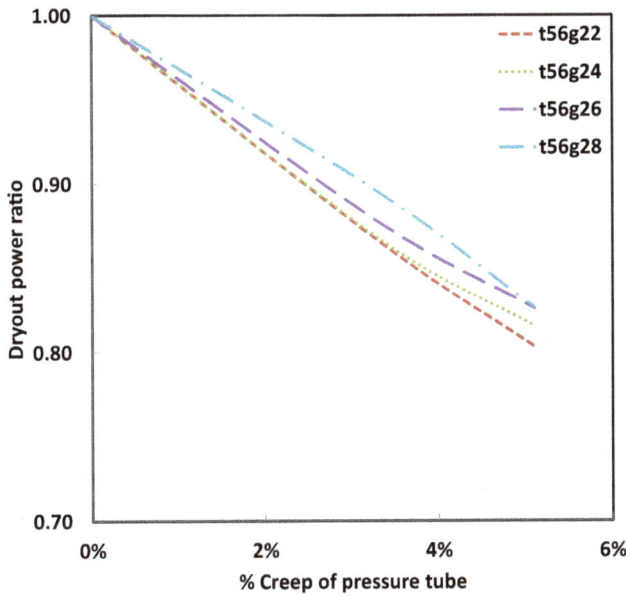

(a) 256°C

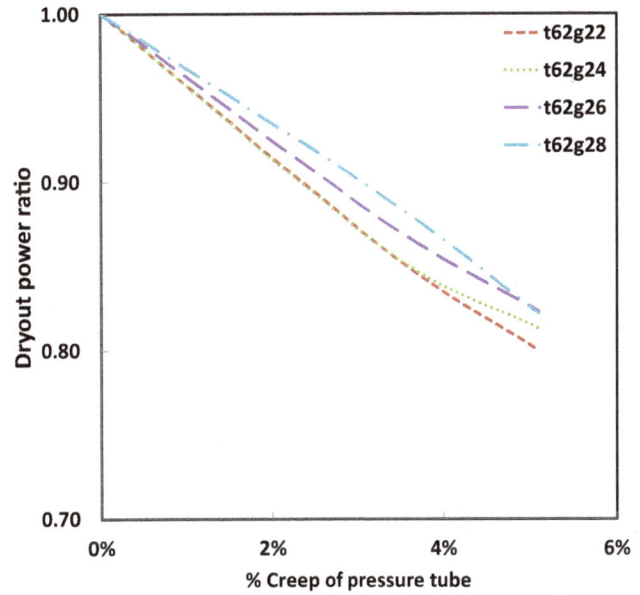

(b) 262°C

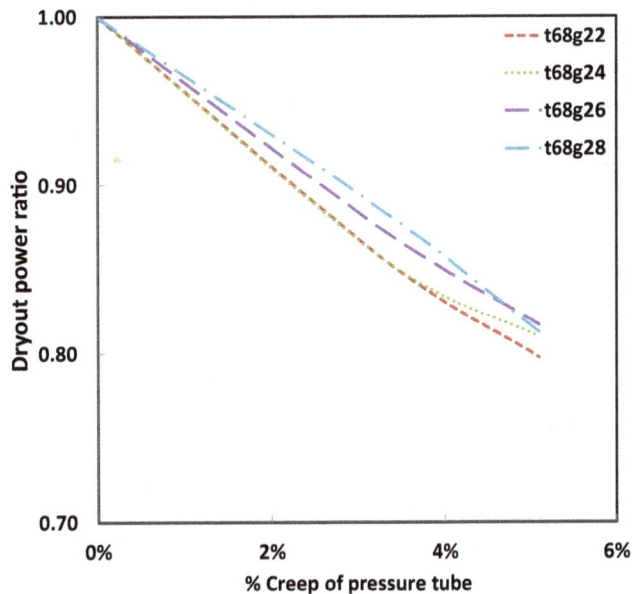

(c) 268°C

Fig. 5. Dryout power ratios for a 37S fuel bundle.

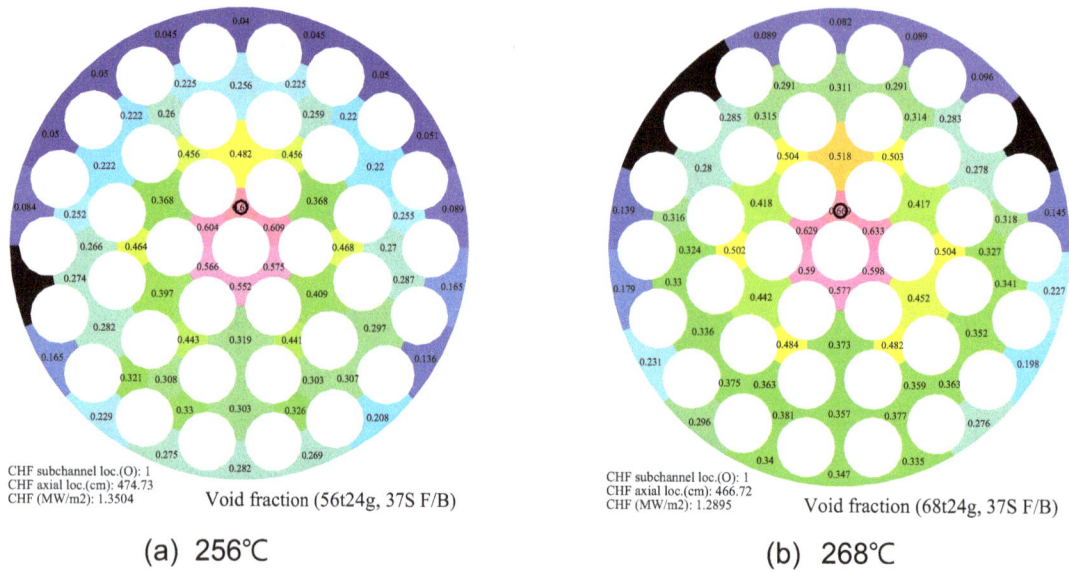

CHF subchannel loc.(O): 1
CHF axial loc.(cm): 474.73
CHF (MW/m2): 1.3504   Void fraction (56t24g, 37S F/B)

CHF subchannel loc.(O): 1
CHF axial loc.(cm): 466.72
CHF (MW/m2): 1.2895   Void fraction (68t24g, 37S F/B)

(a) 256°C   (b) 268°C

Fig. 6. Void fraction of a 37S fuel bundle for the uncrept pressure tube at 24 kg/s.

where 'MCHFLO' and 'MCHFUP' are the lower and upper bounds, respectively, for the target minimum CHFR (MCHFR), and 'MCHFR' is the minimum CHF ratio and is defined as:

$$\text{MCHFR} = \min\left(\frac{q'_{cr}}{q'}\right), \quad (2)$$

where q'_{cr} is the CHF, and q' is the zonal heat flux. 'ODVTOL' is the relative convergence tolerance on the iteration parameter, which is defined as follows:

$$\left|\frac{\Psi_n - \Psi_{n-1}}{\Psi_{n-1}}\right| \leq \text{ODVTOL}, \quad (3)$$

where Ψ is the iteration parameter and n is the iteration number. Ψ and n are given as 1.00004 and 20, respectively, for the present calculation.

Generally, the subchannel can be defined by the hypothetical line connected from one rod center to an adjacent rod center. Hence, the subchannels of a 37S fuel bundle are composed of three types, i.e., triangular, rectangular, and wall subchannels. Those subchannel numbers in Figure 2 are as follows:

– rectangular: 11, 13, 15, 17, 19, 23, 27, 31, 35, 39;
– wall: 43 to 60;
– triangular: the remainder.

The number of rods and subchannels are 37 and 60, respectively. For the present subchannel analysis, the full subchannel geometry was considered.

For a CANDU-6 reactor, the coolant temperature at the reactor inlet header and the coolant pressure at reactor outlet header were designed as 262 °C and 10.0 MPa respectively under D₂O condition and it was limited to 268 °C during the lifetime [10]. If the temperature of the reactor inlet header approaches the limited value, the steam generator should be generally cleaned to lower the reactor

inlet head temperature. And the reference flow rate in the fuel channel was designed as 24 kg/s and the maximum flow rate of the fuel channel can be estimated to be about 28 kg/s [10]. Hence, the present subchannel analysis was performed using the boundary conditions, which are three inlet temperatures, i.e., 256 °C, 262 °C, and 268 °C, four mass flows, 22 kg/s, 24 kg/s, 26 kg/s, and 28 kg/s and the same outlet pressure condition, 10.0 MPa with heavy water coolant to consider the actual reactor operating conditions.

3 Results and discussions

3.1 Pressure tube creep effect on dryout power of a 37S fuel bundle

The subchannel analysis for a 37S fuel bundle was performed to investigate the dryout power according to the increase of the creep rates of the pressure tube from 0% to 3.3% and 5.1% using the ASSERT code with a CHF look-up table [11]. For comparison of the dryout powers of the crept pressure tubes with those of the uncrept pressure tube, the dryout power ratio for a 37S fuel bundle, $r_{DP,37S}$, was defined as follows:

$$r_{DP,37S} = \frac{\text{Dryout Power}_{\text{crept PT,37S fuel bundle}}}{\text{Dryout Power}_{\text{uncrept PT,37S fuel bundle}}}.$$

The results of the dryout power ratios for a 37S fuel bundle, $r_{DP,37S}$, were plotted in Figure 5. As shown in Figure 5, $r_{DP,37S}$ decreases with an increase in the creep rates of the pressure tube for all flow conditions while $r_{DP,37S}$ increases with an increase in the flow rate as expected. The minimum $r_{DP,37S}$ was found to be 0.80 at 22 kg/s of the low flow condition. It means that the dryout power for 5.1% crept pressure tube and 22 kg/s of mass flow condition was about 20% lower than that for the uncrept

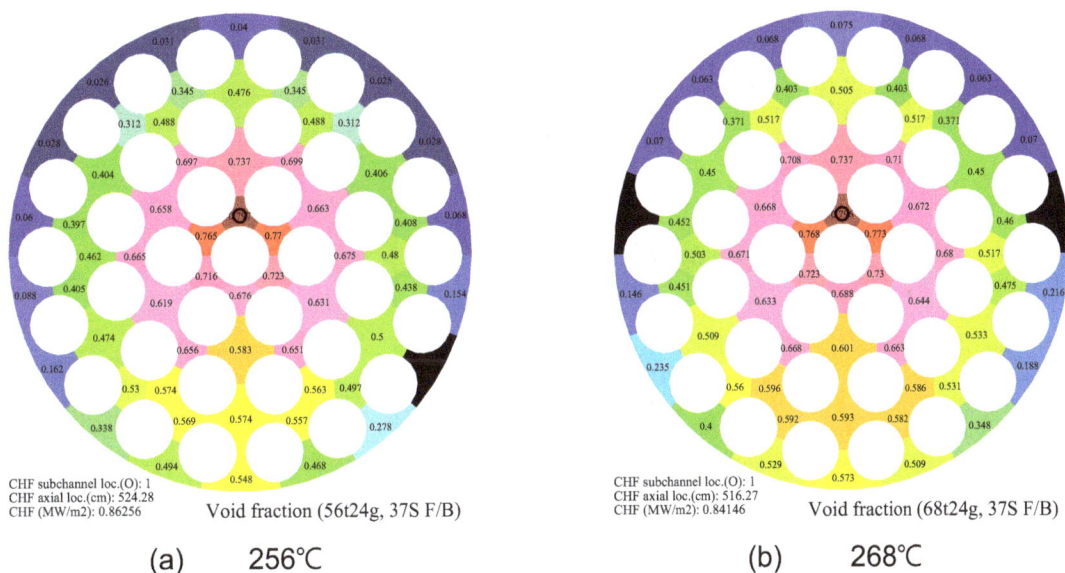

Fig. 7. Void fraction of a 37S fuel bundle for the 3.3% crept pressure tube at 24 kg/s.

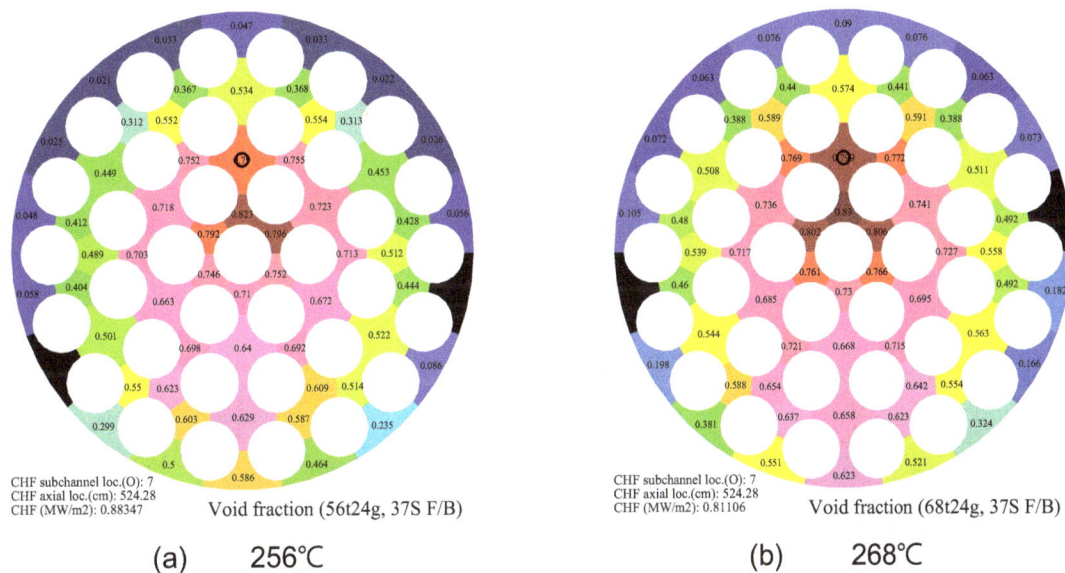

Fig. 8. Void fraction of a 37S fuel bundle for the 5.1% crept pressure tube at 24 kg/s.

pressure tube due to the by-pass flow at upper section of the fuel bundle as described in the above. Also, it can be noted that the high flow rate in the fuel channel makes the coolant have a higher mixing among the subchannels, and the effects of the flow area distortion factor, ξ_d, on the dryout power become less significant for the high flow rate conditions. And the variations of the dryout power ratio for different inlet temperatures were not significant as shown in Figure 5.

Figures 6, 7 and 8 show the void fraction distributions of the subchannels at the first CHF location for the 0%, 3.3%, and 5.1% crept pressure tubes, respectively. As shown in Figure 6, the first CHF occurred at the subchannel #1 under 256 °C and 268 °C inlet temperature conditions. The void fractions at the CHF location or subchannel #1 are

0.650 and 0.669 for 256 °C and 268 °C inlet temperatures respectively and those values are the highest among all the subchannels.

On the other hand, the first CHFs for the 3.3% crept pressure tube occurred at the same location as for the uncrept pressure tube, but the void fraction for 256 °C inlet temperature condition is 0.797 and very close to that for 268 °C inlet temperature condition as shown in Figure 7. For the 5.1% crept pressure tube, the first CHFs for both inlet temperature conditions were occurred at subchannel #7, although the void fraction at the subchannel #7 was lower than that of subchannel #1. Also, the axial CHF locations for both inlet temperature conditions were the same, 524.98 mm which was the location just after the middle bearing pad of the 11th fuel bundle.

In addition, the void fractions of the outer subchannels were lower than those of other subchannels and it was caused by non-heated wall effect of the pressure tube. And the void fractions of the higher inlet temperature condition are higher than those of the lower inlet temperature at the first CHF location.

3.2 Pressure tube creep effect on dryout power of a 37A fuel bundle

The subchannel analysis was performed for a 37A fuel bundle for the 3.3% and 5.1% crept pressure tubes as well as the 0% crept pressure tube. It is focused on examining the diameter increase effect of the pressure tube caused by the irradiation creep. For a comparison of the dryout powers of the 37A and 37S fuel bundles, the dryout power ratio for a 37A fuel bundle, $r_{DP,37A}$, was defined as follows:

$$r_{DP,37A} = \frac{\text{Dryout Power}_{37A \text{ fuel bundle}}}{\text{Dryout Power}_{37S \text{ fuel bundle}}}.$$

The results were plotted in Figures 9, 10 and 11 for uncrept, 3.3%, and 5.1% pressure tubes, respectively. As shown in Figure 9, $r_{DP,37A}$ for the 0% crept pressure tube under 256 °C of the inlet temperature condition is increasing up to 15.18 mm of the inner pitch length, and decreasing for further increases of the inner pitch length. The maximum $r_{DP,37A}$ was found to be 1.057 at 15.18 mm of the inner pitch length under 28 kg/s of the highest flow condition. The behaviors of $r_{DP,37A}$ for all inlet temperature conditions are similar but the dependencies of $r_{DP,37A}$ on the mass flows are a little significant.

For the 3.3% crept pressure tube or 106.79 mm of its diameter, the $r_{DP,37A}$ for each inlet temperature and mass flow is shown in Figure 10. The maximum $r_{DP,37A}$ for 24 kg/s of the mass flow appeared at 15.28 mm of the inner pitch length, while the maximum $r_{DP,37A}$ for 26 kg/s and 28 kg/s of the mass flows were found at 15.18 mm of the inner pitch length. It means that the inner pitch length to give the maximum $r_{DP,37A}$ may tend to be decreased as increasing the mass flow. This trend can be found more distinctly at the case of the 5.1% crept pressure tube as shown in Figure 11. And the maximum $r_{DP,37A}$ was 1.07 for the case of 15.28 mm of the inner pitch length under 24 kg/s and 268 °C of the flow conditions as shown in Figure 10c. The effects of the inner pitch length on $r_{DP,37A}$ for the 3.3% crept pressure tube were more significant than those for the 0% crept pressure tube. It is noted that the modification of the inner pitch length can be more effective as increasing the pressure tube diameter.

For the 5.1% crept pressure tube, 108.65 mm of its diameter, the $r_{DP,37A}$ for each inlet temperature and mass flow is shown in Figure 11. The maximum $r_{DP,37A}$ appeared at the higher inner pitch length than the 0% or 3.3% crept cases for all flow conditions, and was 1.065 at 15.28 mm for 28 kg/s of the highest flow conditions, as shown in Figure 11c. However, $r_{DP,37A}$ for the lower flow conditions such as 22 kg/s and 24 kg/s was monotonically increased by increasing the inner pitch length. In addition, the $r_{DP,37A}$ for all conditions was increased with an increase of the mass flow. The 15.38 mm of the inner pitch length is the maximum allowable

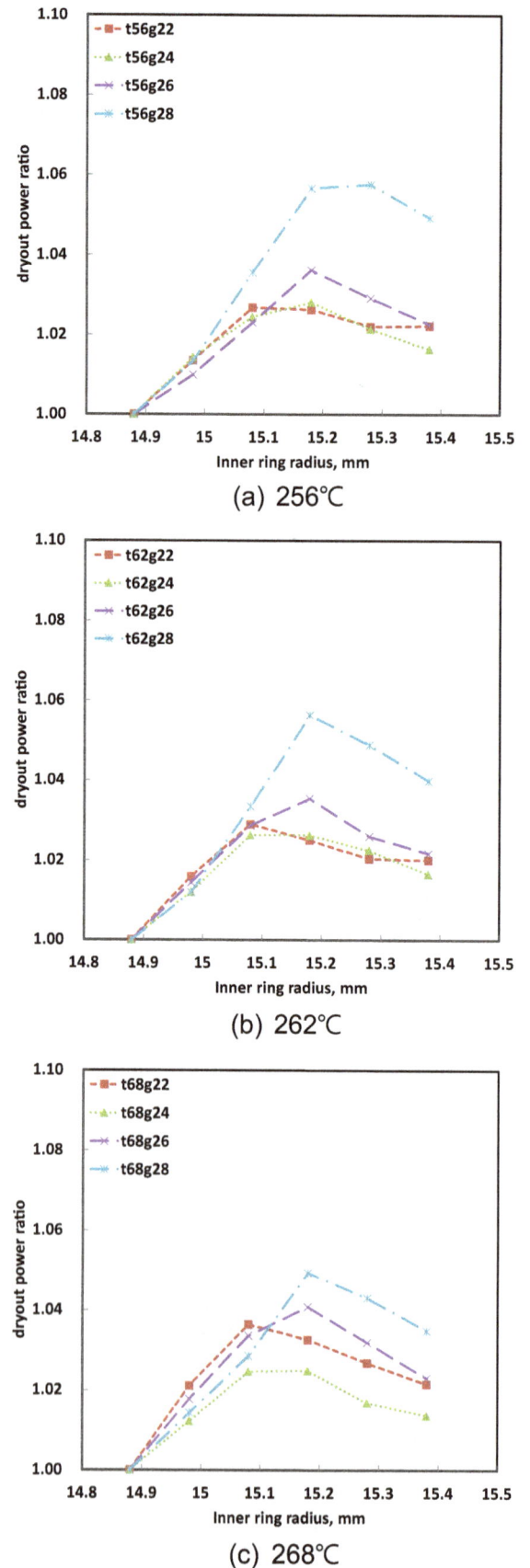

(a) 256°C

(b) 262°C

(c) 268°C

Fig. 9. Dryout power ratio of a 37A fuel bundle for 0% crept pressure tube.

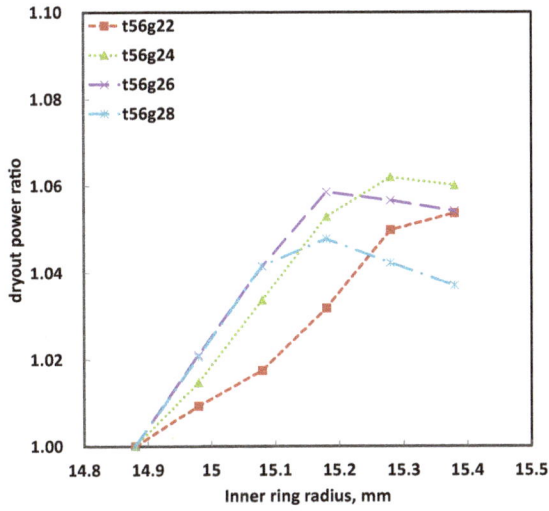

(a) 256℃

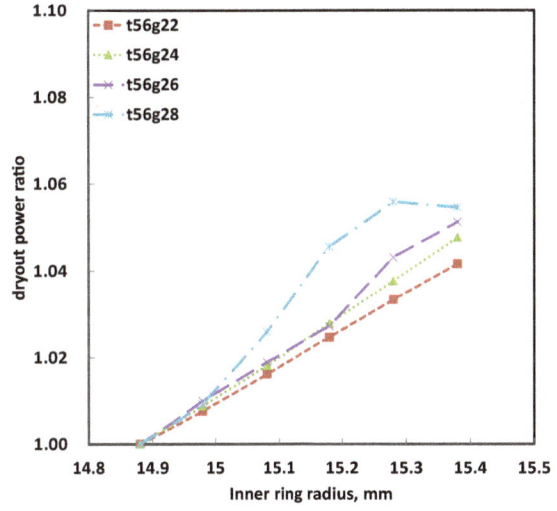

(a) 256℃

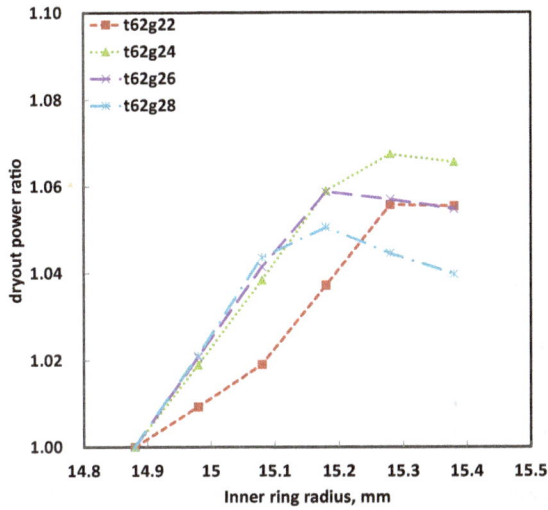

(b) 262℃

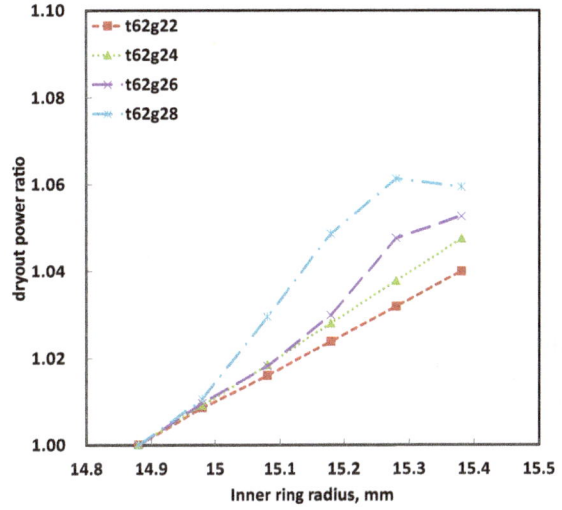

(b) 262℃

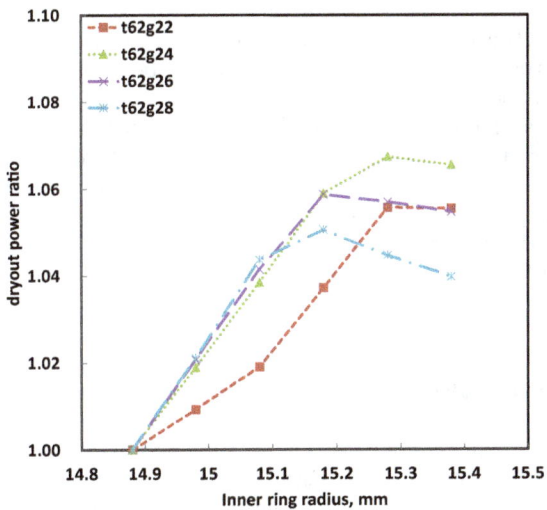

(c) 268℃

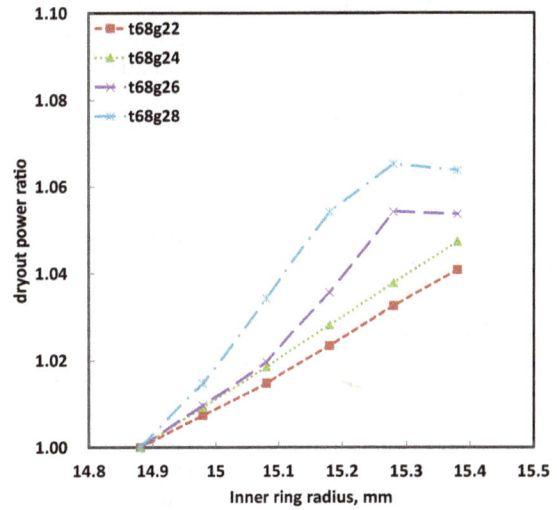

(c) 268℃

Fig. 10. Dryout power ratio of a 37A fuel bundle for 3.3% crept pressure tube.

Fig. 11. Dryout power ratio of a 37A fuel bundle for 5.1% crept pressure tube.

Table 2. Subchannel location of the 1st CHF occurrence for 3.3% crept pressure tube.

Inet temp., °C	256				262				268			
Mass flow, kg	22	24	26	28	22	24	26	28	22	24	26	28
Inner pitch length, mm												
14.88 (37S fuel bundle)	1	1	1	1	1	1	1	1	1	1	1	1
14.98	1	1	1	1	1	1	1	1	1	1	1	1
15.08	1	1	1	1	1	1	1	1	1	1	1	1
15.18	1	1	33	10	1	1	32	10	1	1	33	32
15.28	1	33	33	10	1	33	33	10	1	33	33	32
15.38	10	33	33	10	10	33	33	10	10	33	33	10

Table 3. Subchannel location of the 1st CHF occurrence for 5.1% crept pressure tube.

Inet temp., °C	256				262				268			
Mass flow, kg	22	24	26	28	22	24	26	28	22	24	26	28
Inner pitch length, mm												
14.88 (37S fuel bundle)	7	7	1	1	7	7	1	1	7	7	1	1
14.98	7	7	1	1	7	7	1	1	7	7	1	1
15.08	7	7	1	1	7	7	1	1	7	7	1	1
15.18	7	7	1	1	7	7	1	1	7	7	1	1
15.28	7	7	1	33	7	7	1	33	7	7	1	33
15.38	7	7	10	33	7	7	10	33	7	7	10	10

because of the limitation of the minimum gap between the inner and middle fuel elements as found in reference [10]. It should be noted that the optimum design of the inner pitch length to achieve the maximum $r_{DP,37A}$ is dependent on not only the creep rates of the pressure tube but the flow conditions. In order to determine the optimum inner pitch length, at first, it should be known which mass flow or inlet temperature is more concerned on overcoming the power derating of a CANDU reactor.

On the other hand, the uncertainty on the above calculation results of the dryout power could exist for both a 37S fuel bundle and its modifications, but it was not considered because of the sensitivity studies for a 37S fuel bundle and its modifications.

3.3 Inner pitch length effect on the CHF location for crept pressure tubes

Since the subchannel locations of the first CHF occurrence for the 0% crept pressure tube were found in reference [5], the present study only discusses the locations of the first CHF occurrence for the crept pressure tubes. The subchannel locations of the first CHF occurrence for the 3.3% and 5.1% crept pressure tubes were found and summarized in Tables 2 and 3, respectively. For a 37S fuel bundle, which has a 14.88 mm inner pitch length, all of the first CHFs for the 3.3% crept case occurred at the center subchannel #1 as those for the 0% crept pressure tube in reference [5].

When the inner pitch length is increasing, the first CHF location moves to the inner or middle subchannels such as #10 or #33 (see Fig. 2 for the subchannel numbers). For the 14.88 mm inner pitch length, the subchannels of the first CHF occurrence for the 5.1% crept case were located at the inner subchannel #7 or the center subchannel #1, which are different from those of the 3.1% crept case. This is caused by the higher by-pass flow at the open top section of the fuel bundle, which has a flow area increase of 91% higher than that of the outer lower subchannel as discussed in Section 2.2.

From the above results, it should be noted that the dryout power should be increased by virtue of moving the subchannel locations of the first CHF occurrence from the center subchannel to the other subchannel, according to enlarging the center subchannel area by increasing the inner pitch length. In addition, it is revealed that the favorable effects of the large center subchannel area on the dryout power become more significant for the higher creep rate of the pressure tube.

4 Conclusions

A subchannel analysis using the ASSERT code was performed for the 37S and 37A fuel bundles with the crept pressure tubes to investigate the dryout power changes in terms of the inner pitch length modification and the pressure tube diameter increase.

It was concluded that the inner pitch length modification of a 37S fuel bundle could make the dryout power of the crept pressure tube to be enhanced more than that of the uncrept pressure tube. The maximum dryout power ratio is obtained at a higher inner pitch length. In addition, it was shown that the favorable effects of the large center subchannel area on the dryout power become more significant for the higher creep rate of the pressure tube, that is, the modification of the inner pitch length can be more effective as increasing the pressure tube diameter.

From the present analysis, it was noted that the dryout power could be enhanced by virtue of moving the center subchannel to the other subchannels of the first CHF occurrence if the center subchannel area could be enlarged by increasing the inner pitch length. And it was shown that the optimum value of the inner pitch length to achieve the maximum dryout power ratio is dependent on not only the creep rates of the pressure tube but the flow conditions. In order to determine the optimum inner pitch length, it should be known which mass flow or inlet temperature is more concerned on overcoming the power de-rating of a CANDU reactor.

This work was supported by the National Research Foundation of Korea (NRF) grant funded by the Korea government (Ministry of Science, ICT, and Future Planning) (No. NRF-2012M2A8A4025960).

References

1. G.C. Dimmick, W.W.R. Inch, J.S. Jun, H.C. Suk, G.I. Hadaller, R.A. Fortman, R.C. Hayes, Full scale water CHF testing of the CANFLEX bundle, in *Proceeding of the 6th International Conference on CANDU fuel, Niagara Falls, Ontario, Canada* (1999), pp. 103–113

2. J.S. Jun, Thermalhydraulic evaluations for CANFLEX bundle with natural or recycled uranium fuel in the uncrept and crept channels of a CANDU-6 reactor, Nucl. Eng. Technol. **35**, 479 (2005)

3. L.K.H. Leung, F.C. Diamayuga, Measurements of critical heat flux in CANDU 37-element bundle with a steep variation in radial power profile, Nucl. Eng. Des. **240**, 290 (2010)

4. A. Tahir, Y. Parlatan, M. Kwee, W. Liauw, G. Hadaller, R. Fortman, Modified 37-element bundle dryout, in *NURETH-14, Hilton Toronto Hotel, Toronto, Ontario, Canada* (2011)

5. J.H. Park, Y.M. Song, The effect of inner ring modification of standard 37-element fuel on CHF enhancement, Ann. Nucl. Energy **70**, 135 (2014)

6. J.H. Park, J.Y. Jung, E.H. Ryu, CHF Enhancement of Advanced 37-element Fuel bundles, Sci. Technol. Nucl. Installations **2015**, 243867 (2015)

7. M.B. Carver, J.C. Kiteley, R.Q.N. Zou, S.V. Junop, D.S. Rowe, Validation of the ASSERT subchannel code; prediction of critical heat flux in standard and nonstandard CANDU bundle geometries, Nucl. Technol. **112**, 299 (1995)

8. C.L. Wheeler et al., COBRA-IV-I: an interim version of COBRA for thermal-hydraulic analysis of rod-bundle nuclear fuel elements and cores, Battelle Pacific Northwest Laboratories Report, BNWL-1962, 1976

9. C.W. Stewart et al., COBRA-IV: The model and the method, Battelle Pacific Northwest Laboratories Report, BNWL-2214, 1977

10. AECL, Fuel Design Manual for CANDU-6 reactors, DM-XX-37000-001, 1989

11. D.C. Groeneveld, L.K.H. Leung, P.L. Kirillov, V.P. Bobkov, I.P. Smogalev, V.N. Vinogradov, X.C. Huang, E. Royer, The 1995 look-up table for critical heat flux in tubes, Nucl. Eng. Des. **163**, 1 (1995)

13

Nuclear core activity reconstruction using heterogeneous instruments with data assimilation

Bertrand Bouriquet[*], Jean-Philippe Argaud, Patrick Erhard, and Angélique Ponçot

Électricité de France, 1 avenue du Général de Gaulle, 92141 Clamart cedex, France

Abstract. Evaluating the neutronic state (neutron flux, power . . .) of the whole nuclear core is a very important topic that has strong implication for nuclear core management and for security monitoring. The core state is evaluated using measurements and calculations. Usually, parts of the measurements are used, and only one kind of instrument is taken into account. However, the core state evaluation should be more accurate when more measurements are collected in the core. But using information from heterogeneous sources is at glance a difficult task. This difficulty can be overcome by Data Assimilation techniques. Such a method allows to combine in a coherent framework the information coming from numerical model and the one coming from various types of observations. Beyond the inner advantage to use heterogeneous instruments, this leads to obtaining a significant increase of the quality of neutronic global state reconstruction with respect to individual use of measures. In order to describe this approach, we introduce here the basic principles of data assimilation (focusing on BLUE, Best Unbiased Linear Estimation). Then we present the configuration of the method within the nuclear core problematic. Finally, we present the results obtained on nuclear measurements coming from various instruments.

1 Introduction

The knowledge of the neutronic state (neutron flux, power . . .) in the core is a fundamental point for the design, the safety and the production process of nuclear reactors. Due to the crucial role of this information, considerable work has been conducted for a long time to accurately estimate the neutronic spatial fields. Spatial distribution of power or activity in the whole core, or hottest point of the core, can be derived from such spatial fields. These information allow mainly to check that the nuclear reactor is working as expected in a very detailed manner, and that it will remain in the operating limits during production.

Two types of information can be used for the neutronic state evaluation.

Firstly, the physical core specifications, including the nuclear fuel description, make it possible to build a numerical simulation of the system. Taking into account neutronic, thermic and hydraulic spatial properties of the nuclear core, such well-known numerical models calculate in particular the reaction rates used for the physical analysis of the core state.

Secondly, various measurements can be obtained from in-core or out-of-core detectors. Some detectors can measure neutron density, either locally or in spatially integrated areas, others can measure temperature of the in-core water at some points. A lot of reliable measures come from periodical flux maps measured in each core reactor, at a periodicity of about one month. Then, all these measurements do not have the same type of physical relation with the neutronic activity, and also not the same accuracy. So it is not easy to take into account simultaneously all these heterogeneous measurements for the experimental evaluation of the neutronic state in the core.

A lot of these measures are local, in determined fuel assemblies, and do not give informations in un-instrumented areas of the core. The activity distribution over the whole core is traditionally obtained through an interpolation procedure, using the calculated fields as first guess (a proxy) of the "real" activity field corresponding to the measurements. In other words, the activity value in un-instrumented areas is calculated as the weighted average of the activity measures, using the calculated activity field to interpolate. The power is then obtained from activity through an observation operator, which depends only on core nominal physical specifications for the periodical flux map measurements. This interpolation procedure gives

---

* e-mail: `bertrand.bouriquet@edf.fr`

already good results, but some drawbacks remain in using only activity measures in a deterministic interpolation procedure.

Both physical core specifications and real measurements are subject to some uncertainties. Moreover, numerical assumptions, required to use the models, add some inaccuracy. All these uncertainties are not used explicitly in the interpolation procedure, but often used to qualify the a posteriori activity field obtained through the procedure. Moreover, the interpolation cannot take into account, for example, heterogeneous instruments, or observed discrepancy of some instruments.

Attempts have been made to overcome these limitations, mainly in two directions. Firstly, studies attempt to combine activity measurements and calculations through least-squares derived methods (for example in Ref. [1]), leading to the most probable activity (or power) field on the whole core. These methods allow to take into account heterogeneous measures, but are difficult to develop because of their extreme sensitivity to the weighting factor in the combination of measures and calculations. Secondly, explicit control of the error, in order to reduce its importance, has been tried through the development of adaptive methods to adjust coefficients in the calculation or the interpolation procedure.

Some of these difficulties can be solved by using data assimilation. This mathematical and numerical framework allows combining, in an optimal and consistent way, values obtained both from experimental measures and from a priori models, including information about their uncertainties. Commonly used in earth sciences as meteorology or oceanography [2], data assimilation has strong links with inverse problems or Bayesian estimation [3,4]. It is specifically tailored to solve such estimation problems through efficient yet powerful procedures such as Kalman filtering or variational assimilation [5,6]. Already introduced in nuclear field [7–10], it can be used both for field reconstruction or for parameter estimation in a unified formalism. In particular, in those papers are detailed effects of number and precision of measurements, as well as effect of instrument localization. Those methods are also used to improve nuclear data evaluation [11,12] as well as nuclear mass [13].

Data assimilation can treat information coming from any type of measure instruments, taking into account the way the measure is related with the objective field to be reconstructed, such as neutronic activity here. Data assimilation can further adapt itself to instrument configuration changes, and for example the removal or the failure of an instrument. Moreover, the method takes natively into account informations on instrumental or model uncertainties, introducing them a priori through the data assimilation procedure, and obtaining a posteriori the reduced uncertainties on the reconstruction solution.

In this paper, we introduce the data assimilation method and how it addresses physical field reconstruction. Then we make a detailed description of the various components that are used in data assimilation, and of the various types of instruments we can use to get in-core neutronic activity measurements. Then we present results with various instrument situations in nuclear core, obtained on a set of true nuclear cores.

2 Data assimilation

We briefly introduce the useful data assimilation key points, to understand their use as applied here. But data assimilation is a wider domain, and these techniques are for example the keys of nowadays meteorological operational forecast. It is through advanced data assimilation methods that long-term forecasting of the weather has been drastically improved in the last 30 years. Forecasting is based on all the available data, such as ground and satellite measurements, as well as sophisticated numerical models. Some interesting information on these approaches can be found in the following basic references [2,5,6].

The ultimate goal of data assimilation methods is to be able to provide a best estimate of the inaccessible "true" value of the system state (denoted $\mathbf{x}^t$, with the t index standing for "true"). The basic idea of data assimilation is to put together information coming from an a priori state of the system (usually called the "background" and denoted $\mathbf{x}^b$), and information coming from measurements (denoted as $\mathbf{y}$). The result of data assimilation is called the analysed state $\mathbf{x}^a$ (or the "analysis"), and it is an estimation of the true state $\mathbf{x}^t$ we want to find. Details on the method can be found in references [4] or [5].

Mathematical relations between all these states need to be defined. As the mathematical spaces of the background and of the observations are not necessarily the same, a bridge between them has to be built. This bridge is called the observation operator H, with its linearisation $\mathbf{H}$, that transforms values from the space of the background state to the space of observations. The reciprocal operator is known as the adjoint of H. In the linear case, the adjoint operator is the transpose $\mathbf{H}^T$ of $\mathbf{H}$.

Two additional pieces of information are needed. The first one is the relationships between observation errors in all the measured points. They are described by the covariance matrix $\mathbf{R}$ of observation errors ε_0, defined by $\varepsilon_0 = \mathbf{y} - H(\mathbf{x}^t)$. It is assumed that the errors are unbiased, so that $E[\varepsilon_0] = 0$, where E is the mathematical expectation. $\mathbf{R}$ can be obtained from the known errors on the unbiased measurements. The second one is similar and describes the relationships between background errors. They are described by the covariance matrix $\mathbf{B}$ of background errors ε_b, defined by $\varepsilon_b = \mathbf{x}^b - \mathbf{x}^t$. This represents the a priori error, assuming it to be also unbiased. There are many ways to obtain this a priori and background error matrices. However, in practice, they are commonly built from the output of a model with an evaluation of its accuracy, and/or the result of expert knowledge.

It can be proved, within this framework, that the analysis $\mathbf{x}^a$ is the Best Linear Unbiased Estimator (BLUE), and is given by the following formula:

$$x^a = x^b + K(y - Hx^b), \qquad (1)$$

where $\mathbf{K}$ is the gain matrix [5]:

$$K = BH^T(HBH^T + R)^{-1}. \qquad (2)$$

Moreover, we can get the analysis error covariance matrix $\mathbf{A}$, characterising the analysis errors $\varepsilon_a = \mathbf{x}^a - \mathbf{x}^t$.

This matrix can be expressed from $\mathbf{K}$ as:

$$A = (I - KH)B, \qquad (3)$$

where $\mathbf{I}$ is the identity matrix.

The detailed demonstrations of those formulas can be found in particular in the reference [5]. We note that, in the case of Gaussian distribution probabilities for the variables, solving equation (1) is equivalent to minimising the following function $J(\mathbf{x})$, $\mathbf{x}^a$ being the optimal solution:

$$J(x) = \left(x - x^b\right)^T B^{-1}\left(x - x^b\right) \\ + (y - Hx)^T R^{-1}(y - Hx). \qquad (4)$$

We can make some enlightening comments concerning this equation (4), and more generally on the data assimilation methodology. If we do extreme assumptions on model and measurements, we notice that these cases are covered by minimising J. Firstly, assuming that the model is completely wrong, then the covariance matrix $\mathbf{B}$ is ∞ (or equivalently $\mathbf{B}^{-1}$ is 0). The minimum of J is then given by $\mathbf{x}^a = \mathbf{H}^{-1}\mathbf{y}$ (denoting by $\mathbf{H}^{-1}$ the inverse of $\mathbf{H}$ in the least square sense). It corresponds directly to information given only by measurements in order reconstruct the physical field. Secondly, on the opposite side, the assumption that measurements are useless implies that $\mathbf{R}$ is ∞. The minimum of J is then evident: $\mathbf{x}^a = \mathbf{x}^b$ and the best estimate of the physical field is then the calculated one. Thus, such an approach covers the whole range of assumptions we can have with respect to models and measurements.

3 Data assimilation method parameters

The framework of the study is the standard configuration of a 1300 electrical MW Pressurized Water Reactor (PWR1300) nuclear core. Our goal is to reconstruct the neutronic fields, such as the activity, in the active part of the nuclear core. For that purpose, we use data assimilation. To implement such methodology, we need both simulation codes and measures. For the simulation code, we use standard EDF calculation code COCCINELLE for nuclear core simulation, in a typical configuration (see Ref. [14] for a general overview). The results are built on a set of 20 experimental neutronic flux maps measured on various PWR1300 nuclear cores. Such measurements are done periodically (about each month) on each nuclear core. These different measurement situations are chosen for their representativeness, in order that statistical results cover a wide range of situations and can have some sort of predictability property.

3.1 The background and the measurements

A standard PWR1300 nuclear core has 193 fuel assemblies within. For the calculation, those assemblies are each considered as homogeneous, and are divided in 38 vertical levels. Thus, the state field $\mathbf{x}$ can be represented as a vector of size $193 \times 38 = 7334$.

The background is built upon neutronic diffusion calculation from operational COCCINELLE code routinely used at EDF. This code produces the fields for neutron flux, power and temperature in the context of real cores.

The measurements come from instruments that can be located on horizontal 2D maps of the core. There are three types of instrument that are usually used to monitor the nuclear power core:

- Mobile Fission Chambers (MFC), which measure neutrons inside the active part of the nuclear core, and for which the locations are presented on Figure 1;
- Thermocouples (TC), which are above the active nuclear core, for which the locations are presented on Figure 2;
- fixed ex-core detector locations.

The data coming from the ex-core detectors are continuous in time and are very efficient for security purpose, which is their main goal. Their purpose is to continuously monitor the core, but not to measure accurately the neutronic activity at each fine flux map. So, their measures are too crude for being interesting on a fine reconstruction of the inner core activity map. Thus, we choose here to not take into account information coming from those ex-core detectors.

All these types of instrumentation (MFC, TC, ex-core) can be found on any power plants. For the purpose of this study, we add artificially an extra type of detector, described as idealized Low Granularity MFC (named here LMFC). The measurement attributed to the LMFC are built artificially from the information given by the MFC. Thus they are replacing the MFC on the given LMFC locations. The evaluation of LMFC response is calculated from the MFC measured neutron flux, assuming a different physic process, and a lower granularity. The lower granularity assumption done on the LMFC induces a partial integration of the results of the MFC over a given area. Of course, the physical process involved to make a measurement being different, the resolution of LMFC will be different from the one from MFC. We take 16 of those instruments. They are located in various area of the core, replacing MFC, to try to make a representative array of measurement as shown on Figure 3.

The main characteristics of the instruments, as their number in the core, the number of considered vertical levels, and the size of the part of the observation vector $\mathbf{y}$ associated with the particular instrument type, are reported in Table 1. The size of the final observation vector is given by summing the size of all the individual $\mathbf{y}$ vector of the instruments used.

3.2 The observation operator H

As the output of the neutronic code COCCINELLE provides results which are equivalent to measurements, the observation operator $\mathbf{H}$ is mainly a selection operator, that picks up the chosen information for an instrument among all the code outputs. A normalisation procedure is added for the measurements that have no absolute value.

In details, the $\mathbf{H}$ observation operator can be built independently for each instrument. Each observation

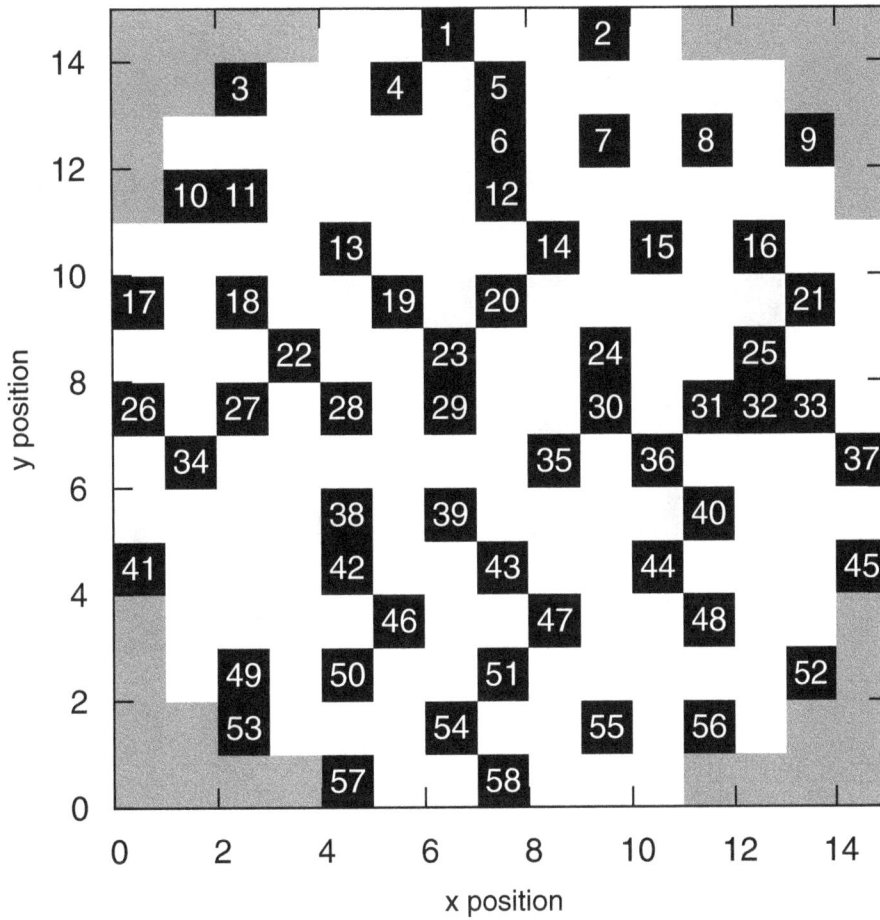

Fig. 1. The Mobile Fission Chambers (MFC) instruments within the nuclear core are localised in assemblies in black, in a horizontal slice of the core. The assemblies without instrument are marked in white, and the reflector is in grey.

operator is then basically a selection matrix, that chooses in the model space a cell that is involved in a measurement in the observation space. In addition, a weight, according to the size of the cell, is affected to the selection. As some experimental data are normalised, this selection matrix is multiplied by a normalisation matrix that represents the effect of the cross normalisation of the data. This observation matrix is a $(7334 \times \sum d_i)$ matrix, where d_i is the size of the part of the observation vector $\mathbf{y}$ for each instrument involved in assimilation, as reported in Table 1.

So there is one individual $\mathbf{H}$ matrix observation operator by instrument type. The complete $\mathbf{H}$ matrix observation operator is the concatenation, as a bloc-diagonal matrix, of all the individual matrix for each instrument.

3.3 The background error covariance matrix B

The $\mathbf{B}$ matrix represents the covariance between the spatialised errors for the background. The $\mathbf{B}$ matrix is estimated as the double-product of a correlation matrix $\mathbf{C}$ by a diagonal scaling matrix containing standard deviation, to set variances.

The correlation $\mathbf{C}$ matrix is built using a positive function that defines the correlations between instruments

with respect to a pseudo-distance in model space. Positive functions allow, through the Bochner theorem, to build symmetric defined positive matrix when they are used as matrix generator (for theoretical insight, see reference documents [15] and [16]). Second Order Auto-Regressive (SOAR) function is used here. In such a function, the amount of correlation depends from the euclidean distance between spatial points in the core. The radial and vertical correlation lengths (denoted L_r and L_z respectively, associated to the radial r coordinate and the vertical z coordinate) have different values, which means we are dealing with a global pseudo euclidean distance. The used function can be expressed as follows:

$$C(r,z) = \left(1 + \frac{r}{L_r}\right)\left(1 + \frac{|z|}{L_z}\right)\exp\left(-\frac{r}{L_r} - \frac{|z|}{L_z}\right). \quad (5)$$

The matrix $\mathbf{C}$ obtained from the above equation (5) is a correlation one. It can be multiplied (on left and right) by a suitable diagonal standard deviation matrix, to get covariance matrix. If the error variance is spatially constant, there is only one coefficient to multiply $\mathbf{C}$. This coefficient is obtained here by a statistical study of difference between the model and the measurements in

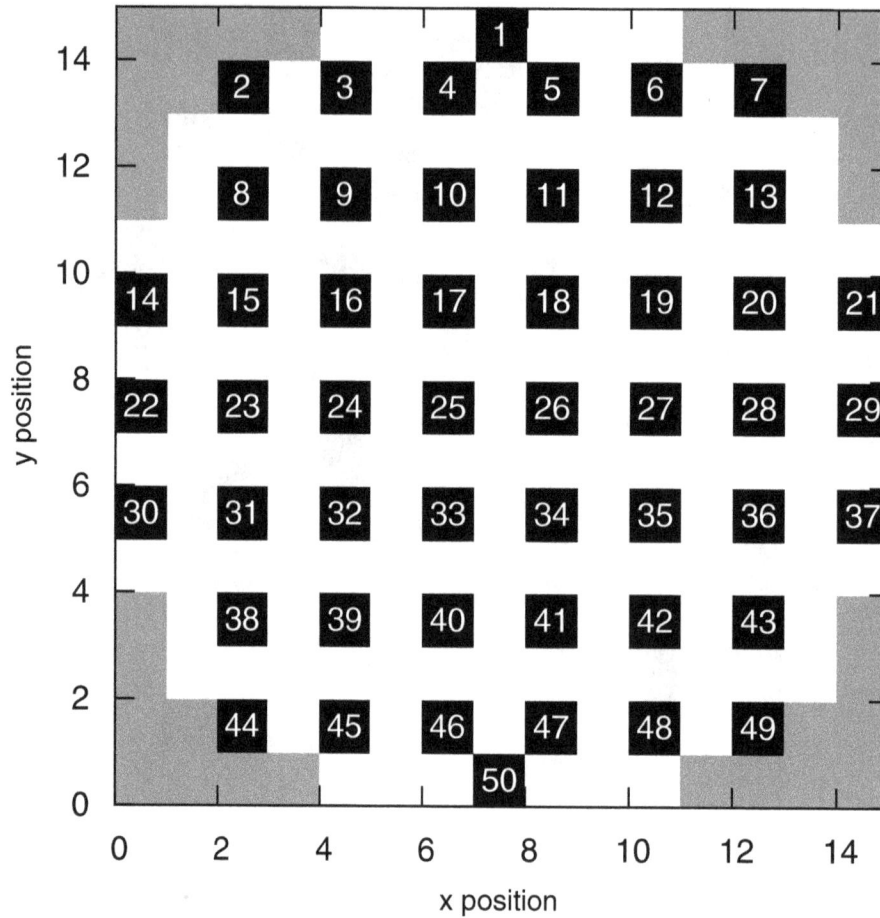

Fig. 2. The Thermocouples (TC) instruments within the nuclear core are localised above assemblies in black, in a horizontal slice of the core. The assemblies without instrument are marked in white, and the reflector is in grey.

real cases. In real cases, this value is set around a few percent.

Globally speaking, the covariance matrix is fully defined by the parameters L_r and L_z that are related to the mean diffusion length of neutrons in the assemblies.

The size of the background error covariance matrix **B** is related to the size of model space, so it is (7334×7334) here.

3.4 The observation error covariance matrix R

The observation error covariance matrix **R** is approximated by a simple diagonal matrix. It means we assume that no significant correlation exists between the measurement errors of all the instruments. A usual modelling consists in taking the diagonal values as a percentage of the observation values. This can be expressed as:

$$R_{jj} = (\alpha y_j)^2, \quad \forall j. \tag{6}$$

The α parameter is fixed according to the accuracy of the measurements and the representative error associated to the instruments. It is the same for all the diagonal coefficients related with one instrument. Its value only

depends on the type of instrument we are dealing with. The α value can be determined by both statistical method and expert opinion about the measurement quality. In the present paper, we will use arbitrary value for the α.

The size of the **R** matrix is related to the size of observation space, so it is $\left(\sum_i d_i \times \sum_i d_i\right)$ where d_i is the size of the observation vector of each instrument i involved in assimilation, as reported in Table 1.

4 Results on data assimilation using only one type of instrument

The first results are showing the quality of the reconstruction as a function of the various types of instruments that are taken into account for reconstructing the activity of the core.

The experimental data are a set of measurements on the 38 levels of all the instrument locations inside of the core. Thus, to evaluate the quality of the reconstruction of the physical fields with one type of instrument, we look for the misfit $(\mathbf{y} - \mathbf{H}\mathbf{x}^a)$ at measurement locations (by other instruments) that are not involved in the assimilation process. The number of locations, where there is a

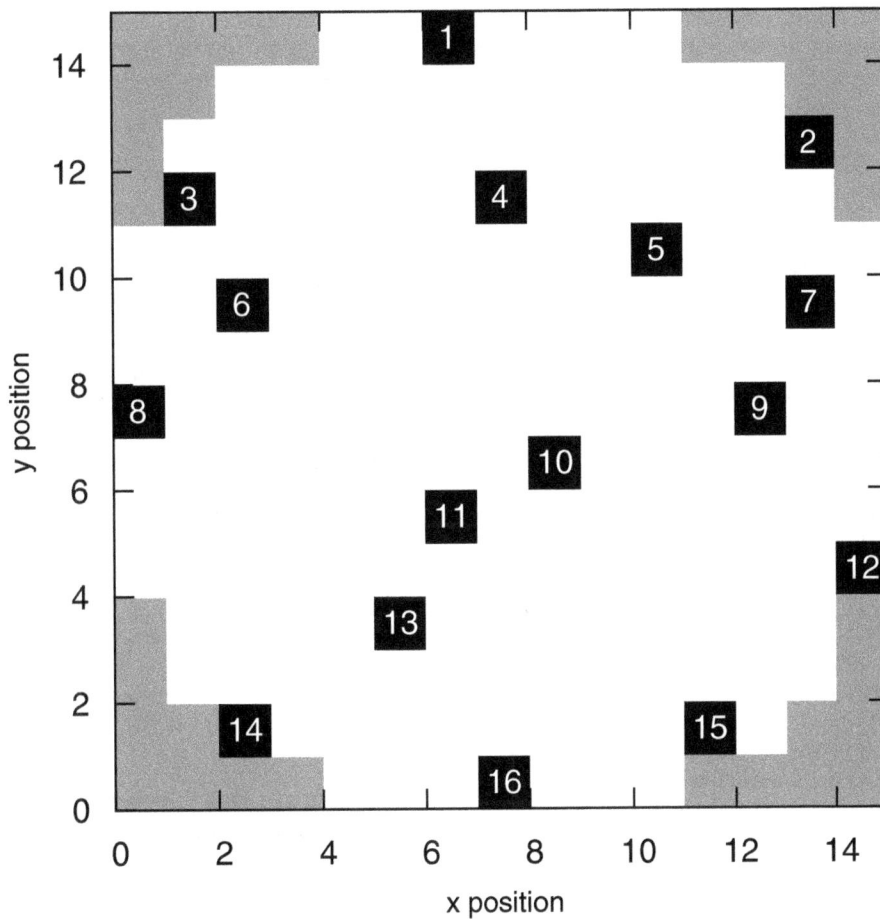

Fig. 3. The idealized Low Granularity MFC (LMFC) instruments within the nuclear core are localised in assemblies in black, in a horizontal slice of the core. The assemblies without instrument are marked in white, and the reflector is in grey.

measurement and that is not involved in data assimilation procedure (so where the misfit $(\mathbf{y} - \mathbf{H}\mathbf{x}^a)$ is calculated) is synthesised in Table 2, as well as the accuracy associated to each instrument through the α parameter of equation (6).

Thermocouples being a fully integral measurement outside of the active core, we can do the misfit $(\mathbf{y} - \mathbf{H}\mathbf{x}^a)$ evaluation of the reconstruction in all the locations of MFC/LMFC.

For each data assimilation procedure associated with an instrument, we calculate the Root Mean Square (RMS, which is the norm) of the misfit $(\mathbf{y} - \mathbf{H}\mathbf{x}^a)$ on all the misfit calculation locations. To synthesize the value for one set of measurement, we take the mean value of the misfit

$(\mathbf{y} - \mathbf{H}\mathbf{x}^a)$ on each of the 38 levels of the core, which leads to a horizontally averaged value of the misfit $(\mathbf{y} - \mathbf{H}\mathbf{x}^a)$. For the sake of more general behaviour, we take 20 sets of flux map measurements, with various settings and ageing of PWR1300 nuclear cores. Then we proceed to the calculation of the mean value on all those sets of measurement.

Globally all the results present strong misfit $(\mathbf{y} - \mathbf{H}\mathbf{x}^a)$ on the upper and lower levels, which is a known effect mainly due to the axial reflector modelling. In addition, the central part of the reactive core in the nuclear plant is also the region where the neutron flux is the most intense, so the hot spot of the activity field in the core is expected to be in this region. Thus, the next plots are restricted to the centre of the core.

Table 1. Main characteristics of the instruments used for data assimilation. These characteristics remain the same, either in mono-instrumented cases or in multi-instrumented ones.

Instrument type	Locations number	Vertical levels	Size of the $\mathbf{y}$ vector part
MFC	42	38	1596
LMFC	16	8	128
TC	50	1	50

Table 2. Number of misfit calculation locations used for each instrument type considered individually for data assimilation, and arbitrary accuracy assumed in the present studies for each type of instrument.

Instrument type	Number of misfit calculations locations	α value (%)
MFC	16	1
LMFC	42	2
TC	58	3

Figure 4 shows the axial misfit measured by the standard RMS of the difference between analysis and measurements, in arbitrary units, on all the data assimilation unused locations for the various types of instrument studied. The oscillating behaviours, that are barely noticeable on all the curves, come from the different materials that are within a core level, and mainly the steel grids that maintain assemblies. Some levels contain a mechanical structure of the core, thus these are more neutron absorption.

We noticed, as expected, that the reconstruction coming from the thermocouples (TC) is the closest to the background, due to their integral measurement property and their lower accuracy. Moreover, an improvement of the accuracy does not improve dramatically the quality of the core state evaluation, mainly due to the integral measurement property.

The Low Granularity MFC (LMFC) are showing a good reconstruction of the physical field in the whole core, despite the not so good accuracy and the limited number of measurements. The increase of the misfit from $(\mathbf{y} - \mathbf{H}\mathbf{x}^b)$ to $(\mathbf{y} - \mathbf{H}\mathbf{x}^a)$ for the lower part of the core is easily explainable by the chosen locations of LMFC that are in core. This part, near the border of the core, does not get enough measurements to be very accurately reconstructed.

The reconstruction using only MFC is, also as expected from accuracy values, the best one. The data assimilation procedure leads to half the misfit $(\mathbf{y} - \mathbf{H}\mathbf{x}^b)$ observed when only using the model.

From results coming from MFC and LMFC, we notice that, within the hypothesis and the chosen modelling of the integration, TC measurement permits only a crude evaluation of the core state.

5 Results on data assimilation with heterogeneous instruments

This section describes results using different instruments together in the data assimilation procedure.

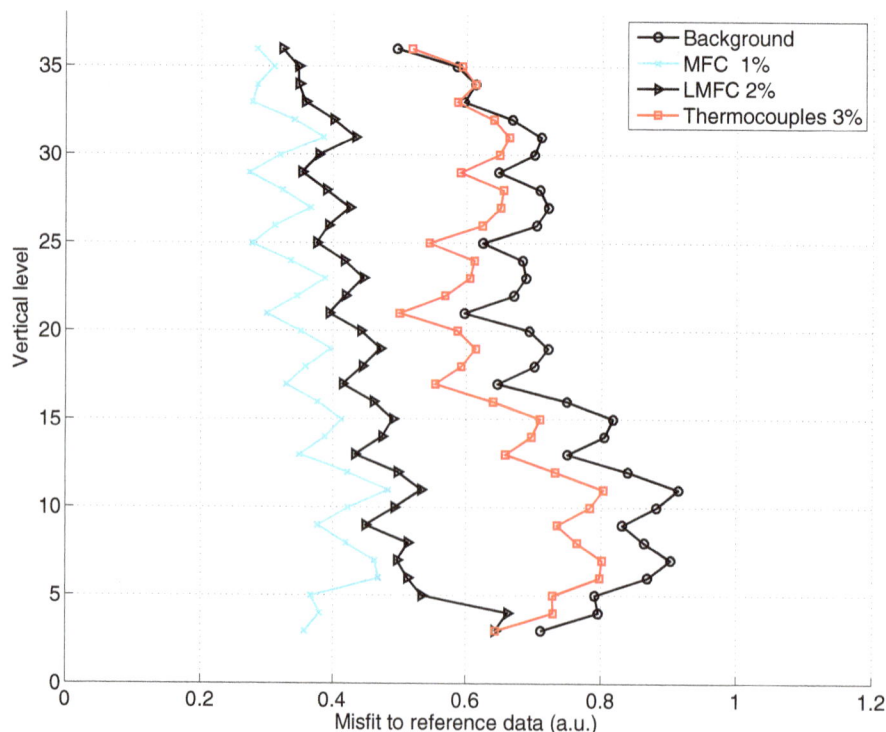

Fig. 4. Vertical misfit for various kind of intruments, measured by the RMS of the differences between the analysis and the measurements at unused locations (see text for details).

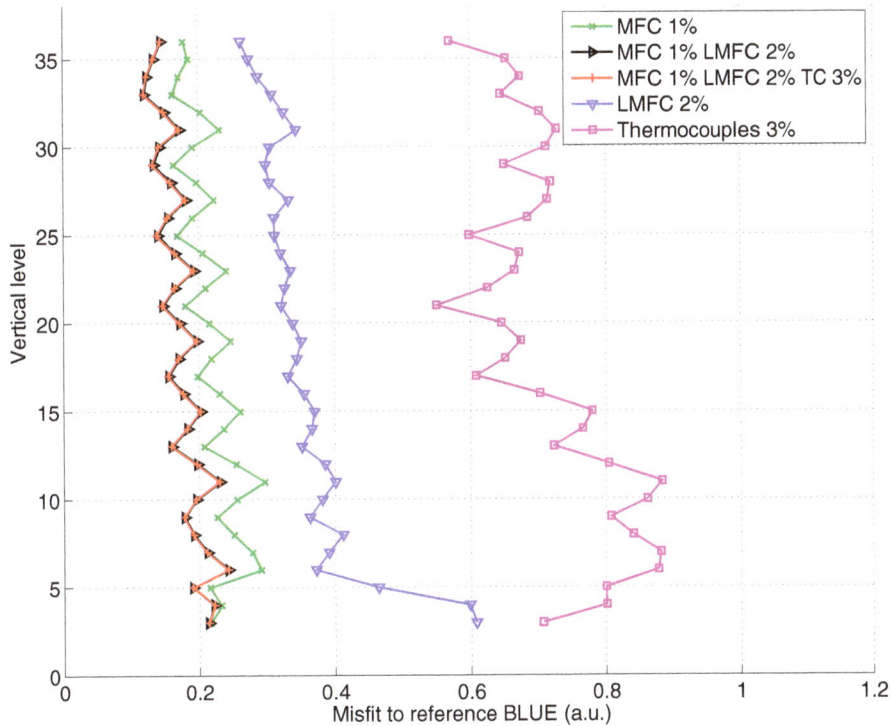

Fig. 5. Axial misfit for various kind of instrumental configurations, measured by the RMS of the difference between the analysis and the reference calculation.

In this case, we cannot take as reference the measures points taken apart, as when we are studying individual instruments as presented in Section 4. Thus, we choose to make an analysis with data assimilation using all the available measures in the core, on the 58 locations, and using a 1% accuracy. Then we evaluate the misfit $(\mathbf{y} - \mathbf{Hx}^a)$ with respect to this reference calculation in all the assemblies where no MFC are present, which means 135 locations (193 total assembly minus 58 instrumented locations). The calculated values at those instrumented locations allow to benchmark the quality of the reconstruction. On this misfit $(\mathbf{y} - \mathbf{Hx}^a)$, we calculate the RMS per horizontal level as previously, and take the average over some 20 selected flux map measurements. The results are presented on Figure 5.

Figure 5 presents RMS per horizontal level of the misfit calculated for various instruments taken alone or in conjunction. We notice that the results have the same behaviour as the one we got in Figure 4. This means that the reference we chose to evaluate the quality of data assimilation using several instrument types can be considered as reliable.

Looking at the successive addition of instrument, we notice that addition of the LMFC has an important effect, the MFC + LMFC configuration presenting an improvement with respect to the configuration with only MFC. However, adding thermocouple to this configuration is not really helpful, and the improvement is not really noticeable on Figure 5.

These results highlight a number of important points on data assimilation methodology. On one hand, when only very few measurements are available, they are very helpful and allow a fairly good improvement of reconstructed physical field. On the other hand, when a lot of measurements are available, adding a few more, or with a lower accuracy as thermocouples, does not change dramatically the result of field reconstruction by data assimilation. On overall, this shows that data assimilation technique is doing the best use of experimental information provided to the procedure. Those results are comforting the ones found on the robustness of the evaluation of the nuclear core by data assimilation, when only MFC is used as presented in reference [9]. Moreover, as expected in data assimilation technique, the use of heterogeneous instruments is integrated within the method.

6 Conclusion

The use of data assimilation has already been proved to be efficient to reconstruct fields in several domains, and recently in neutronic activity field reconstruction for nuclear core. The present paper demonstrates that, within the data assimilation framework, information coming from heterogeneous sources can be used without making any adjustment to the method.

Looking at the various types of instruments we have (MFC, LMFC and TC), we notice that the influence they have on reconstruction depends on three parameters:

- the granularity of each type of instrument, that is the density of instrument, their integral measurement property and their repartition all over the core;
- the accuracy of each instrument, possibly with respect to the accuracy of the others;

– and the global instrumentation configuration, that is the complex repartition of all instruments.

Those conclusions arise from the comparison of the various instruments, using them individually or together, in various instrumental configurations.

Data assimilation gives a very efficient and adaptable framework in order to take the best from both experimental data and model. Moreover, this can be done without heterogeneities constraints between instruments. This technique is used here in a very elementary way, but it opens the door to many developments, for example in systematic data analysis, models comparison, in dynamic modelling, etc. for nuclear reactors.

References

1. H. Ezure, Estimation of most probable power distribution in BWRs by least squares method using in-core measurements, J. Nucl. Sci. Technol. **25**, 731 (1988)

2. E. Kalnay, *Atmospheric modeling, data assimilation and predictability* (Cambridge University Press, 2003)

3. S.M. Uppala et al., The ERA-40 re-analysis, Q. J. R. Meteorol. Soc. **131**, 2961 (2005)

4. A. Tarantola, *Inverse problem: theory methods for data fitting and parameter estimation* (Elsevier, 1987)

5. F. Bouttier, P. Courtier, Data assimilation concepts and methods, Meteorological training course lecture series, ECMWF, 1999

6. O. Talagrand, Assimilation of observations, an introduction, J. Meteorol. Soc. Jpn **75**, 191 (1997)

7. S. Massart, S. Buis, P. Erhard, G. Gacon, Use of 3DVAR and Kalman filter approaches for neutronic state and parameter estimation in nuclear reactors, Nucl. Sci. Eng. **155**, 409 (2007)

8. B. Bouriquet, J.-P. Argaud, P. Erhard, S. Massart, A. Ponçot, S. Ricci, O. Thual, Differential influence of instruments in nuclear core activity evaluation by data assimilation, Nucl. Instrum. Methods Phys. Res. A **626–627**, 97 (2011)

9. B. Bouriquet, J.-P. Argaud, P. Erhard, S. Massart, A. Ponçot, S. Ricci, O. Thual, Robustness of nuclear core activity reconstruction by data assimilation, Nucl. Instrum. Methods Phys. Res. A **629**, 282 (2011)

10. B. Bouriquet, J.-P. Argaud, R. Cugnart, Optimal design of measurement network for neautronic activity field reconstruction by data assimilation, Nucl. Instrum. Methods Phys. Res. A **664**, 117 (2012)

11. P. Blaise, D. Bernard, C. de Saint Jean, N. dos Santos, P. Leconte, A. Santamarina, J. Tommasi, J.-F. Vidal, The uncertainty quantification for complex neutronics code packages: issues on modern nuclear data assimilation and transposition to future nuclear systems, in *ASME 2011 PVP-MF Symposium on risk-informed monitoring, modeling, and operation of aging systems Baltimore, Maryland, USA, July 17–22, 2011* (2011)

12. D.W. Muir, The contribution of individual correlated parameters to the uncertainty of integral quantities, Nucl. Instrum. Methods Phys. Res. A **644**, 55 (2011)

13. B. Bouriquet, J.-P. Argaud, Best linear unbiased estimation of the nuclear masses, Ann. Nucl. Energy **38**, 1863 (2011)

14. A. Santamarina, C. Collignon, C. Garat, French calculation schemes for light water reactor analysis, in *PHYSOR 2004 - The Physics of Fuel Cycles and Advanced Nuclear Systems: Global Developments Chicago, Illinois, April 25–29, 2004, on CD-ROM, American Nuclear Society, Lagrange Park, IL, 2004* (2004)

15. G. Matheron, La théorie des variables régionalisées et ses applications, Cahiers du Centre de Morphologie Mathématique de l'ENSMP, Fontainebleau, Fascicule 5, 1970

16. D. Marcotte, Géologie et géostatistique minières (lecture notes), 2008

14

Thermal hydraulic simulations of the Angra 2 PWR

Javier González-Mantecón, Antonella Lombardi Costa, Maria Auxiliadora Fortini Veloso, Claubia Pereira, Patrícia Amélia de Lima Reis, Adolfo R. Hamers[*], and Maria Elizabeth Scari

Departamento de Engenharia Nuclear, Universidade Federal de Minas Gerais Av. Antônio Carlos, 6627, Escola de Engenharia, Pampulha CEP 31270-901, Belo Horizonte, Brazil

Abstract. Angra 2, the second Brazilian nuclear power plant, began the commercial operation in 2001. The plant is a pressurized water reactor (PWR) type with electrical output of about 1350 MW. In the present work, the thermal hydraulic RELAP5-3D code was used to develop a model of this reactor. The model was performed using geometrical and material data from the Angra 2 final safety analysis report (FSAR). Simulations of the reactor behavior during steady state and loss of coolant accident were performed. Results of temperature distribution within the core, inlet and outlet coolant temperatures, coolant mass flow, and other parameters have been compared with the reference data and demonstrated to be in good agreement with each other. This study demonstrates that the developed RELAP5-3D model is capable of reproducing the thermal hydraulic behavior of the Angra 2 PWR and it can contribute to the process of the plant safety analysis.

1 Introduction

As the global population increases, the demand for energy and the benefits that it provides also grow up. With the worldwide concern over global warming, it is necessary to use clean sources, which do not cause the greenhouse effect. Nuclear energy is increasingly considered an attractive energy source that can bring an answer to this increasing demand, but safety of nuclear power reactors is one of the most important public worries.

For many years, the main research area in the nuclear field has been focused on the performance of nuclear power plants (NPPs) during accident conditions. In order to simulate the behavior of water-cooled reactors, the nuclear engineering community developed several complex thermal hydraulic codes systems. RELAP5-3D [1], developed by the Idaho National Laboratory, is one of the most used best-estimate thermal hydraulic codes. It is capable of performing steady state, transient and postulated accident simulations, including loss of coolant accidents (LOCAs) and a several types of transients in light water reactors (LWRs).

The aim of this work is to simulate the Angra 2 nuclear reactor behavior during steady state condition and for a postulated Small Break Loss of Coolant Accident (SBLOCA) in the primary circuit, using the thermal hydraulic computer code RELAP5-3D [2–4]. The accident simulated consists of a total break (200 cm^2) in the cold-leg piping of one of the reactor loops. A variety of break sizes in the cold-leg and hot-leg piping and other parts of the reactor coolant system representing a typical range and locations of small- and medium-break LOCAs are described and studied in the final safety analysis report of Angra 2 [5]. Hence, that document is taken as reference for the development of the present work.

2 Model description

2.1 Angra 2 plant description

In June 1975, it was signed a cooperation agreement for the peaceful uses of nuclear energy between Brazil and the Federal Republic of Germany. Under this agreement, Brazil acquired two nuclear plants, Angra 2 and 3, from the German company Siemens/KWU, currently Areva ANP. The Almirante Álvaro Alberto NPP – Unit 2 (Angra 2) is located on the Atlantic Coast in a bay at the western extremity of the Brazilian state of Rio de Janeiro.

The second Brazilian nuclear power plant began commercial operation in 2001. The plant is equipped with a pressurized water reactor with an electrical output of about 1350 MWe, which uses light water as both reactor coolant and moderator. The PWR is designed as a four-loop plant, which is based on the proven technology of other four-loop plants. Some technical data of the plant are shown in Table 1 [5].

---

*e-mail: adolforhamers@gmail.com

Table 1. Some technical data of the Angra 2 NPP.

Reactor power	
Reactor thermal power	3771.4 MW
Gross electrical	1350 MW
Thermal yield	35.8%
Reactor core	
Fuel material	Enriched uranium – UO_2
Number of fuel elements	193
Number of fuel rods	236
per assembly	
Cladding material	Zircaloy 4 (Zr)
Thickness	0.72 mm
Mean linear power density	20.7 kW/m
Mean temperature rise	34 °C
in core	
Plant systems	
Primary system description	4 Loops
Number of pumps of the	4
primary system	
Primary system pressure	15.7 MPa
Average temperature	308.6 °C
Steam generator	
Type	Vertical U-Tubes
Material	20 Mn Mo Ni 55
Tubes	Incoloy 800

2.2 RELAP5-3D nodalization

The structure of the RELAP5-3D nodalization is simple (Fig. 1) and it is based on the component design and operating data. The model contains 162 hydraulic components and 14 heat structures (HSs). All the four coolant loops are independently modeled. The loops were simulated symmetrically except for the differences due to the location of the pressurizer in loop 1. All loops have steam generators (SGs) that include both the primary and secondary sides with heat exchange structures.

Both the SG inlet and outlet plena are modelled as a separated branch. The SG tubes are represented by a pipe consisting of an "up" (hot) leg and a "down" (cold) leg, and each one is represented by eight volumes. The secondary side nodalization is limited to the SG riser and downcomer, the SG dome and the main steam line. Both the main feedwater system (MFW) and the emergency feedwater system (EFW) are modelled by a separate time-dependent volume (TDV) and time-dependent junction (TDJ) for each steam generator. The main steam relief valve (MSRV) is modelled by a trip valve. The four reactor coolant pumps (RCPs) included in this system are identical for each loop and contain realistic characteristics.

The coolant flow area through the core was divided into 10 regions (600–609) representing the same number of thermal hydraulic channels, and heat structures were associated to each one. The effective flow rate for heat transfer in the core is 17,672 kg/s. A non-heated channel

represents the bypass (550). Thermal hydraulic channels and its connected heat structures were subdivided axially into 34 volumes. The axial power distribution was calculated considering a cosine profile. The 3D capability of RELAP5-3D code for conducting neutronic calculations was not used. In a 3D reconstruction, it is possible to define exactly the position of the fuel element in the core to perform more realistically the transient evolution [6]. The point reactor kinetics model was used to compute the transient behavior of the neutron fission power in the nuclear reactor. Appropriate factors were defined to account for the fraction of the thermal power produced in the fuel rods and the one released to the coolant. The ANS79-3 decay heat model was selected to calculate the fission product decay.

The emergency core cooling system (ECCS) is also modelled, including accumulators (146, 246, 346 and 446) and safety injection (SI) pumps. The SI pumps are represented by TDV (142, 242, 342 and 442) to set-up the boundary conditions of the injected water (temperature and pressure), and a TDJ (144, 244, 344 and 444) to impose the mass flow rate. Values of geometrical dimensions, set-up points and initial conditions used are given in the reference document [5].

Control variables are included in the RELAP5-3D model to simulate the main control logic of primary system and the ECCS actuation during accident scenario.

2.3 Accident description

The transient analyzed is the postulated 200-cm^2 break in the cold-leg of loop 2. This size of break falls into the category of SBLOCA, in which the secondary side is always required for heat removal from the reactor coolant system (RCS). The accident initial and boundary conditions are described in the FSAR. The simulation of this accident was performed by incorporating the logic of operation of the reactor protection system. The conditions considered are:

– Reactor power – 100% nominal power.
– Reactor trip from RCS pressure (P_{RCS}) < 13.2 MPa.
– Offsite power is not available.
– Reactor coolant pumps coast down.
– ECCS criteria, P_{RCS} < 11.0 MPa and pressurizer level (L_{PRZ}) < 2.28 m.
– Emergency feedwater supply to the SG secondary side, SG level (L_{SG}) < 5 m.

Shutdown of the reactor is performed by inserting the control rods, which is assumed to be initiated from the second trip signal – P_{RCS} < 13.2 MPa – to be issued during the course of the accident. To ensure a conservative delay time for the actuation of the SI pumps, it is assumed that offsite power is not available and that loss of offsite power occurs at the same time as reactor trip. For a conservative availability of the ECCS, it is assumed that the diesel engines of loops 3 and 4 are not available because of single failure and repair, respectively (see Tab. 2). Failure and repair criteria are adopted for the ECCS components in order to verify the system operation in carrying out its

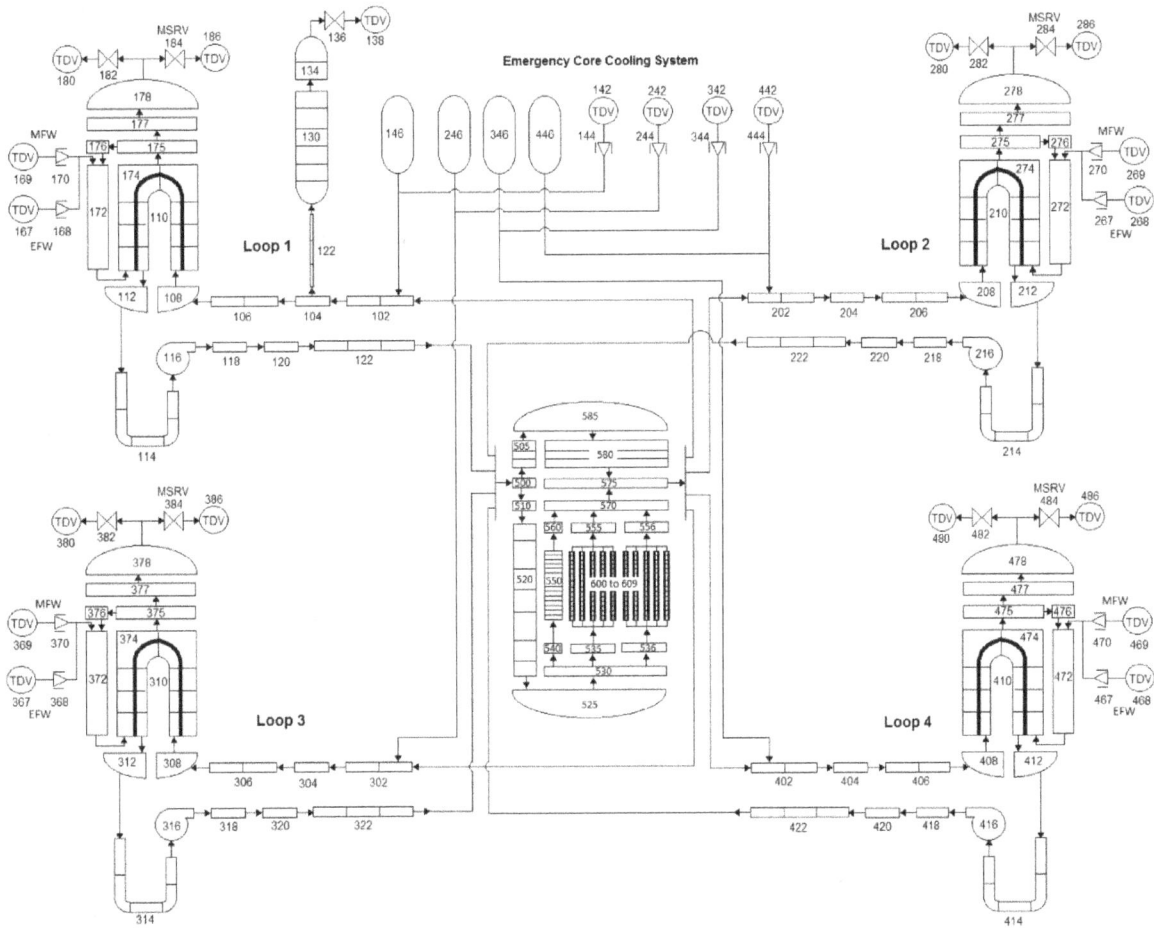

Fig. 1. RELAP5-3D nodalization for the Angra 2 PWR.

function as expected to preserve the integrity of the reactor core and to guarantee its cooling.

The reactor coolant pumps coast down due to the postulated loss of offsite power. If offsite power was available, the RCPs would be switched off by the reactor protection system when the ECC criteria are met or when the pressurizer level criteria – $L_{PRZ} < 2.28$ m – is met.

When the ECC criteria – $P_{RCS} < 11.0$ MPa and $L_{PRZ} < 2.28$ m – are met, the SI pumps are started. When the RCS pressure decrease to 2.6 MPa, the initial pressure of the accumulators, the accumulators start to inject borated water into the reactor coolant system.

Because of the assumed loss of offsite power and postulated unavailability of loops 3 and 4 diesel engines,

and assuming that the startup and shutdown pumps are electrically connected to these diesel engines, the startup and shutdown pump are not available to provide water to the secondary side of the steam generators. Water is therefore supplied by the emergency feedwater pumps, which will be started when the secondary level drops below 5 m and injects water at 36 kg/s per steam generator.

According to FSAR, for this transient, it must be demonstrated that the following acceptance criteria are met under best-estimate conditions:

– Cladding temperature < 1200 °C.
– Local cladding oxidation < 17%.
– No fuel centerline melting.

Table 2. Availability of the ECCS components [5].

ECCS components	Injection							
	Loop 1		Loop 2		Loop 3		Loop 4	
	Hot	Cold	Hot	Cold	Hot	Cold	Hot	Cold
Safety injection pumps	1	–	1	–	SF	–	RC	–
Accumulators	1	–	1	–	1	–	1	–

SF: single failure of diesel engine; RC: diesel engine down for repair.

Table 3. Comparison between the steady state thermal hydraulic parameters calculated by RELAP5-3D code and FSAR data.

Parameter	Nominal value	RELAP5-3D	Error*
Reactor coolant system side			
Reactor thermal power	3771.4 MW	3771.4 MW	0.0%
Coolant temperature			
RPV inlet	292.1 °C	293.45 °C	0.46%
RPV outlet	326.1 °C	328.40 °C	0.71%
Coolant pressure			
RPV inlet	16.05 MPa	16.19 MPa	0.87%
RPV outlet	15.7 MPa	15.59 MPa	0.70%
Coolant mass flow rates			
Total loop flow rate	4700 kg/s	4675.28 kg/s	0.53%
Total RPV flow rate	18,800 kg/s	18,701.23 kg/s	0.53%
Secondary side			
Pressure at SG exit	6.29 MPa	6.25 MPa	0.64%
Feedwater temperature	218.9 °C	217.85 °C	0.48%
Main steam mass flow rate	2068.4 kg/s	2086 kg/s	0.85%

*Error = 100 × |(Nominal value − calculated value)/Nominal value|.

Additionally, the core must remain amenable to cooling during and after the event. These criteria were established to provide significant margin in the emergency core cooling system performance following a LOCA.

3 Results

3.1 Steady state simulation

To verify a RELAP5 nodalization, the model must reproduce the steady state conditions of the simulated system with acceptable margins. Moreover, the nodalization must have a geometric fidelity with the system and to reproduce satisfactorily the time evolution conditions [7].

RELAP5-3D steady state calculations were performed for Angra 2 PWR operating at 3771.4 MWt. The steady state parameters calculated are presented in Table 3 and are compared with the nominal values provided in FSAR. The results show good agreement with the reference data and the calculated errors are in correspondence with the usual criteria for quantification of the steady state prediction quality that have been adopted [8,9]. This means that the model reproduces with good approximation the steady state thermal hydraulic behavior of the reactor.

In most reactor designs, 200 s null transient is typically sufficient time to achieve stable steady state conditions [5,8]. The dynamic behavior of the models is satisfactory and most of the equilibrium values were reached, or their rates of change were small after first 200 s of calculations.

The time evolution of the coolant temperature and pressure at inlet and outlet of the reactor pressure vessel (RPV) are shown in Figure 2. As can be seen, temperatures reach stable condition in approximately 50 s of simulation. It is also possible to conclude that the pressure drop in the vessel predicted by the code is approximately 0.6 MPa.

Figure 3a presents the time evolution for the heat structure 6050 (HS-605) fuel centerline temperature at four different axial levels. In addition, Figure 3b shows the fuel centerline and cladding temperature evolution for the heat structure 6050 associated to the channel 605 at mid high (level 18). As it can be observed, these temperatures are completely stables and are within the expected range [5].

As for the axial power distribution, also the axial fuel temperature distribution follows the cosine-shaped profile, reaching higher temperatures in central part of the element as shown in Figure 4 for the case of the HS-6050. As it was expected, the coolant temperature increases as the fluid movement along the heated length. The results are in agreement with theoretical predictions [10].

3.2 Transient analysis

The break is modelled adding a trip valve (800), which connects the reactor coolant line with the containment. The valve is opened after 300 s of steady state simulation, and its area is equal to the break area. On the side of the reactor coolant line, the valve is attached to the center or the respective hydraulic volume. Figure 5 gives a nodalization diagram of the break in the cold-leg of loop 2. The containment pressure is established by a time-dependent volume (802).

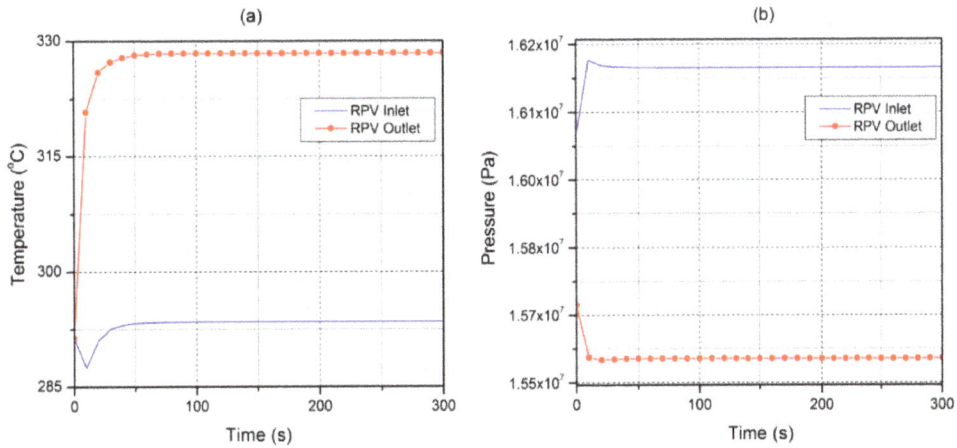

Fig. 2. Time evolution of coolant temperature (a) and pressure (b) at inlet and outlet of the RPV.

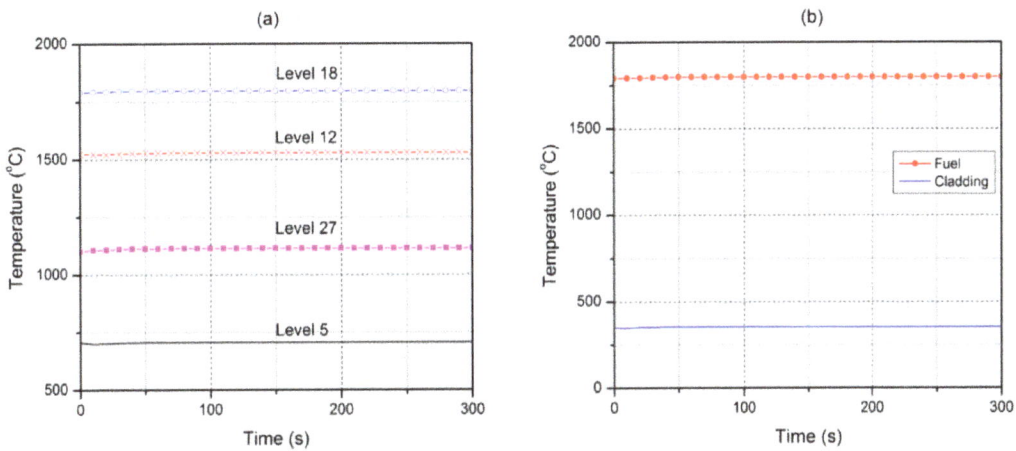

Fig. 3. Time evolution of fuel temperature at different axial positions in HS-6050 (a). Fuel and cladding temperature at level 18 (b).

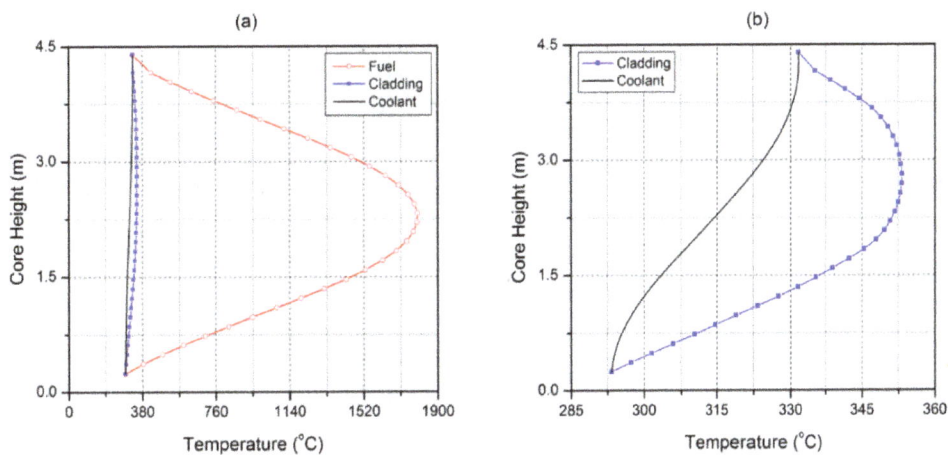

Fig. 4. Axial distribution of fuel, cladding and coolant temperatures (a). Comparison of coolant and cladding temperatures (b).

Because of the initial break flow rate and the incompressibility of the subcooled reactor coolant, the pressure on the primary side drops rapidly. At the same time, the loss of external power is assumed, which results in the RCPs coast down. The sequence of events is summarized in Table 4.

Figure 6a shows the mass flow rate through the break. The accident beginning is characterized by a sudden discharge of subcooled water into the containment. A fast depressurization of the primary system also occurs. The coolant pressure at inlet and outlet of the RPV can be seen in the same figure (Fig. 6b).

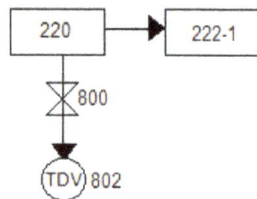

Fig. 5. RELAP5-3D nodalization diagram of the break.

Table 4. Sequence of events in the accident evolution.

Event	Time (s)
1 – Break initiation	0
2 – Reactor trip from RCS pressure ($P_{RCS} < 13.2$ MPa)	16
3 – ECC criteria met ($P_{RCS} < 11.1$ MPa and $L_{PRZ} < 2.28$ m)	47
4 – Safety injection pumps start	47
5 – Accumulators injection start	82
6 – Steam generators level recovered	950
7 – Calculation terminated	1200

At 16 s, the RCS pressure < 13.2 MPa and reactor protection system signal is generated. The control rods are inserted, beginning the fast reactor shutdown; 31 s later, the SI pumps start to deliver ECC water. The initial break mass flow rate is much higher than the injection rate of the SI pumps, then, the coolant inventory of the RCS is reduced continuously, and thus, the collapsed liquid levels in the core and the steam generator tubes gradually decrease. Because of the assumed loss of offsite power and the postulated unavailability of loops 3 and 4 diesel engines, the startup and shutdown pumps are not available to provide water to the secondary side of the SGs. Water is therefore supplied by the emergency feedwater pumps only, which are started when the secondary side water level drops below 5 m. Steam generators level is recovered about 950 s after the transient beginning (see Fig. 7b). The reactor total power evolution during the transient is represented in Figure 7a.

Approximately 1000 s after transient started, the volume of liquid injected by the ECCS is sufficient to compensate the loss of coolant through the break, as can be observed in Figure 8a. At the beginning of the transient, the cladding temperature starts to rise, reaching a peak of 752.34 °C at 18 s. With the quick actuation of control and

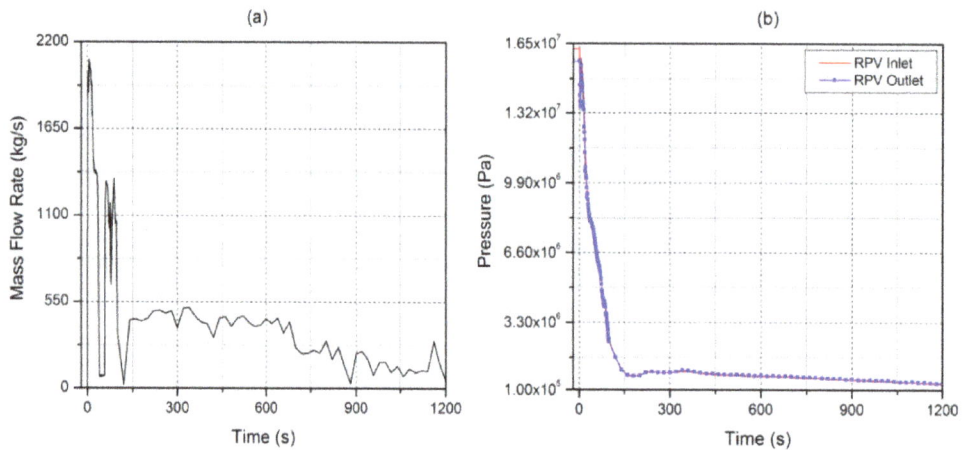

Fig. 6. Total break mass flow rate (a). Coolant pressure at inlet and outlet of the RPV (b).

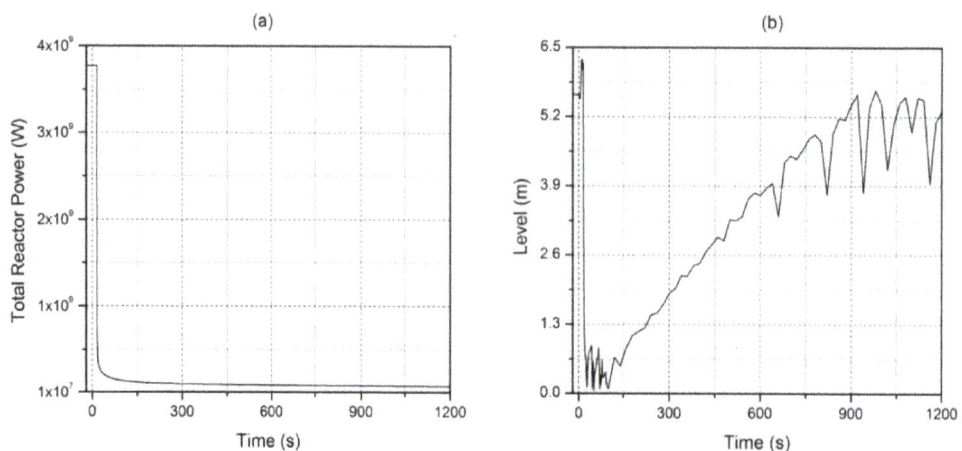

Fig. 7. Total reactor power (a). Steam generator 2 secondary side water level (b).

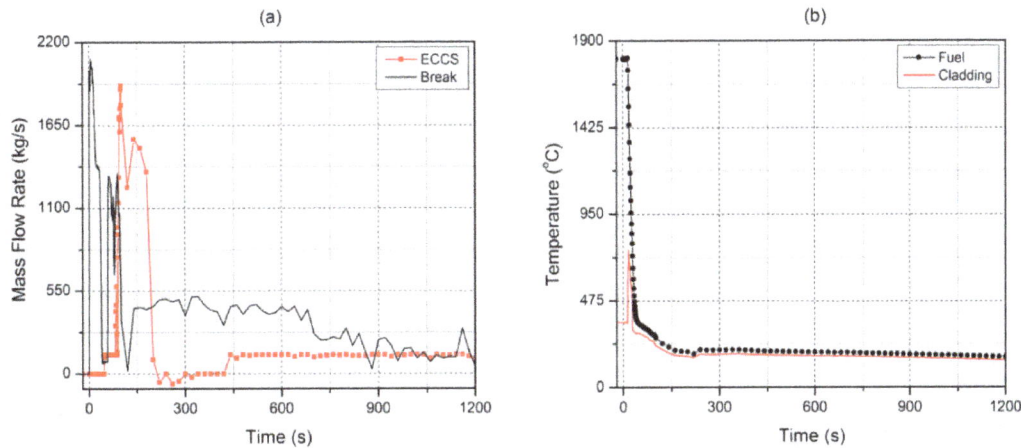

Fig. 8. Total mass flow rate injected and removed from RCS (a). Fuel and cladding temperature at level 18 of HS-605 (b).

protection system, this temperature increment does not reach values out of the allowed limits and, therefore, the reactor core integrity is guaranteed (see Fig. 8b).

4 Conclusions

A RELAP5-3D model of the Angra 2 PWR was developed using geometrical and material data from the final safety analysis report. Simulations of the reactor behavior during steady state and loss of coolant accident were performed. The analyzed parameters for the simulated cases demonstrated that the model with the control and protection system could successfully describe the reactor performance as in steady state as in transient operation conditions.

The analysis of the 200-cm^2 break between RCP and RPV demonstrates that the ECCS can provide sufficient cooling to prevent threat to the core. In the long term, the ECCS keeps the reactor coolant system filled and the decay heat is removed partly by the break flow. Taking the results of FSAR as a reference, the results obtained in this study show qualitatively similar behavior.

The next step of this work will be to insert others safety dispositive in the model and to observe how they can mitigate consequences after a severe transient. Moreover, the neutronic parameters will be inserted in the RELAP5-3D model, through the NESTLE code, to realistically reproduce 3D transient conditions with coupled thermal hydraulic/neutron kinetics effects.

The authors are grateful to CAPES, FAPEMIG and CNPq for the support. Thanks also to Idaho National Laboratory for the license to use the RELAP5-3D computer software.

Nomenclature

FSAR	Final safety analysis report
NPP	Nuclear power plant
LOCA	Loss of coolant accident
SBLOCA	Small break loss of coolant accident
RCP	Reactor coolant pump
RCS	Reactor coolant system
RPV	Reactor pressure vessel
TDV	Time-dependent volume
TDJ	Time-dependent junction
PRZ	Pressurizer
HS	Heat structure
ECCS	Emergency core cooling system
EFW	Emergency feedwater system
MFW	Main feedwater system

References

1. *THE RELAP5-3D$^{©}$ CODE DEVELOPMENT TEAM, RELAP5-3D$^{©}$ User's Manual* (Idaho National Laboratory, Idaho Falls, 2009)
2. G. Sabundjian, D.A. Andrade, A. Belchior Jr. et al., The behavior of Angra 2 nuclear power plant core for a small break LOCA simulated with RELAP5 Code, AIP Conference Proceedings **1529**, 151 (2013)
3. M.S. Rocha, G. Sabundjian, A. Belchior Jr. et al., Angra 2 Small Break LOCA Flow Regime Identification Through RELAP5 Code, in *Proceedings of ENCIT 2012, 14th Brazilian Congress of Thermal Sciences and Engineering, Rio de Janeiro, Brazil* (2012)
4. T. Crook, R. Vaghetto, A. Vanni, Y.A. Hassan, Sensitivity analysis of a PWR response during a loss of coolant accident under a hypothetical core blockage scenario using RELAP5-3D, in *Proceedings of the 2014, 22nd International Conference on Nuclear Engineering, ICONE22, Prague, Czech Republic* (2014)
5. Eletrobrás Eletronuclear, *Final Safety Analysis Report – Central Nuclear Almirante Álvaro Alberto – Unit 2*, Eletrobrás Termonuclear S.A., Rev. **13** (2013)
6. K. Ivanov, A. Olson, E. Sartori, OECD/NRC BWR turbine trip transient benchmark as a basis for comprehensive qualification and studying best-estimate coupled codes, Nucl. Sci. Eng. **148**, 195 (2004)
7. F. D'Auria, M. Frogheri, W. Giannoti, *RELAP5/MOD3.2 Post Test Analysis and Accuracy Quantification of Lobi Test BL-44*, International Agreement Report, NUREG/IA-0153 (1999)

8. T. Bajs, D. Grgić, V. Sêgon, L. Oriani, L.E. Conway, Development of a RELAP5 Nodalization for IRIS Non-LOCA Transient Analyses, in *Nuclear Mathematical and Computational Sciences: A Century in Review, A Century Anew* (Gatlinburg, United States, 2003)

9. A. Petruzzi, F. D'Auria, Thermal hydraulic system codes in nuclear reactor safety and qualification procedures, Sci. Technol. Nucl. Install. **2008**, 460795 (2008)

10. J.J. Duderstadt, L.J. Hamilton, *Nuclear Reactor Analysis* (John Wiley & Sons Inc., 1976)

15

An attempt for a unified description of mechanical testing on Zircaloy-4 cladding subjected to simulated LOCA transient

Jean Desquines[*], Doris Drouan, Elodie Torres, Séverine Guilbert, and Pauline Lacote

PSN-RES/SEREX/LE2M, IRSN, Bâtiment 327, BP 3, 13115 Saint-Paul-Lez-Durance, France

Abstract. During a Loss Of Coolant Accident (LOCA), an important safety requirement is that the reflooding of the core by the emergency core cooling system should not lead to a complete rupture of the fuel rods. Several types of mechanical tests are usually performed in the industry to determine the degree of cladding embrittlement, such as ring compression tests or four-point bending of rodlets. Many other tests can be found in the open literature. However, there is presently no real intrinsic understanding of the failure conditions in these tests which would allow translation of the results from one kind of mechanical testing to another. The present study is an attempt to provide a unified description of the failure not directly depending on the tested geometry. This effort aims at providing a better understanding of the link between several existing safety criteria relying on very different mechanical testing. To achieve this objective, the failure mechanisms of pre-oxidized and pre-hydrided cladding samples are characterized by comparing the behavior of two different mechanical tests: Axial Tensile (AT) test and "C"-shaped Ring Compression Test (CCT). The failure of samples in both cases can be described by usual linear elastic fracture mechanics theory. Using interrupted mechanical tests, metallographic examinations have evidenced that a set of parallel cracks are nucleated at the inner and outer surface of the samples just before failure, crossing both the oxide layer and the oxygen rich alpha layer. The stress intensity factors for multiple crack geometry are determined for both AT and CCT samples using finite element calculations. After each mechanical test performed on high temperature steam oxidized samples, metallography is then used to individually determine the crack depth and crack spacing. Using these two important parameters and considering the applied load at fracture, the stress intensity factor at failure is derived for each tested sample. This procedure provides an assessment scheme to determine experimentally the fracture toughness of the prior-β region in the mid-wall of the oxidized samples. The obtained fracture toughness for CCT and AT samples are thus compared, confirming that the linear elastic fracture mechanics is a relevant tool to describe the strength of LOCA embrittled cladding alloys.

1 Introduction

Two main types of tests have been developed to check the degree of material embrittlement after high temperature oxidation under steam environment, some combining the influence of applied load and oxidation-induced embrittlement and others relying on the mechanical testing of separately oxidized samples.

A very high degree of prototypicality is obtained using the integral thermal shock test originally developed by JAEA [1–7]. The test sample is a pressurized rodlet with inserted alumina pellets. The sample is inserted in a quartz tube and heated using an infrared lamp furnace. The quartz tube allows injection of steam environment, and at the end

of the test, the rod is quenched by water injection from the bottom end of the quartz tube. An axial tensile (AT) device is also connected to the upper plug of the cladding tube to apply an axial load during the quench. This test reproduces the main phases expected during a LOCA: cladding creep, ballooning and burst, high temperature oxidation under steam environment and finally water quench under applied load. The axial load results from partial restraint of thermal contraction of the rod during the quench under actual LOCA conditions. JAEA performed a large set of such integral thermal shock tests providing a rather simple result at the end of the test: failure or integrity of the rod. The complexity of the phenomena activated during such tests cannot offer a straightforward interpretation of the fuel rod failure conditions for modeling purpose.

In many laboratories, the LOCA consequences on the fuel cladding are rather studied by a sequential testing

* e-mail: jean.desquines@irsn.fr

Table 1. Ingot chemical composition of the tested SRA Zry-4.

Sn (wt%)	Fe (wt%)	Cr (wt%)	O (wt%)	H (wppm)
1.30	0.21	0.11	0.14	7

procedure: high temperature steam oxidation of a cladding sample or a fuel rod followed by a low temperature mechanical testing. The steam oxidation is sometimes performed using pressurized rodlets with pellets [8,9] or more simply on cladding samples in many other labs. The mechanical tests are usually: ring compression, AT loading [8], three- or four-point bending [8,10]. Such sequential tests can provide results of interest to clarify the failure mode of the rod and expectedly modeling data. However, there is currently no robust procedure to extrapolate the results from one mechanical test to another one.

In the present paper, two different sample geometries, corresponding to extremely different mechanical tests, are subjected to high temperature steam oxidation at 900 °C followed by mechanical testing at room temperature. The AT test is governed by rather uniform axial stress whereas the "C"-shaped Ring Compression Test (CCT) test is governed by hoop bending load. The failure mode is studied by performing interrupted tests, just before sample failure to better understand the influencing parameters on the oxidized cladding embrittlement. After this, the failure of AT samples is analyzed and compared to the failure conditions of CCT. A unified understanding of the failure conditions of these tests is provided and discussed.

2 Experiments

2.1 Materials

Stress Relieved Annealed (SRA) low-tin Zry-4 cladding tube with nominal chemical composition described in Table 1 is used in this study. The alloy was manufactured by CEZUS. The outer diameter of the tubes is 9.5 mm and the cladding thickness 0.57 mm.

2.2 Testing protocol

The first step of the testing protocol consists in low temperature oxidation and machining of the test samples. Some of the samples were not oxidized but directly machined. The low temperature (470 °C) oxidation under wet-air environment induces several hundred wppm hydrogen charging in the samples and a thin zirconia layer is formed at the inner and outer diameter of the sample.

Two sample geometries were considered in this preliminary step of the testing protocol:

– 20 mm long ring samples;
– AT samples with geometry specified in Figure 1.

The ring samples were obtained by diamond saw cutting whereas the AT samples were spark machined in order to avoid residual stress deposition in the gage sections.

After this step, the samples were high temperature oxidized, at both inner and outer surfaces, in a vertical furnace heated at 900 °C under steam environment. This temperature is relevant for small-break LOCA studies. The water quench was simulated by dropping the sample at 900 °C into a water bath. The steam oxidation protocol and its qualification are detailed in references [11–13]. The samples were weighted before and after 900 °C steam oxidation to determine a measured Equivalent Cladding Reacted (ECR). The ECR is a key parameter influencing the material embrittlement. It is defined as the ratio between the weight gain versus its maximum possible value corresponding to full oxidation of the sample.

The 20 mm long ring samples were then divided into two 10-mm long rings and one edge of the ring samples was then cut, using a diamond wire saw, to form "C"-shaped samples.

After the high temperature oxidation, the AT samples and CCT samples were subjected to mechanical testing, using an INSTRON 5566 electromechanical test device, to determine the degree of material embrittlement. The CCT test consists in compressing the sample as illustrated in Figure 2 at a displacement rate of 1 mm/min (see Refs. [14,15] for complementary details). After mechanical testing, upper and lower parts of AT samples were re-used to machine 10 mm long "C"-shaped ring compression samples. Post-test analyses were additionally performed to clarify the material failure mode, such as: metallography and hydrogen content measurements. Hydrogen contents are measured by sample fusion using a Brücker ONH mat 286 device.

2.3 Test matrixes

As explained above, some of the samples were hydrogen charged using wet-air oxidation at 470 °C. A thin oxidation layer was also formed during this period. This protocol leads to about 10% uncertainty on hydrogen content and about one micrometer variation in oxide layer thickness.

Three main test matrixes were defined, dedicated to the mechanical behavior of 900 °C steam oxidized cladding samples:

– test matrix #1: two pairs of CCT tests on steam oxidized samples, each pair corresponds to the samples originating from the same 20 mm long steam oxidized ring, one is loaded to failure and the second is interrupted just before failure load;
– test matrix #2: six CCT tests on steam oxidized cladding samples with hydrogen content below 1000 wppm;
– test matrix #3: eight AT samples, steam oxidized, were tested with hydrogen contents ranging between as-received content up to about 1000 wppm.

The first test matrix aims at clarifying the failure mode of oxidized cladding samples. The second and the third test matrixes provide respectively determination of the failure behavior of CCT samples and AT samples for comparison.

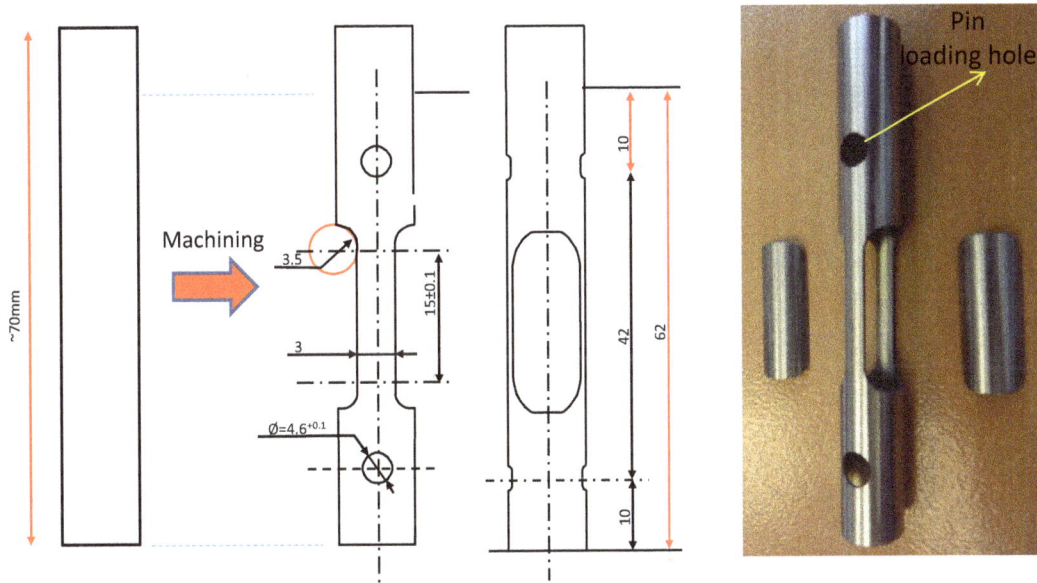

Fig. 1. Axial tensile sample geometry.

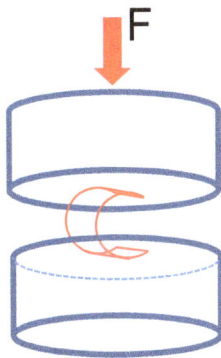

Fig. 2. Principle of "C"-shaped ring Compression Test (CCT) between two flat surfaces.

3 Test results

3.1 Oxidation results

The main oxidation results associated to each mechanical test are summarized in Table 2. The oxide layer thickness and the oxygen stabilized $\alpha(O)$ layer thickness are measured on post-test metallography after both low temperature oxidation and steam oxidation. A key parameter influencing material embrittlement is the sample hydrogen content ([H]), the measured value is provided with two standard deviations uncertainty after high temperature steam oxidation. The protective effect of pre-existing oxide layer explains some very low ECR values and sometimes zero weight gain values.

3.2 Mechanical test results

The key mechanical test results are summarized in Table 3. A failure mode is determined for each sample: ductile or brittle. A ductile failure is associated to samples developing macroscopic plastic deformation or stable crack growth prior to failure, all others are considered as brittle.

The CCT test load is normalized by sample axial length (L). During CCT testing, the hoop bending stress is heterogeneous and its maximum value is obtained at outer diameter and equatorial location. This maximum hoop stress value is determined from the normalized applied load in the elastic range according to references [14,15]:

$$\sigma_{max}(MPa) = \frac{1}{1.2987 \times 10^{-2}} \frac{F}{L} (N/mm). \qquad (1)$$

The CCT sample failure is expected to be governed by this stress value in the brittle range.

For AT testing, the axial stress (σ_{max} in Tab. 3) is assumed to be homogeneous and can be determined straightforward as the ratio between applied load and gage cross-section. However, it is clear that there are material properties heterogeneities in the various layers of the oxidized samples. But it is reasonable to consider that the equivalent homogeneous material properties are acceptably

Table 2. Low temperature oxidation conditions and 900 °C oxidation results for the test matrix.

| Test Matrix (#) | Sample (#) | 470 °C wet-air oxidation | | 900 °C steam oxydation | | | | |
		Time (days)	ZrO$_2$ layer thickness $\pm 1\,\mu$m (μm)	Time (min)	ZrO$_2$ layer thickness $\pm 1\,\mu$m (μm)	α(O) layer thickness $\pm 1\,\mu$m (μm)	[H] (wppm)	Measured ECR (%)
1	CCT1	15	8	15	8	28	430 ± 50	1.2
	CCT2	15	8	30	10	22	590 ± 60	3.2
	CCT3	15	8	15	8	28	430 ± 40	1.2
	CCT4	15	8	30	10	30	610 ± 80	3.2
2	CCT5	15	7.5	5	7.5	19	370 ± 50	0.0
	CCT6	15	7.5	15	9.5	31	630 ± 100	1.2
	CCT7	15	6.5	30	9.5	36	700 ± 90	3.4
	CCT8	20	9.5	60	22	26	970 ± 110	5.5
	CCT9	20	9.5	60	22	27	1060 ± 110	5.5
	CCT10	20	9.5	60	24.5	27	1030 ± 100	5.5
3	AT1	20	9.5	60	22	30	970 ± 120	5.5
	AT2	20	9.5				1060 ± 110	5.5
	AT3	20	10	60	27.5	26	1070 ± 120	5.5
	AT4	15	8	5	7.5	12	370 ± 50	0.0
	AT5	15	8	15	9.5	19	630 ± 100	1.2
	AT6	15	8	30	9.5	36	700 ± 160	3.4
	AT7	0	0	15	13	21	40 ± 20	3.8
	AT8	0	0	30	16.5	20	40 ± 10	4.8

represented by the most resistant phase property which is not cracked and is consequently affected by the strongest stress levels. This is a common assumption for materials affected by a cracked brittle surface layer. The failure axial load is extrapolated to the expected value for an entire fuel cladding (F/rod) without any machined gage section.

Before mechanical testing, some incipient cracks were observed in the oxygen stabilized α(O) layer. Similar cracks were observed by Nagase and Fuketa [4] and Kim et al. [16] and most probably form during the water quench. When comparing CCT3 interrupted test (Fig. 3) and CCT1 twin sample loaded up to failure (Fig. 4) many nucleated cracks propagating through the zirconia layer and the oxygen stabilized α(O) layer were observed along the sample outer surface. The observed cracks never propagated into α(O) inclusions in the $\alpha +$ prior-β region of the cladding sample. These two samples confirm that the brittle failure of CCT samples is governed by crack instability. The relevant parameter to describe the intensity of the stress singularity at the crack tip is the stress intensity factor (K_I). The critical value of this parameter under plane strain loading is the sample fracture toughness (K_{Ic}). Similar conclusions were obtained comparing CCT2 and CCT4 interrupted test on twin material. The procedure to determine the stress intensity factor is described in the following.

It is well known that multiple crack nucleation has a shadowing effect [2,17] limiting the stress intensity factor

at the crack tips. It was thus decided to measure the crack spacing on metallographic samples. For CCT tests, the crack spacing was considered to be the largest average value (left and right) associated to each observed crack. The measured crack spacing for each CCT sample is reported in Table 3. It is normally considered under uniform applied stress, that the largest stress intensity factor is obtained on the edge of multiple nucleated crack set. Considering CCT sample, the situation is different because the applied stress (assuming no crack nucleated) at the sample surface decreases when moving away from equator. For this reason, the maximum stress intensity factor is considered to be obtained within the nucleated crack set.

Considering now, the AT test with uniform applied stress, the worst location associated with maximum stress intensity factor corresponds to the edge of the multiple nucleated cracks set. The measured crack spacing for AT tests is also reported in Table 3.

The mechanical test results are analyzed in the next paragraph after calculation of the stress intensity factor associated to each sample geometry. The Cast3m finite element code[1] developed by CEA was used for the calculations.

---

[1] http://www-cast3m.cea.fr/

Table 3. Mechanical test results and determination of the sample fracture toughness.

Test Matrix	Sample	Failure (y/n)	Mode	F/L (N/mm)	F/rod (N)	$\sigma_{\max}$ (MPa)	λ: crack spacing (±1 μm) (μm)	a: crack depth (±1 μm) (μm)	K_{Ic} (MPa√m)
1	CCT1	y	Brittle	15.5		1190	50	36	7.2
	CCT2	y	Brittle	11.2		865	85	32	6.3
	CCT3	n		14.2		1090	50	36	6.6>
	CCT4	n		9.9		765	185	40	7.4>
2	CCT5	y	Ductile	63.8		4910	30	27	
	CCT6	y	Brittle	17.3		135	62	40	8.7
	CCT7	y	Brittle	17.0		1310	100	45	10.3
	CCT8	y	Brittle	9.8		755	80	48	5.4
	CCT9	y	Brittle	11.1		855	90	51	6.5
	CCT10	y	Brittle	9.1		700	56	51	4.2
3	AT1	y	Brittle		4860	295	>1000	52	4.2
	AT2[a]	y	Brittle						
	AT3	y	Brittle		4890	295	800	54	4.2
	AT4	y	Ductile		13790	855	49	20	
	AT5	y	Brittle		9660	600	65	28	5.2
	AT6	y	Brittle		7600	470	>1000	45	5.0
	AT7[b]	y	Ductile		11450	705	18	21	
	AT8[b]	y	Ductile		11930	730	20	24	

[a] Sample failure when mounting.
[b] Oxide layer spalling during the mechanical test.

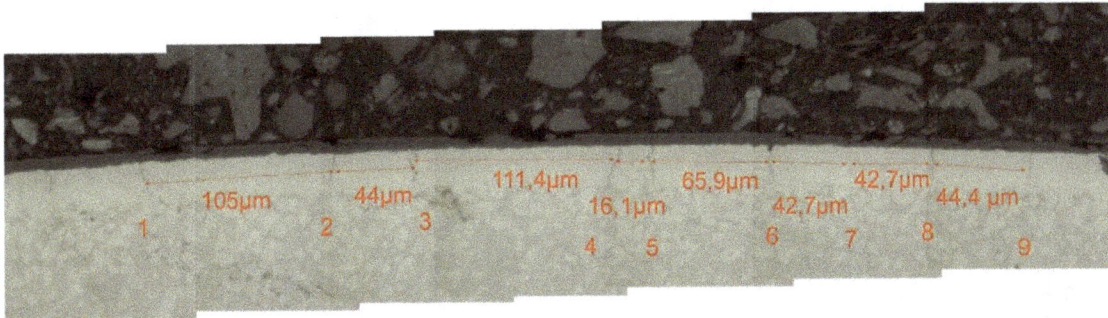

Fig. 3. Crack nucleation and crack spacing at outer diameter equatorial location of CCT3 interrupted CCT test at 92% of the failure load (comparison to CCT1 test twining sample).

4 Analysis of the mechanical test results

4.1 Modeling CCT test with multiple crack nucleation at outer diameter

A two-dimensional finite element model describing the elastic behavior of a CCT sample with a set of regularly spaced cracks nucleated at its outer surface was developed. The nominal geometry of the cladding is 9.5 mm outer diameter and 0.57 mm cladding thickness. The crack spacing and crack depth are parameters of this model. The contact conditions between crack lips are incorporated as boundary conditions.

The shape function at equatorial location, $\Phi\left(\frac{\lambda}{e}, \frac{a}{e}\right)$, describing the geometry influence on the stress intensity factor, was tabulated for various crack set geometries and the consistency with single crack nucleated $\left(\frac{\lambda}{e} = +\infty\right)$ was checked. A correlation is established providing the Φ value for all possible crack geometries (see Fig. 5). This shape function was used to determine the stress intensity factor value for CCT tests in Table 3. Figure 5 confirms that the stress intensity factor increases with increasing crack spacing. This phenomenon is known as shadowing effect of neighbor parallel cracks. A sensitivity study that is not reported in the present paper showed that the stress intensity factor decreases considering cracks away from equatorial location.

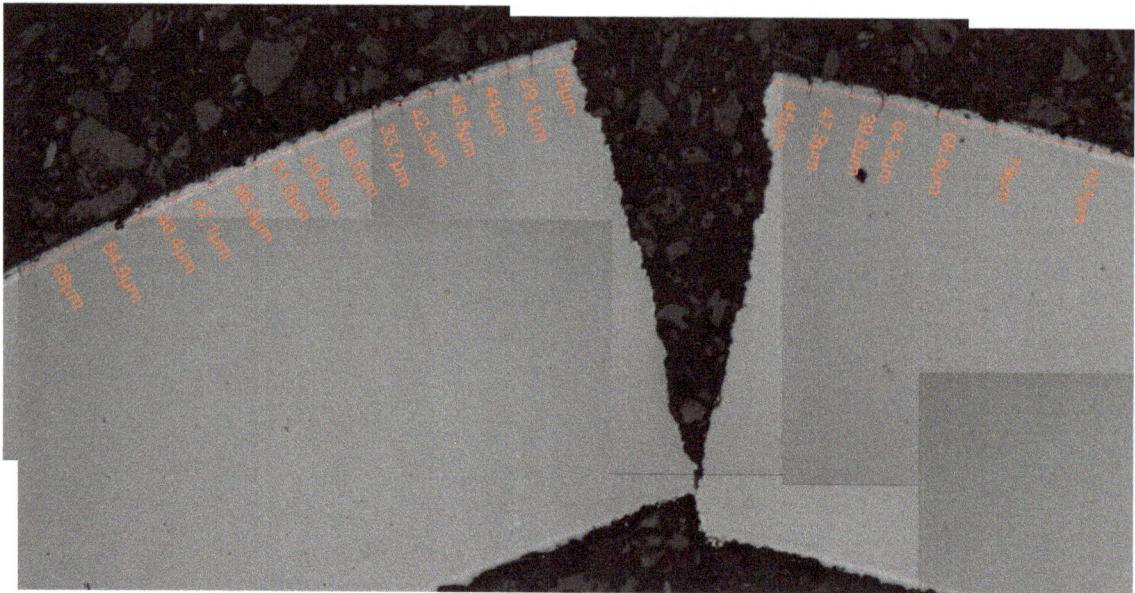

Fig. 4. Crack nucleation and crack spacing at outer diameter equatorial location of CCT1 after failure.

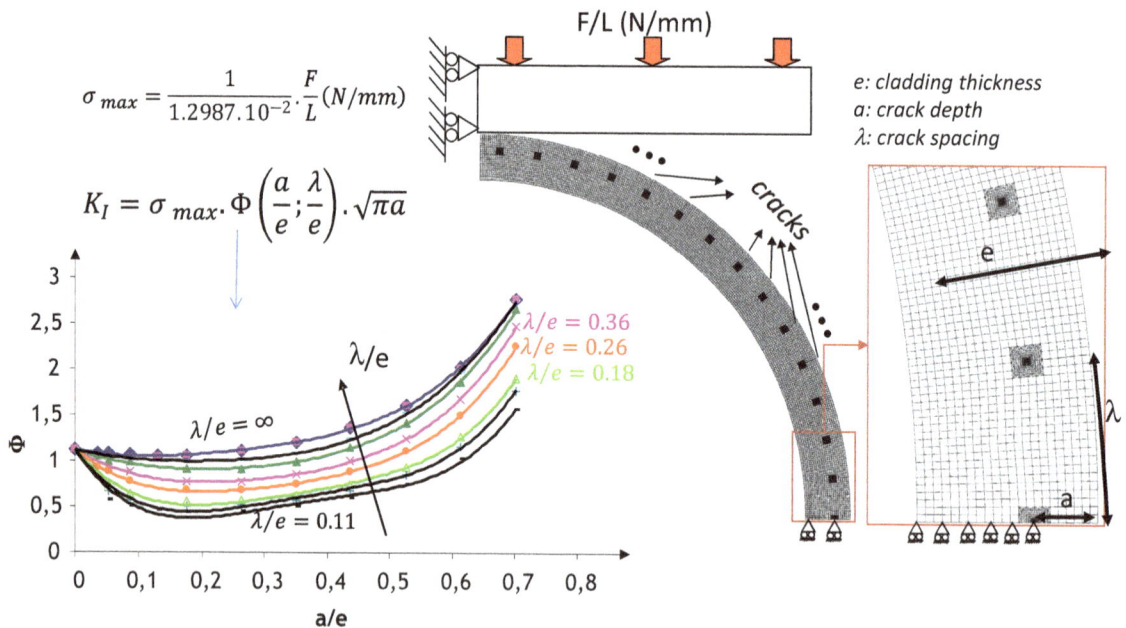

Fig. 5. Stress intensity factor at outer diameter equatorial location during CCT with multiple crack nucleation (marks: calculated values; lines: correlated values).

4.2 Modeling axial tensile tests with multiple crack nucleation at outer diameter

Using a similar approach, the multiple crack nucleation along the legs of the AT sample was studied using finite element simulations. The largest value of the stress intensity factor (see Fig. 6) is obtained at the edge of the array of parallel cracks. Consequently, the sample failure is expected at this location. In order to describe accurately multiple crack influence, a set of nine regularly spaced cracks subjected to tension was modeled using a 2D finite element model considering the sample symmetries, as illustrated in Figure 6. Murakami [2] shows that the crack number influence is close to saturated above about five cracks. There is consequently no need for additional cracks. The shadowing effect also disappears at large crack spacing, this can be easily described using the nine-crack model. The crack number is considered as sufficiently large to have no influence on the stress intensity factor of the edge crack. For crack clusters with varying crack spacing, the stress intensity factor at the edge is rather governed by the spacing between the two edge cracks. The crack depth and crack spacing are the key parameters of the modeling.

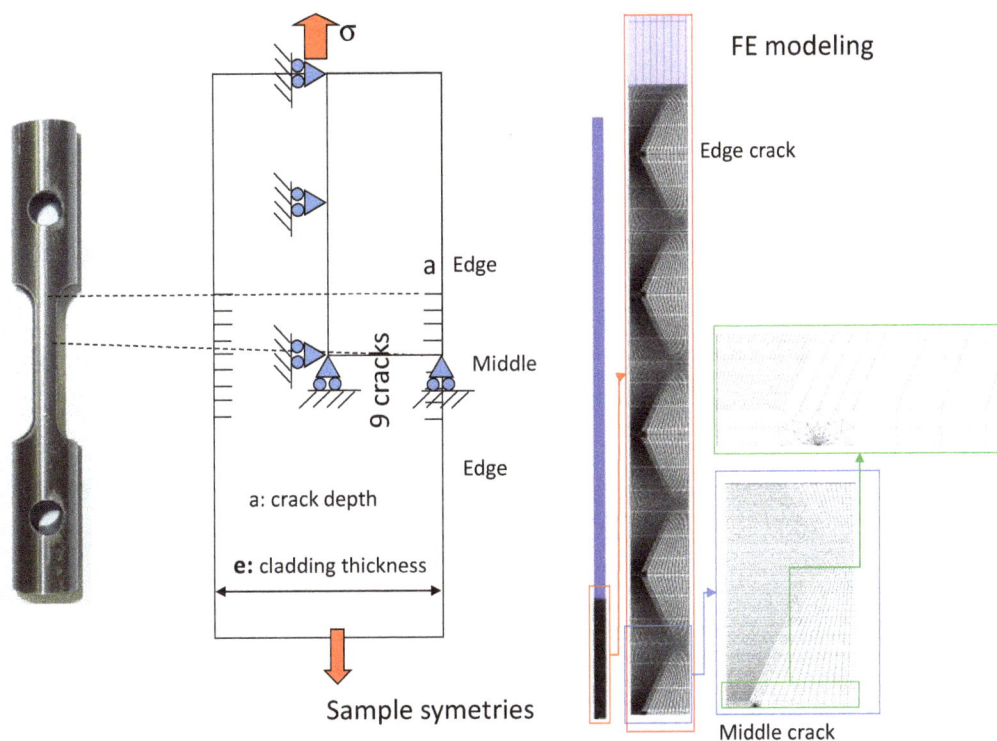

Fig. 6. Periodic crack array modeling to simulate an axial tensile test with multiple crack nucleation at its inner and outer surface.

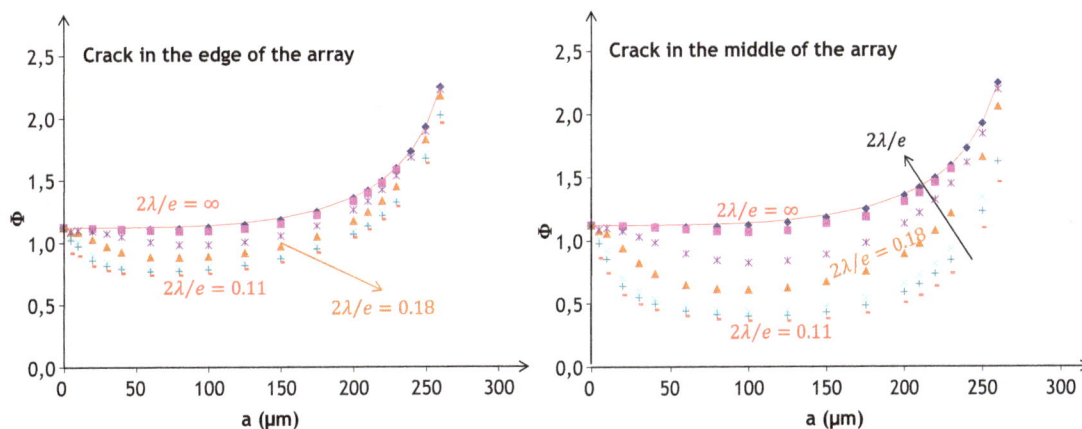

Fig. 7. Shape function values at the edge and at the middle location of the crack array (marks: calculated values; lines: correlated values).

The stress intensity factor is determined as previously done for CCT by incorporating a shape function describing the combined influence of normalized crack depth and crack spacing: $K_I = \sigma \Phi\left(\frac{2a}{e}; \frac{2\lambda}{e}\right)\sqrt{\pi a}$.

The shape function was tabulated at edge crack and middle crack as illustrated in Figure 7. The values at both locations are comparable for large crack spacing and significantly higher at the edge of the array, as expected, for close cracks. In other words, when a dense array of cracks is nucleated, the fracture is expected at the edge of the AT sample gage section. On the contrary, when only few cracks with large crack spacing are nucleated, the stress intensity is expected to be comparable at any crack tip of the array, the failure is expected anywhere in the gage section. The edge crack shape function was used to determine the stress intensity factors values reported in Table 3 for AT tests.

4.3 Evaluation of CCT and AT fracture toughness values

The fracture toughness values are valid only for brittle samples, other parameters are required to describe the ductile failures. The obtained values using the above-described procedure are reported in Figure 8. The values are rather low considering the usual values obtained for metals and are close to expected range for ceramics, especially at large hydrogen content.

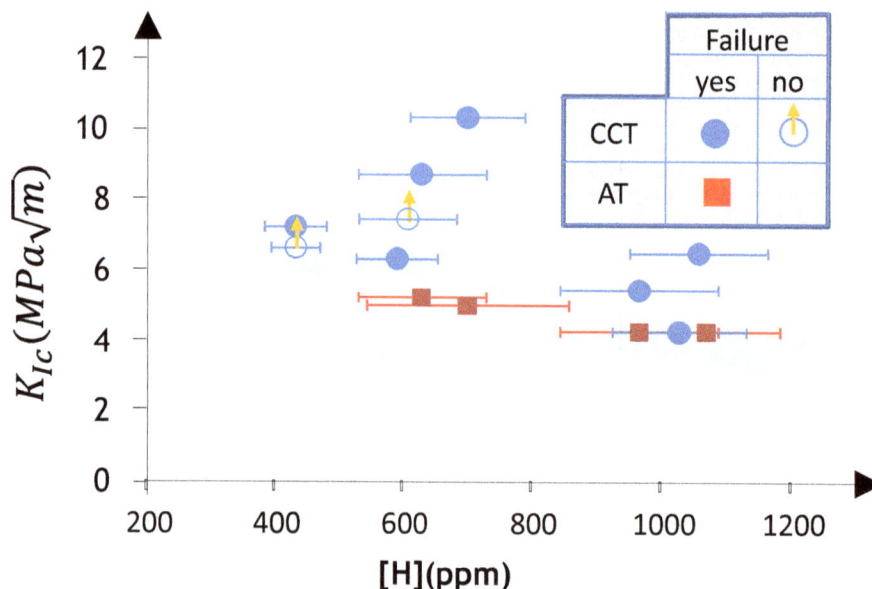

Fig. 8. Influence of sample hydrogen content on the evaluated cladding fracture toughness.

The fracture toughness of the AT tests corresponds to the lower trend of CCT tests. This might be linked to the limited volume of material exposed to the maximum stress value during a CCT compared to the gage volume of an AT test. If the crack instability happens slightly away from equatorial location, the stress intensity is clearly over-estimated using the value at equator. However, there is an acceptable consistency between the obtained fracture toughness values after 900 °C steam oxidation. An average value representing the average trend of AT tests and the lower trend of CCT tests is plotted in blue in Figure 8, this curve represents the expected value of material fracture toughness for large scale samples.

5 Conclusion

The room temperature mechanical behavior of cladding samples exposed to high temperature steam oxidation has been analyzed in the present paper. The mechanical behavior of the brittle samples appears consistent with linear elastic fracture mechanics. The cracks responsible for material embrittlement are nucleated in the oxygen stabilized $\alpha(O)$ layer during the sample quench and some of them propagate through the oxide layer during the mechanical test. A set of parallel crack forms at the oxidized sample surface. The sample failure was obtained for K_{Ic} (critical stress intensity factor or material fracture toughness) that is comparable during CCT and AT tests. However, the fracture toughness was slightly lower during AT tests. This was attributed to scale effect associated to very different gage volumes exposed to the maximum stress in the two kinds of tests. A large number of CCT tests would then be required to determine the lower fracture toughness expected on large scale samples.

The crack spacing is an influencing parameter for brittle samples with significant strength (close to the ductile-brittle transition), but for samples subjected to strong embrittlement a single crack nucleation is expected before sample failure.

The fracture toughness appears as an interesting parameter to describe the post-quench cladding embrittlement after a LOCA. The quantitative comparison between AT and CCT test results is possible using this material parameter.

To the authors knowledge, linear elastic fracture mechanics was surprisingly never evaluated to describe post-LOCA cladding embrittlement and appears as a powerful approach to model the cladding failure. Further extrapolation at higher temperature (1200 °C) and for a large set of ECR is now required for full validation of the proposed methodology. Some difficulties are expected when transposing this approach to Nb-bearing alloys for which the boundary between the oxygen stabilized α-layer and the prior β-layer is not well defined.

References

1. K. Honma, S. Doi, M. Ozawa, S. Urata, T. Sato, Thermal-shock behavior of PWR high burnup fuel cladding under simulated LOCA conditions, in *ANS 2001 Annual Meeting, Milwaukee, Wisconsin, USA, June 17–21* (2001)
2. Y. Murakami, *Stress intensity factors handbook* (Pergamon Press, 1987), Vols. 1, 2
3. F. Nagase, T. Fuketa, Effect of pre-hydriding on thermal-shock resistance of Zircaloy-4 cladding under simulated loss-of-coolant conditions, J. Nucl. Sci. Technol. **41**, 723 (2004)
4. F. Nagase, T. Fuketa, Behavior of pre-hydrided Zircaloy-4 cladding under simulated LOCA conditions, J. Nucl. Sci. Technol. **42**, 209 (2005)
5. F. Nagase, Fracture behavior of irradiated Zircaloy-4 cladding under simulated LOCA conditions, J. Nucl. Sci. Technol. **43**, 1114 (2006)
6. F. Nagase, T. Chuto, T. Fuketa, Behavior of high burnup fuel cladding under LOCA conditions, J. Nucl. Sci. Technol. **46**, 763 (2009)

7. F. Nagase, Status and plan of LOCA studies at JAEA, in *Fuel Safety Research Meeting, May 19–20, 2010, Tokai, Japan* (2010)

8. M.C. Billone, Assessment of current test methods for post-LOCA cladding behavior, NUREG Report, CR-7139, 2012

9. J. Stuckert, M. Grosse, C. Rössger, M. Klimenkov, M. Steinbrück, M. Walter, QUENCH-LOCA program at KIT on secondary hydriding and results of the commissioning bundle test QUENCH-L0, Nucl. Eng. Des. **255**, 185 (2013)

10. J.-C. Brachet, J. Pelchat, D. Hamon, R. Maury, P. Jacques, J.-P. Mardon, Mechanical behavior at room temperature and metallurgical study of low-tin Zy-4 and M5TM (Zr-NbO) alloys after oxidation at 1100 °C and quenching, in *Proceedings of TCM on Fuel behavior under transient and LOCA conditions, Organized by IAEA, Halden, September 10–14, 2001* (2001)

11. C. Duriez, S. Guilbert, A. Stern, C. Grandjean, L. Belovsky, J. Desquines, Characterization of oxygen distribution in LOCA situations, J. ASTM Int. **8** (2011). DOI:10.1520/JAI103156

12. S. Guilbert, C. Duriez, C. Grandjean, Influence of a pre-oxide layer on oxygen diffusion and on post-quench mechanical properties of Zircaloy-4 after steam oxidation at 900 °C, in *Proceedings of 2010 LWR Fuel Performance/TopFuel/WRFPM, Orlando, Florida, USA, September 26–29, 2010* (2010), Paper 121

13. S. Guilbert, P. Lacote, G. Montigny, C. Duriez, J. Desquines, C. Grandjean, Effect of Pre-oxide on Zircaloy-4 high-temperature steam oxidation and post-quench mechanical properties, ASTM STP 1523 (2013)

14. J. Desquines, D. Drouan, P. March, S. Fourgeaud, C. Getrey, V. Elbaz, M. Philippe, Characterization of radial hydride precipitation in Zy-4 using "C"-Shaped samples, in *LWR Fuel Performance Meeting Topfuel 2013, Charlotte, North Carolina* (2013)

15. J. Desquines, D. Drouan, M. Billone, M.P. Puls, P. March, S. Fourgeaud, C. Getrey, V. Elbaz, M. Philippe, Influence of temperature and hydrogen content on stress-induced radial hydride precipitation in Zircaloy-4 cladding, J. Nucl. Mater. **453**, 131 (2014)

16. J.H. Kim, B.K. Choi, J.H. Baek, Y.H. Jeong, Effects of oxide and hydrogen on the behavior of Zircaloy-4 cladding during the loss of the coolant accident (LOCA), Nucl. Eng. Des. **236**, 2383 (2006)

17. Z.D. Jiang, A. Zeghloul, G. Bezine, J. Petit, Stress intensity factors of parallel cracks in a finite width sheet, Eng. Fract. Mech. **35**, 1073 (1990)

16

Monte Carlo MSM correction factors for control rod worth estimates in subcritical and near-critical fast neutron reactors

Jean-Luc Lecouey[1*], Anatoly Kochetkov[2], Antonin Krása[2], Peter Baeten[2], Vicente Bécares[3], Annick Billebaud[4], Sébastien Chabod[4], Thibault Chevret[1], Xavier Doligez[5], François-René Lecolley[1], Grégory Lehaut[1], Nathalie Marie[1], Frédéric Mellier[6], Wim Uyttenhove[2], David Villamarin[3], Guido Vittiglio[2], and Jan Wagemans[2]

[1] Laboratoire de Physique Corpusculaire de Caen, ENSICAEN/Université de Caen/CNRS-IN2P3, 14050 Caen, France
[2] SCK·CEN, Belgian Nuclear Research Centre, Boeretang 200, 2400 Mol, Belgium
[3] Nuclear Fission Division, CIEMAT, Madrid, Spain
[4] Laboratoire de Physique Subatomique et de Cosmologie, Université Grenoble-Alpes, CNRS/IN2P3, 53, rue des Martyrs, 38026 Grenoble Cedex, France
[5] Institut de Physique Nucléaire d'Orsay, CNRS-IN2P3/Université Paris Sud, Orsay, France
[6] Commissariat à l'Énergie Atomique et aux Énergies Alternatives, DEN, DER/SPEX, 13108 Saint-Paul-lez-Durance, France

Abstract. The GUINEVERE project was launched in 2006, within the 6th Euratom Framework Program IP-EUROTRANS, in order to study the feasibility of transmutation in Accelerator Driven subcritical Systems (ADS). This zero-power facility hosted at the SCK·CEN site in Mol (Belgium) couples the fast subcritical lead reactor VENUS-F with an external neutron source provided by interaction of deuterons delivered by the GENEPI-3C accelerator and a tritiated target located at the reactor core center. In order to test on-line subcriticality monitoring techniques, the reactivity of all the VENUS-F configurations used must be known beforehand to serve as benchmark values. That is why the Modified Source Multiplication Method (MSM) is under consideration to estimate the reactivity worth of the control rods when the reactor is largely subcritical as well as near-critical. The MSM method appears to be a technique well adapted to measure control rod worth over a large range of subcriticality levels. The MSM factors which are required to account for spatial effects in the reactor can be successfully calculated using a Monte Carlo neutron transport code.

1 Introduction

The GUINEVERE (Generator of Uninterrupted Intense NEutrons at the lead VEnus REactor) project [1] was launched in 2006, within the 6th Euratom Framework Program IP-EUROTRANS [2], in order to study the feasibility of transmutation in Accelerator Driven subcritical Systems (ADS). This facility hosted at the SCK·CEN site in Mol (Belgium) is presently used in the follow-up FREYA project (7th European FP) [3]. It couples the fast subcritical lead-moderated reactor VENUS-F with an external neutron source provided by the deuteron accelerator GENEPI-3C via T(d,n)[4]He fusion reactions occurring at the reactor core center (Fig. 1). It is partially dedicated to the investigation of techniques of on-line subcriticality monitoring.

The VENUS-F reactor core is very modular and its reactivity can range from deep subcritical to critical by varying the number of fuel assemblies loaded in the core. It is

also equipped with two boron carbide control rods which allow for a finer tuning of the reactivity. Fission chambers, spread throughout the reactor, allow recording count rates during either steady-state or time-dependent measurements.

In order to test on-line subcriticality monitoring techniques, the reactivity of all the VENUS-F configurations used must be known beforehand to serve as benchmark values. Thus, the reactivity worth of the control rods must be known as accurately as possible so that the reactivity of every new reactor configuration created by moving the control rods be estimated correctly.

Although the reactor asymptotic period measurement is a usual technique to determine the reactivity worth of control rods, it is limited to a small reactivity range (from ≈ -0.3 \$ to $+0.3$ \$) [4]. Consequently, it does not always allow measuring the total reactivity worth of the control rods. Furthermore, it is obviously inapplicable to control rod worth measurement in deep subcritical reactors.

This is the reason why the Modified Source Multiplication Method (MSM) [5] is under consideration to be used as an

*e-mail: lecouey@lpccaen.in2p3.fr

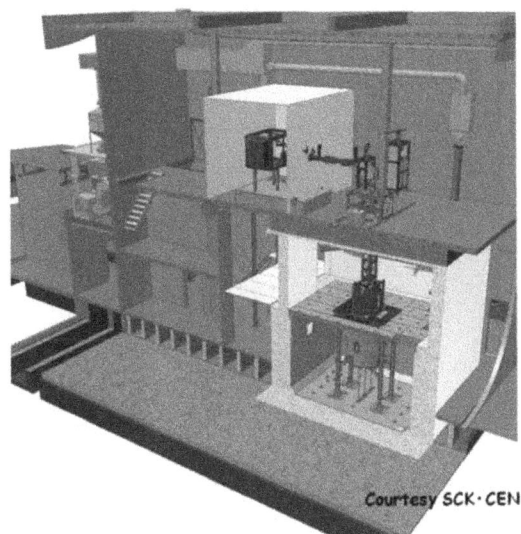

Fig. 1. Overview of the GUINEVERE facility at SCK·CEN.

alternative method for estimating the reactivity worth of the VENUS-F control rods when the reactor is largely subcritical as well as near-critical. In this technique, the unknown reactivity is determined by comparing detector count rates driven by an external neutron source in the configuration of interest (in this paper it will be a new configuration obtained by moving the control rods) with those obtained with the same neutron source in another subcritical configuration whose reactivity is already known (reference configuration). However, to account for the flux shape differences between the two reactor configurations, some position-dependent correction factors (the so-called MSM factors) must be calculated using a neutron transport code.

In this paper, we first present the GUINEVERE facility and the various configurations of the VENUS-F reactor studied. Then the principle of the MSM method is briefly exposed. The results of MSM factor calculations performed with the Monte Carlo neutron transport code MCNP are also shown. They were carried out in support to MSM experiments dedicated to the measurement of the VENUS-F control rod worth when the reactor was either subcritical or near critical. In the former case, the GENEPI-3C was used to generate the neutron external source. In the latter case, an Am-Be neutron source was inserted in the reactor.

General trends in the MSM factor behavior which depend on the neutron source and detector locations, as well as on the reactor subcriticality level are outlined. Finally the calculated MSM correction factors are applied to the detector count rates measured during the MSM experiments. The consistency between the reactivity values given by the detectors is discussed.

2 The GUINEVERE facility

2.1 The VENUS-F reactor

The VENUS-F fast reactor is contained in a cylindrical vessel of approximately 80 cm in radius and 140 cm in

height. A 12×12 grid surrounded by a square stainless steel casing can receive up to 144 elements of $\approx 8 \times 8$ cm^2 in section which can be fuel assemblies, lead assemblies or specific elements for accommodating detectors or absorbent rods. The remaining room in the vessel is filled with semi-circular lead plates, which act as a radial neutron reflector. In addition, the core is equipped with top and bottom 40 cm-thick lead reflectors. Each fuel assembly (FA) contains a 5×5 pattern, filled with 9 fuel rodlets and 16 lead bars, surrounded by lead plates. The fuel is 30 wt.% enriched metallic uranium provided by CEA. Among the set of FAs, six are actually safety rods (SR) made of boron carbide and fuel followers with the absorbent part retracted from the core in normal operation. Two control rods (CR) made of natural boron carbide square cuboids can be positioned at various locations in the 12×12 grid. They can be moved vertically from 0 mm (fully inserted in the core) to 600 mm (fully retracted). Another absorbent rod, whose reactivity worth is very small, called PEAR (Pellet Absorber Rod) rod, is available for performing rod drop experiments.

Various configurations of the reactor in terms of reactivity can be studied thanks to the modular shape of the core. In this paper, since we are interested in measuring the reactivity worth of the set of two CRs, all the reactor configurations studied were obtained from either a near-critical reactor configuration called CR0↓ or a subcritical configuration named SC1↓, by moving the two CRs together at various heights. Since the reactivities of the CR0↓ and SC1↓ configurations had been measured during previous experiments [6], they could serve as reference values for applying the MSM method.

The so-called CR0↓ configuration is represented in Figure 2. Ninety-seven FAs (in blue for the regular ones, in light blue for the SRs with fuel followers) are arranged in a way to create a pseudo-cylindrical core. The two boron-carbide CRs (in red) are located at the core periphery and retracted at approximately 515 mm in height. The CR0↓ configuration was created from a critical one by dropping the PEAR rod (in green). After analyzing the rod drop experiments using Inverse Point Kinetics, the reactivity of CR0↓ was found to be −136(2) pcm [6]. As shown in Figure 2, the reactor was equipped with 9 fission chambers (FCs) working in pulse mode. Three different types of FCs were used, either Photonis CFUL01 and CFUM21[1], or GE Reuter-Stokes (RS), whose specifications are listed in Table 1. In order to help localizing the various assemblies and detectors, an arbitrary coordinate system is used in the 12×12 grid: the upper left corner is labeled (−6,6) and the lower right one (6,−6), there is no (0,0) element. Outside the 12×12 grid, six cylindrical cavities bored in the outer reflector can receive experimental devices. They are labeled, from left to right: A1, B1, C1, A2, B2, and C2.

The so-called SC1↓ configuration is shown in Figure 3. It is derived from the CR0↓ configuration by removing the four central FAs. This removal also permits the insertion of the accelerator thimble inside the VENUS-F core.

[1]http://www.photonis.com/nuclear/products/fission-chambers-for-out-of-core-use/

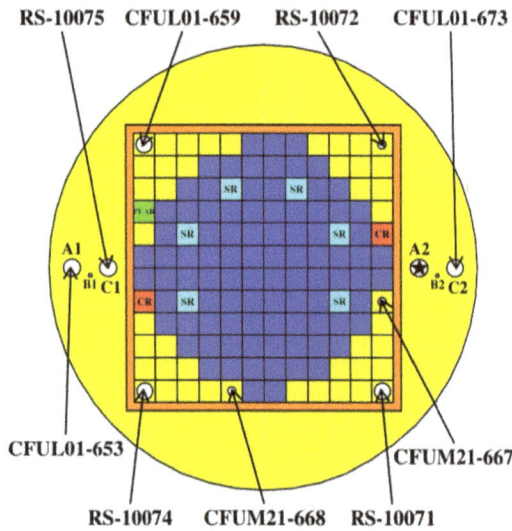

Fig. 2. Schematic view of the CR0↓ configuration. The black star shows the position of the external neutron source (Am-Be). Control rods (CR) are in red.

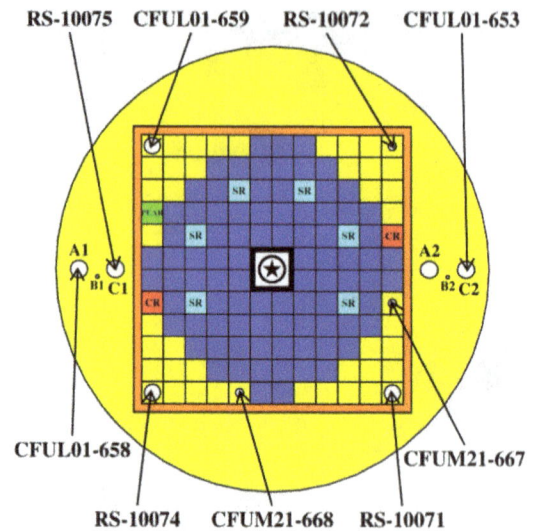

Fig. 3. Schematic view of the SC1↓ configuration. The black star shows the position of the external neutron source (GENEPI-3C). Control rods (CR) are in red.

Compared to CR0↓, some additional minor differences are itemized below:

- the CRs are slightly more inserted inside the core (CR height is 479 mm instead of 515 mm);
- the detector set is slightly different: the CFUL01-673 detector is replaced by the CFUL01-653 FC. The latter is replaced in the A1 location by the CFUL01-658 FC which is identical to CFUL01-659 and CFUL01-653 FCs. The reactivity of SC1↓ was measured using the MSM method and was found to be −3824(96) pcm [6].

2.2 External neutron sources

The external neutron source used for performing the MSM experiments was different depending on whether the reference configuration was CR0↓ or SC1↓.

Table 1. Fission chambers used in configurations CR0↓ and SC1↓.

Name	Main deposit	Approximate mass (mg)	Location in CR0↓	Location in SC1↓
CFUL01-653	^{235}U	1000	A1	C2
CFUL01-658	^{235}U	1000	None	A1
CFUL01-659	^{235}U	1000	(−6,6)	(−6,6)
CFUL01-673	^{238}U	1000	C2	None
RS-10071	^{235}U	100	(6,−6)	(6,−6)
RS-10072	^{235}U	100	(6,6)	(6,6)
RS-10074	^{235}U	100	(−6,−6)	(−6,−6)
RS-10075	^{235}U	100	C1	C1
CFUM21-667	^{235}U	10	(6,−2)	(6,−2)
CFUM21-668	^{235}U	10	(−2,−6)	(−2,−6)

In the latter configuration, the external source was created at the center of the VENUS-F core by deuterons interacting with a tritiated titanium target. The deuteron ions were accelerated up to an energy of 220 keV by the GENEPI-3C particle accelerator [7] built by a collaboration of CNRS-IN2P3 laboratories. The fusion reactions at core mid-plane generate a quasi-isotropic field of ∼14-MeV neutrons. The GENEPI-3C can operate in pulsed mode, in continuous mode, and also in continuous mode with short beam interruptions. During the MSM experiments reported here, GENEPI-3C delivered a continuous deuteron beam whose intensity ranged from ∼400 μA to ∼500 μA.

However, the intensity of the external neutron source created by the accelerator had to be monitored directly. Indeed, the tritium release and the beam tuning variations over time prevent the direct calculation of the neutron source intensity from that of the beam on target. This is the reason why the accelerator is equipped with two Si detectors which can detect either alpha particles from T$(d,n)^4$He reactions or protons from D(d,p)T reactions. The detection of α particles allows one to quantify the amount of 14-MeV neutrons produced whereas the detection of protons allows estimating the parasitic production of 2.5-MeV neutrons by D$(d,n)^3$He due to the implantation of deuterons in the target. During the MSM experiments reported here, the neutron source intensity varied from ∼1×10^9 to ∼3×10^9 14-MeV neutrons/s.

In the CR0↓ configuration, the external neutron source was an Am-Be source inserted in the outer reflector slot A2 (denoted by a star in Fig. 2) which emitted only 2.2×10^6 neutrons/s. Thus the Am-Be neutron source intensity is lower than that induced by the GENEPI-3C by three orders of magnitude. Furthermore, mainly because the Am-Be source is off-centered, its importance is approximately eight times lower than that of GENEPI-3C. In terms of detector count rates, these source dissimilarities are only (very) partially compensated by the difference in reactivity between the two CR0↓ and SC1↓

configurations. Therefore, since the two reference configurations are very dissimilar both in terms of reactivity and of source location, interesting differences in the results of the MSM experiments can be anticipated.

3 The MSM method

3.1 Principle

The MSM (Modified Source Multiplication) method is a technique for estimating the unknown reactivity of a subcritical configuration by comparing detector count rates driven by an external neutron source in this configuration with those obtained in another subcritical configuration whose reactivity is known.

The inhomogeneous transport equation associated with a subcritical configuration of a reactor driven by an external neutron source reads:

$$A\Phi = P\Phi + S \qquad (1)$$

where P is the neutron production operator (by fission or (n, xn) reactions), A is the migration and loss operator and S is the external neutron source intensity. Φ is the neutron flux which is present inside the reactor when the external neutron source is inserted.

This transport equation can be made homogeneous by introducing the neutron multiplication coefficient k_{eff}:

$$A\varphi = \frac{1}{k_{eff}}P\varphi. \qquad (2)$$

In that case, φ is the fundamental mode corresponding to the associated critical reactor. $\lambda = 1/k_{eff}$ is also an eigenvalue of the adjoint homogeneous equation:

$$A^{\dagger}\varphi^{\dagger} = \frac{1}{k_{eff}}P^{\dagger}\varphi^{\dagger}. \qquad (3)$$

where $A^{\dagger}$ and $P^{\dagger}$ are the adjoint operators of A and P, respectively. $\varphi^{\dagger}$ is the adjoint flux, also called neutron importance function.

Multiplying the adjoint homogeneous equation (3) by Φ and integrating over space, angle and energy, one gets:

$$\rho = \frac{\left\langle\Phi, \left(P^{\dagger} - A^{\dagger}\right)\varphi^{\dagger}\right\rangle}{\left\langle\Phi, P^{\dagger}\varphi^{\dagger}\right\rangle} = \frac{\left\langle\varphi^{\dagger}, (P - A)\Phi\right\rangle}{\left\langle\varphi^{\dagger}, P\Phi\right\rangle} \qquad (4)$$

where $\langle\rangle$ denotes such an integration.

Then, multiplying the inhomogeneous equation (1) by $\varphi^{\dagger}$ and integrating over space, angle and energy leads to:

$$\left\langle\varphi^{\dagger}, (P - A)\Phi\right\rangle = -\left\langle\varphi^{\dagger}, S\right\rangle \qquad (5)$$

and combining equation (4) and equation (5), one gets:

$$\rho = -\frac{\left\langle\varphi^{\dagger}, S\right\rangle}{\left\langle\varphi^{\dagger}, P\Phi\right\rangle}. \qquad (6)$$

As in reference [5], we introduce the reaction rate in the detector $C = \langle\Sigma_d, \Phi\rangle$, where Σ_d is the macroscopic reaction cross-section of the detector, and rewrite equation (6):

$$\rho = -\left\langle\varphi^{\dagger}, S\right\rangle \times \frac{\langle\Sigma_d, \Phi\rangle}{\langle\varphi^{\dagger}, P\Phi\rangle} \times \frac{1}{\langle\Sigma_d, \Phi\rangle}$$

$$= -S_{eff} \times \varepsilon \times \frac{1}{C} \qquad (7)$$

where $S_{eff} = \langle\varphi^{\dagger}, S\rangle$ is called the effective neutron source and $\varepsilon = \langle\Sigma_d, \Phi\rangle/\langle\varphi^{\dagger}, P\Phi\rangle$ the detector efficiency.

Now let us consider two subcritical configurations. Let configuration 0 be the subcritical configuration of known reactivity ρ_0 and configuration 1 be that of unknown reactivity ρ_1. Assuming that the neutron external source and the detectors utilised are the same in both configurations, equation (7) can be used to find a relationship between ρ_0, ρ_1, and the detector count rates C_0 and C_1 in configurations 0 and 1:

$$\frac{\rho_1}{\rho_0} = \frac{S_{eff,1}\varepsilon_1}{S_{eff,0}\varepsilon_0} \times \frac{C_0}{C_1} = f_{MSM} \times \frac{C_0}{C_1} \qquad (8)$$

where f_{MSM} is the MSM correction factor. One can also introduce the source importance φ which is defined as the ratio of the average importance of external source neutrons to the average importance of fissions in the reactor [8,9]:

$$\varphi = \left(\frac{\langle\varphi^{\dagger}, S\rangle}{\langle S\rangle}\right)\left(\frac{\langle\varphi^{\dagger}, P\Phi\rangle}{\langle P\Phi\rangle}\right)^{-1}. \qquad (9)$$

If one introduces the source multiplication coefficient k_s as [9]:

$$k_s = \frac{\langle P\Phi\rangle + \langle S\rangle}{\langle S\rangle} \qquad (10)$$

the source importance appears as the ratio of the neutron gain with the external neutron source to a hypothetical gain which would be obtained with a stabilized fission source in the same reactor:

$$\varphi = \left(\frac{k_s}{1-k_s}\right)\left(\frac{1-k_{eff}}{k_{eff}}\right). \qquad (11)$$

Then the MSM factor can be rewritten with the source importance of the two configurations:

$$f_{MSM} = \frac{\varphi_1}{\varphi_0} \times \frac{e_1}{e_0} \qquad (12)$$

where φ_i and $k_{s,i}$ are respectively the source importance and the source multiplication coefficient in configuration i. The parameter e_i is defined as:

$$e_i = \frac{\langle\Sigma_d, \Phi_i\rangle}{\langle P_i\Phi_i\rangle}. \qquad (13)$$

It represents the ratio of the reaction rate in the detector to the total rate of neutron produced in the reactor. Thus, formula (12) shows that the MSM factor accounts for the differences in neutron and source importance as well as in flux shapes between the two configurations considered.

However, if configurations 0 and 1 are very similar, such differences may vanish and formula (8) reduces to the Approximate Source Method (ASM) formula:

$$\frac{\rho_1}{\rho_0} = \frac{k_{eff,0}}{k_{eff,1}}\frac{C_0}{C_1} \approx \frac{C_0}{C_1} \tag{14}$$

where the approximation $k_{eff,0}/k_{eff,1} \approx 1$ is often made.

The MSM correction factors must be calculated using a transport code, either deterministic or stochastic. It is worth mentioning that the value of the MSM correction factor is expected to depend on the detector location. Indeed, any difference in the flux shape between the two configurations will result in position-dependent ratios in the f_{MSM} formula.

3.2 Calculation of MSM factors

Starting from equation (8), the MSM factor reads:

$$f_{MSM} = \frac{\rho_1}{\rho_0} \times \frac{C_1}{C_0} \tag{15}$$

where the reactivity of configurations 0 and 1, ρ_0 and ρ_1, as well as the detector count rates in configurations 0 and 1, C_0 and C_1, can be calculated using a neutron transport code.

Although the use of deterministic codes is largely reported in literature, MSM factors can also be calculated using stochastic neutron transport codes. On one hand, the use of a Monte Carlo code advantageously allows one to transport neutrons in the reactor theoretically without any geometry simplification (to the extent that the reactor geometry be accurately known) and with pointwise energy dependent cross-sections. On the other hand, Monte Carlo calculations are much more computer-time-consuming than deterministic ones and provide as results only statistical estimates of quantities of interest. In this paper, the Monte Carlo simulation code MCNP 5 [10] was employed, together with ZZ ALEPH-LIB-JEFF3.1.1, a continuous energy multi-temperature library created at SCK·CEN and based on JEFF3.1.1 [11]. Once the geometry as well as the material composition of the various elements constituting the reactor have been described in an MCNP input file, the corresponding multiplication factor (hence the reactivity) can be estimated using a generation-based, iterative fission neutron source whose spatial distribution converges towards the fundamental mode of the reactor (the so-called "kcode" source). On the other hand, standard fixed-source calculations can provide estimates of reaction rates anywhere in the reactor. So, for the calculation of MSM factors, four Monte Carlo simulations must be run: two fixed-source simulations for calculating the source driven reaction rates C_0 and C_1 in the fission chambers for configurations 0 and 1, and two "kcode" simulations for estimating the reactivity of the same two configurations, ρ_0 and ρ_1.

As a first step towards the calculation of MSM factors, MCNP input files had to be built for the configurations CR0↓ and SC1↓ as well as their variants created by moving the CRs. In order to save computing times (a factor of ~4.5 was gained), it was decided to use a simplified reactor geometry. Indeed, the MSM method bears interest only if the calculation of MSM factors turns out to be rather insensitive to the details and errors on the reactor geometry, as well as to uncertainties on material compositions and on nuclear data: since MSM experiments are carried out to estimate the unknown reactivity of a reactor configuration, one can imagine that the reactor itself could be not very well known either. Fortunately, this robustness of MSM factor calculations has already been observed for previous MSM experiments at the VENUS-F reactor and can be understood by recalling that MSM factors are double ratios of quantities: one can expect that any reasonable difference between the calculated reactivity values and the real ones will be at least partially compensated by corresponding differences between the calculated reaction rates and the measured ones [6].

Since the control and safety rods are nearly homogeneous, the principal source of geometrical simplification was the homogenization of the fuel assemblies. Additionally, some details of the bottom reactor reflector geometry were not considered. Also, the GENEPI-3C accelerator was not modelled. Instead, a 14-MeV point source was placed in vacuum at the core center. For the Am-Be source, the average source energy of 5 MeV was used. Finally, the FCs were not modelled at all. Instead, use was made of the next-event estimator MCNP tally F5 (point detector) to estimate the fission rates of the FC deposits, at the center of each detector location.

One MCNP input file was created for each CR height selected for the MSM experiments (from 0 to 600 mm by step of 60 mm around the reference CR position of the SC1↓ configuration and by step of 50 mm around the one of CR0↓). Then, prior to calculating the four terms of formula (15), the reactivity scale of the MCNP models of VENUS-F configurations had to be adjusted so that the calculated reactivities of CR0↓ and SC1↓ be approximately equal to their measured values of –136 pcm and –3824 pcm, respectively. This allowed an overall consistency between experimental results concerning the configurations used as references and the subsequent calculations. This was achieved by multiplying the average number of neutrons per fission ν used inside the MCNP code by a factor of 1.001071. This slightly modified value of ν was then used for calculating the reactivity of all the other configuration variants obtained by changing the CR heights.

3.3 Results of MSM factor calculations

Figure 4 shows the evolution of MSM factors as a function of the new height of the CRs after moving them away from their position associated with the reference configuration SC1↓ (479 mm). Error bars were calculated using the quadratic sum of the uncertainties on the four terms of formula (15). The relative uncertainty is basically

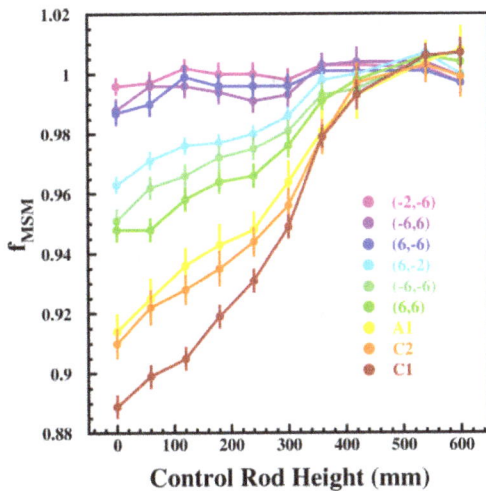

Fig. 4. MSM factors as a function of FC position and CR height using SC1↓ as reference.

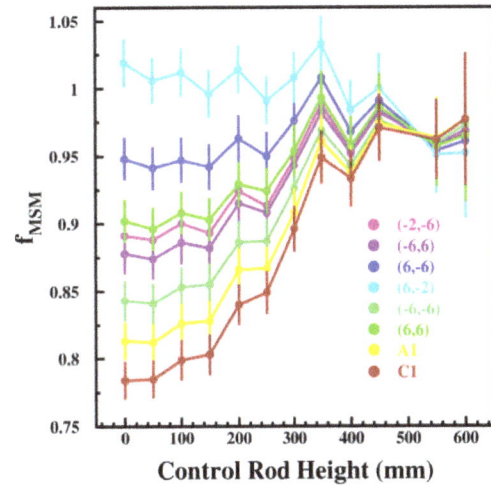

Fig. 5. MSM factors as a function of FC position and CR height using CR0↓ as reference.

dominated by that on the fission reaction rates for which a precision of less than 1% could be achieved in a reasonable computing time.

The first observation to be made is that, as expected, regardless of the detector position, there is roughly no MSM correction to consider when the CRs do not move much around 479 mm. However, as the amplitude of the CR motion and thus the dissimilarity between the neutron flux shapes increases, the MSM factors tend to deviate more and more from 1. Regarding the evolution of the MSM factors, the detectors seem to fall into three or four different groups. In the first group (positions (−2,−6), (−6,6) and (6,−6)), the MSM factors do not deviate much from 1, even for the largest CR motion. This can be explained by the fact that those three detectors are rather far from the source and far from the CRs and therefore rather protected from the modifications of the flux shape caused by the CR motion. It is less and less the case as we move from the first group to the second one (detectors in (6,−2), (−6,−6), (6,6)) and then to the third one (detectors in A1, C2 and C1).

In conclusion, in the case of the SC1↓ configuration, it seems possible to estimate the CR worth using a simple ASM approach without any calculated correction factor as long as the detectors are carefully selected.

Figures 5 and 6 deal in the same way with the MSM factors for the CRs moving when the reactor is almost at critical, that is when the reference configuration used for the MSM calculation is CR0↓ (CR height at 515 mm). The color code for detector positions is identical to that of Figure 4. First of all, it is worth mentioning that the statistical error bars are significantly larger than those shown in Figure 4. Indeed, since CR0↓ is almost critical, the reactivity values are rather close to zero and the relative uncertainties on the calculated reactivity tend to be much larger than those calculated for the configurations based on SC1↓. Furthermore, some MCNP fixed-source calculations needed for estimating detector fission rates can become very computer-time-consuming as the CR height increases, and hence as the multiplication factor k_{eff} becomes very close to 1. To quantify this evolution, one can make use of the

Figure of Merit (FOM) [10] which is defined as:

$$FOM = \frac{1}{R^2 T}$$

where T is the computing time and R the relative statistical uncertainty on the quantity of interest (here the detector reaction rates). For instance, between fixed-source calculations performed at 0 mm and at 600 mm, the Figure of Merit drops by a factor of ∼300.

As in the case where the reference configuration is SC1↓, the MSM factors tend to deviate more and more from 1 as the amplitude of the CR motion increases, as expected by the associated larger perturbation of the reference flux. However, compared to Figure 4, two main differences are visible in Figures 5 and 6.

The most striking one concerns the FC located in C2 (Fig. 6). The behavior of the associated MSM factor is so different from the others that it had to be shown in a separate figure. This is due to the extremely short distance from the Am-Be source (13.5 cm) combined to the high energy

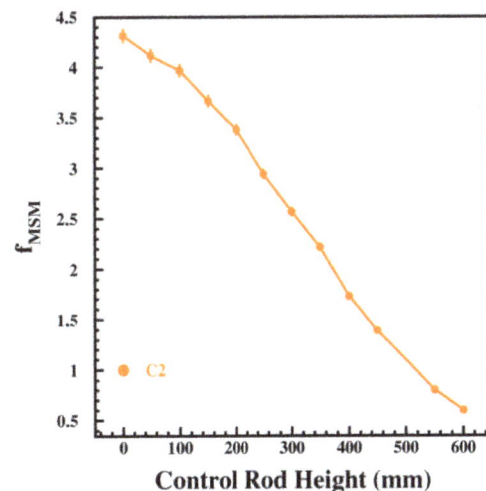

Fig. 6. MSM factors associated with detector CFUL01-673 in C2 as a function of CR height using CR0↓ as reference.

threshold of ^{238}U (~1 MeV) which represents 99.965% of the CFUL01-673 deposit mass. Thus, on the one hand, CFUL01-673 is proportionally much more sensitive than the other FCs to the fast neutrons originating directly from the Am-Be source and, on the other hand, it is much less sensitive to the regular neutron multiplication in the core.

When looking at Figure 5, it also appears that the clear division of detectors in groups proposed for Figure 4 does not hold any longer. Although some detectors very close to one CR exhibit strong correction factors, such as RS-10075, it is not the case for the FC located in (6,−2). On the other hand, the detector located in (−6,6) is very far away from the CRs and from the Am-Be source and still, its MSM correction factor is far from remaining close to 1 when the CRs are moved.

To understand why these MSM factors exhibit a much more complex behavior than in the case of those calculated for SC1↓ and its variants, it is worth using formula (12). The latter relates the MSM factors to the ratio of the source importance in the configuration of interest to that in the reference configuration. The source importance can be easily calculated by combining the results of MCNP kcode and fixed-source calculations.

As already mentioned herein above, the Am-Be source importance is much smaller (~0.3) than the GENEPI-3C one (~2.5), mainly because of the difference in the source location. Hence, it is more fruitful to compare the source importance variations (compared to the value of φ^* taken arbitrarily at 0 mm) as a function of CR height for the variants of the configurations based on CR0↓ and for those based on SC1↓. Results (in %) are shown in Figure 7.

The difference in behavior of the source importance between the two sets of reactor configurations is striking. On one hand, in the case of SC1↓-derived configurations (in red), the variation of source importance as the CR height changes, if any, is very small. On the other hand, when considering CR0↓-derived configurations (in black), the source importance appears to increase significantly as the CRs are raised. This is obviously due to the very short distance between the Am-Be source and one of the two CRs,

which makes the source multiplication very sensitive to the motion of the neutron absorbent (while the external neutron source was at the core center in the case of the variants based on SC1↓). Hence, whereas the MSM factor evolution seems to be mainly explained by the flux shape modification occurring around the CRs when they are moved from the 479 mm position for SC1↓, the MSM factor variation with the CR motion around CR0↓ seems to be more complex. In this case, the source importance and the flux shape are both modified.

In short, the behavior of the MSM correction factors as a function of the detector position appears to be much more complex when using the Am-Be source instead of the external neutron source provided by means of the GENEPI-3C accelerator. This difference is likely to originate from the off-centered position, close to one CR, of the Am-Be source.

4 Application to MSM experiments

It is beyond the scope of this paper to apply the MSM factors calculations presented herein to the numerous measurements performed at the VENUS-F reactor. Instead we present hereinafter a few results which illustrate the most the performances of the MSM method with Monte Carlo simulations.

4.1 MSM experiments with the SC1↓ configuration

As shown in Figure 4, the farther from that of the reference configuration (479 mm) the CR height is, the larger the MSM correction factor is. This is the reason why the results corresponding to the CR heights settled at 0 mm have been selected and are shown in Table 2.

First the ASM reactivity ρ_{ASM} of the VENUS-F reactor when the CRs are positioned at 0 mm was calculated from each detector count rate $\tilde{R}$ as follows:

$$\rho_{ASM}(0\,\text{mm}) = \frac{\tilde{R}(479\,\text{mm})}{\tilde{R}(0\,\text{mm})}\rho(SC1↓) \qquad (16)$$

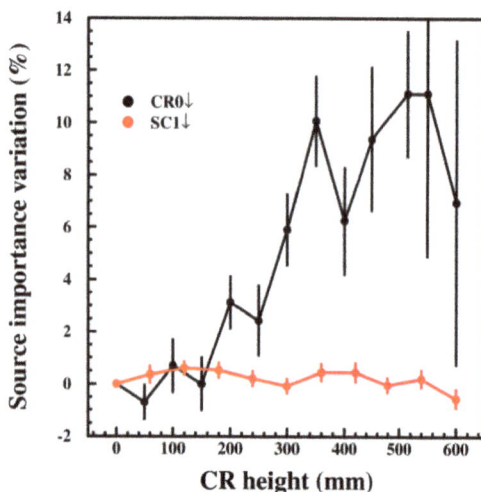

Fig. 7. Source importance variations as a function of the CR height.

Table 2. ASM reactivity, MSM factor and MSM reactivity when CRs are lowered to 0 mm (from SC1↓) for each of the nine FCs used.

Name	Location	ρ_{ASM}(pcm)	f_{MSM}	ρ_{MSM}(pcm)
CFUL01-658	A1	−4884(123)	0.914(6)	−4464(116)
CFUL01-659	(−6,6)	−4573(115)	0.988(4)	−4518(115)
CFUL01-653	C2	−4874(122)	0.910(5)	−4436(114)
RS-10071	(6,−6)	−4597(116)	0.987(4)	−4537(115)
RS-10072	(6,6)	−4792(120)	0.948(4)	−4543(116)
RS-10074	(−6,−6)	−4772(120)	0.951(4)	−4539(116)
RS-10075	C1	−5090(128)	0.889(4)	−4525(115)
CFUM21-667	(6,−2)	−4718(119)	0.963(3)	−4544(115)
CFUM21-668	(−2,−6)	−4576(115)	0.996(3)	−4558(115)

where $\rho(SC1\downarrow)$ is equal to -3824 ± 96 pcm. Since the neutron external source was created by means of the GENEPI-3C accelerator, a specific normalization had to be applied to the detector count rates, $\tilde{R}(479\,\text{mm})$ and $\tilde{R}(0\,\text{mm})$, measured respectively in the reference configuration (SC1↓, CRs at 479 mm) and in the other selected configuration (SC1↓, CRs at 0 mm). In order to take into account the fluctuations of the neutron production in the tritiated target, the total numbers of counts in the detectors were normalized per alpha particle detected in the dedicated Si detector, instead of per second.

As can be seen in Table 2, the ASM reactivity varies by about 550 pcm, depending on the detector considered. Some spatial effects are definitely at work. Now, applying the MSM factors calculated with MCNP simulations which are shown in Figure 4 and also listed in Table 2, the MSM reactivity can be built for each detector as follows:

$$\rho_{MSM}(0\text{mm}) = f_{MSM} \times \rho_{ASM}(0\text{mm}). \quad (17)$$

After multiplication by the MSM factors, the dispersion of the detector reactivity values is successfully reduced to ~120 pcm.

To finish with the SC1↓-related experiments, we make use of one interesting result of the MSM factor calculations. In Figure 4, one can see that the MSM factors for three FCs (positions $(-2,-6)$, $(-6,6)$ and $(6,-6)$) are very close to 1. This suggests that the simple ASM method applied to these three detectors might give the same results as the MSM method does. The comparison of the reactivity results of the ASM method applied to the three detectors with the MSM ones using the full set of FCs, as a function of CR height, is presented in Figure 8. For each CR position and each method, the reactivity was calculated as the weighted mean of the values given by the two different sets of selected detectors (3 and 9 FCS for the ASM and MSM methods, respectively):

$$\langle\rho\rangle = \frac{\sum_i \rho_i/s_i^2}{\sum_i 1/s_i^2} \quad (18)$$

where s_i is the uncertainty on the reactivity given by detector i. The uncertainty associated with the average reactivity was conservatively calculated by assuming that correlation was at maximum between all the detectors considered:

$$s = \sqrt{\frac{\sum_i \sum_j 1/(s_i s_j)}{\left(\sum_i 1/s_i^2\right)^2}}. \quad (19)$$

As can be seen in Figure 8, the agreement between the two methods is excellent over the whole range of CR heights. Not only it validates the use of MSM factor calculations to choose the right detector subset to use an ASM approach but it also shows that the measurements made with all the other detectors can be well corrected. It is also remarkable that, once the kcode MCNP reactivity values are properly scaled by adjusting the calculated reactivity of SC1↓ to the measured one, they (in blue) also are in very good agreement with the experimental results over the whole range of CR height. In addition, it is worthwhile to mention that some dynamical reactivity measurements carried out in pulsed neutron source experiments [12] or in experiments with programmed interruptions of a continuous beam [13], with the CR heights at 0, 240, 479 and 600 mm, gave results consistent with those presented here.

4.2 MSM experiments with the CR0↓ configuration

Since, as in the case of the SC1↓ experiments, the MSM corrections to the ASM reactivity values are expected to be the strongest at 0 mm for the CR0↓ experiments, the results obtained for this height are gathered in Table 3. The spread observed among the ASM reactivity values is much more dramatic (~850 pcm) because of the presence of the CFUL01-673 FC with ^{238}U as main deposit (Sect. 3.3). However, once the MSM factors have been applied, the dispersion of the results drops down to ~80 pcm. Even though the CFUL01-673 final reactivity value seems to be

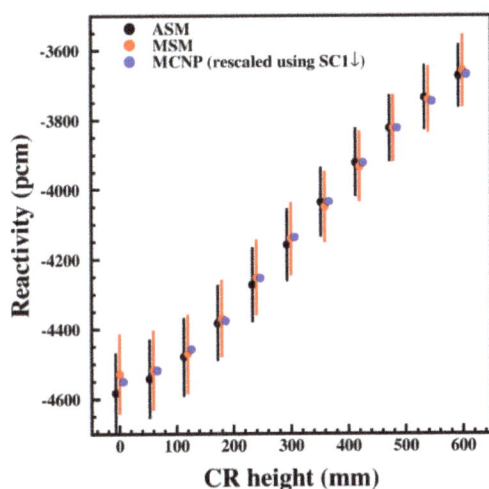

Fig. 8. Reactivity of VENUS-F as a function of CR height when CRs are moved together from 0 to 600 mm, around the SC1↓ configuration.

Table 3. ASM reactivity, MSM factor and MSM reactivity when CRs are lowered to 0 mm (from CR0↓) for each of the nine FCs used.

Name	Location	ρ_{ASM} (pcm)	f_{MSM}	ρ_{MSM} (pcm)
CFUL01-653	A1	−1013(13)	0.813(14)	−824(18)
CFUL01-659	(−6,6)	−950(12)	0.878(15)	−834(18)
CFUL01-673	C2	−210(5)	4.316(72)	−905(27)
RS-10071	(6,−6)	−876(12)	0.948(16)	−830(18)
RS-10072	(6,6)	−924(13)	0.902(16)	−833(19)
RS-10074	(−6,−6)	−987(14)	0.843(15)	−832(19)
RS-10075	C1	−1058(15)	0.784(14)	−829(19)
CFUM21-667	(6,−2)	−858(21)	1.019(18)	−874(27)
CFUM21-668	(−2,−6)	−937(23)	0.891(15)	−835(25)

slightly overestimated (in absolute value), one has to keep in mind that the MSM correction is a tour-de-force for this detector considering the amplitude of the correction to be applied.

5 Conclusions

The so-called Modified Source Multiplication Method (MSM) technique consists in determining the unknown reactivity of a reactor configuration by comparing detector count rates driven by an external neutron source in the configuration of interest with those obtained in another subcritical configuration whose reactivity is already known (reference configuration). This method can be used as an alternative method to the asymptotic period measurement for determining control rod worth.

This paper focused on the use of the Monte Carlo neutron transport code MCNP to calculate position-dependent MSM correction factors needed to account for the flux shape differences between the reference reactor configuration and the configuration whose reactivity is to be measured.

A comparison was made between the MSM factors obtained for a set of nine detectors spread in the lead-moderated fast neutron reactor VENUS-F, for a largely sub-critical configuration and a near-critical one. It was found that the MSM factors exhibited some common trends but also that the behavior of these factors was much more difficult to explain simply in the near-critical case because of the specific location of the external neutron source.

However, in both cases, the MSM factors calculated with the MCNP Monte Carlo code were successfully applied to ASM reactivity values obtained experimentally: the reactivity spread among the detectors was strongly reduced by the MSM correction.

In conclusion, the MSM method seems to be a technique well adapted to measure control rod worth over a large range of subcriticality levels. The required MSM factors can be easily calculated using a Monte Carlo neutron transport code, although the computing time can become very large when the studied reactor configurations are very close to criticality. Consequently, appropriate variance reduction techniques remain to be investigated.

This work was partially supported by the 6th and 7th Framework Programs of the European Commission (EURATOM) through the EUROTRANS-IP contract # FI6W-CT-2005-516520 and FREYA contract # 269665, and the French PACEN and NEEDS programs of CNRS. The authors want to thank the VENUS reactor and GENEPI-3C accelerator technical teams for their help and support during experiments.

References

1. A. Billebaud et al., in *Proceedings of Global 2009* (Paris, France, 2009)
2. J. Knebel et al., in *Proceedings of the International Conference on Research and Training in Reactor Systems (FISA 2006), Luxembourg, 2006*
3. A. Kochetkov et al., in *Proceedings of the International Conference on Technology and Components of Accelerator Driven Systems (TCADS-2), Nantes, France, 2013*
4. K.O. Ott, R.J. Neuhold, *Introductory Nuclear Reactor Dynamics*, (American Nuclear Society, 1985)
5. P. Blaise, F. Mellier, P. Fougeras, IEEE Trans. Nucl. Sci. **58**, 1166 (2011)
6. J.L. Lecouey et al., Ann. Nucl. Energ. **83**, 65 (2015)
7. M. Baylac et al., in *Proceedings of the International Topical Meeting on Nuclear Research Applications and Utilization of Accelerators (AccApp '09), Vienna, Austria, 2009*
8. M. Salvatores et al., Nucl. Sci. Eng. **126**, 333 (1997)
9. P. Seltborg et al., Nucl. Sci. Eng. **145**, 390 (2003)
10. Los Alamos National Laboratory Report LA-ORNL, RSICC LA-UR-03-1987 (2003)
11. W. Haeck, B. Verboomen, Technical Report NEA/JEFF/DOC-1125, OECD/NEA (2006)
12. N. Marie et al., in *Proceedings of the International Conference on Technology and Components of Accelerator Driven Systems (TCADS-2), Nantes, France, 2013*
13. T. Chevret et al., The Role of Reactor Physics Toward a Sustainable Future, in *PHYSOR 2014, Kyoto, Japan, 2014*

17

Modelling of as-fabricated porosity in UO₂ fuel by MFPR code

Vladimir I. Tarasov[*] and Mikhail S. Veshchunov

Nuclear Safety Institute (IBRAE), Russian Academy of Sciences, 52, B. Tulskaya, 115191, Moscow, Russia

Abstract. For consistent modelling of behaviour of as-fabricated porosity in UO₂ fuel irradiated under various conditions of in-pile and out-of-pile tests as well as under normal and abnormal conditions of nuclear reactor operation, the additional analysis of experimental observations and critical assessment of available models are presented. On this base, the mechanistic MFPR code, including physically-grounded models for the fuel porosity evolution in UO₂ fuel under various irradiation and thermal regimes, is refined. These modifications complete the consistent description of the fuel porosity evolution in the MFPR code and result in a notable improvement of the code predictions.

1 Introduction

The in-pile dimensional behaviour of oxide fuels in nuclear reactors is a well-known phenomenon of great technological interest. It is generally established that at the beginning of irradiation the fuel densifies due to shrinking of the as-fabricated pores remaining from the fuel sintering process with a wide distribution of their sizes [1,2]. The densification is most pronounced in low density fuel, especially in the case of fine-dispersed porosity with pores typically less than one micron diameter.

Re-sintering in the furnace can be generally understood and described analytically by thermal diffusion processes, but not so in-pile densification: it was additionally assumed by Stehle and Assmann [3] that in-pile densification is a mixed athermal/thermal process, including the thermal evaporation of vacancies from pores (which dominates at relatively high temperatures above ≈1200 °C), and the athermal atomization of pores into lattice vacancies by fission spikes.

For consistent modelling of porosity behaviour in UO₂ fuel irradiated under various conditions of in-pile and out-of-pile tests as well as under normal and abnormal conditions of nuclear reactor operation, the critical assessment of available models, their modification and development of more advanced models for implementation in the mechanistic codes, become rather an important task. The code MFPR (Module for Fission Products Release) was developed for analysis of fission products (FP) release from

irradiated UO₂ fuel in collaboration between IBRAE and IRSN (Cadarache, France) [4,5]. The mechanistic approach applied in this code allows the realistic consideration of fuel porosity evolution, self-consistently with analysis of FP release, based on physically-grounded parameters and mechanisms.

Some important modifications of the existing models of MFPR and development of new models for the fuel porosity evolution in UO₂ fuel under various irradiation and thermal regimes are presented in this paper.

2 Initial fuel porosity

Optical microscopy reveals that the majority of the internal cavities are located on grain boundaries [6]; the pores are generally non-spherical in shape. In the current analysis, the pores are considered as intergranular lenticular voids with the dihedral angle $\theta = 50°$. Their volume, V, and surface area, S, are [7]:

$$V = \frac{4\pi}{3}\left(1 - \frac{3}{2}\cos\theta + \frac{1}{2}\cos^3\theta\right)R^3 \equiv \frac{4\pi}{3}f_V R^3,$$
$$S = 4\pi(1 - \cos\theta)R^2 \equiv 4\pi f_S R^2, \tag{1}$$

where the pore curvature radius, R, relates to the experimentally measured median radius, ρ (which below will be simply referenced to as 'radius') as:

$$\rho \frac{\sin\theta}{\xi} R, \tag{2}$$

where $\xi \approx 1.29$ (see Appendix A of Ref. [8]).

* e-mail: tarasov@ibrae.ac.ru

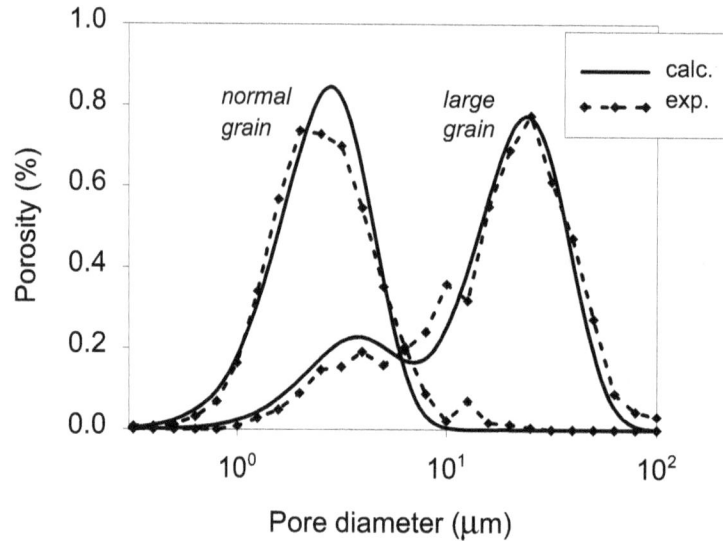

Fig. 1. Initial pore size distributions in fuel samples of the Harada and Doi test [9] and their approximations.

In the typical fresh fuel, the pores sizes are distributed within the interval 0.1–10 μm, their density distribution function can be satisfactory approximated as [8]:

$$C(\rho) = \frac{C_0}{\overline{\rho}} e^{-\rho/\overline{\rho}}, \qquad (3)$$

where C_0 is the total pore concentration and $\overline{\rho}$ is the mean radius. The maximum contribution to the total porosity makes pores with $\rho = 3\overline{\rho}$.

Figure 1 illustrates approximations of the experimental distributions observed in reference [9] for normal grain (mean grain diameter $d_{gr} = 8\,\mu$m) and large grain (23 μm) samples. The normal grain data are approximated by equation (3) with the mean radius of 0.35 μm and total porosity of 4.7%. Note that pores with ρ from 0.5 to 2 μm contribute near 75% to the total porosity. The large grain data are approximated by a superposition of two exponents corresponding to two pore populations, P1 and P2, with the mean radii of 0.45 and 3.0 μm, the partial porosities being 1.2 and 4.3% respectively.

3 Mechanisms of pore size relaxation

If the grain boundary self-diffusion is the rate controlling mechanism of the thermal pore relaxation, the pore volume change is described by the equation [10,11]:

$$\left(\frac{d}{dt}V\right)_{therm} = 4\pi D_{gb} w \frac{\Omega}{kT} \delta P F_{SB}(\varphi), \qquad (4)$$

where D_{gb} is the grain boundary diffusivity, $w \approx 0.5$ nm is the thickness of the grain boundary layer, $\Omega = 4.09 \times 10^{-29}$ m^3 is the atomic volume, k is the Boltzmann constant, T is the temperature. The pressure difference, δP, is given by equation:

$$\delta P = \frac{N_p kT}{V - N_p B} - \frac{2\gamma}{R} - P_h, \qquad (5)$$

where N_p is the number of gas atoms in the pore, P_h is the external pressure, B is the van der Waals constant, φ is the fractional coverage of the grain boundary by pores.

The factor F_{SB} was derived in reference [10] for the case of small identical pores uniformly distributed over an infinitely large grain boundary, if the vacancy diffusion in the grain boundary is rate controlling:

$$F_{SB}(\varphi) = -\left(\ln\varphi + \frac{1}{2}(1-\varphi)(3-\varphi)\right)^{-1}. \qquad (6)$$

This function is of order of 1 for typical φ values of 10–20%, however it has logarithmic singularity at $\varphi \to 0$ and cubic singularity at $\varphi \to 1$. Moreover, applicability of equation (6) is unclear in the case of ensemble of different pores as well as in the case of large pores, which size is comparable with inter-pore distance or with grain face size. Therefore, for simplification it was assumed in this paper that $F_{SB} = 1$.

As for the grain boundary diffusivity, considerable uncertainty still exists in the literature. It was shown in reference [8] that the best fit to the re-sintering data of references [9] and [12] is provided by the Arrhenius correlation for the diffusivity with parameters of Reynolds and Burton [13]. For instance, simulations of the re-sintering conditions (24 h at 1700 °C) in the Harada and Doi test resulted in the density change of 1.15% for the normal grain and 0.175% for the large grain samples (including reduction by 0.173% for population P1 and 0.002% for P2), which should be compared with the experimental values of 1.08% and 0.19% [9].

Dollins and Nichols [10], following Stehle and Assmann [3], concluded that the thermal vacancy emission alone is not sufficient to explain the healing of pores under irradiation, especially at low temperatures. This is illustrated in Figure 2 where the results are presented of simulation of porosity evolution with equation (4) under steady irradiation conditions in the Harada and Doi test (the line denoted as 'thermal'). In these calculations, the mean pellet temperature was supposed to be equal to 1100 K, in accordance with reference [9].

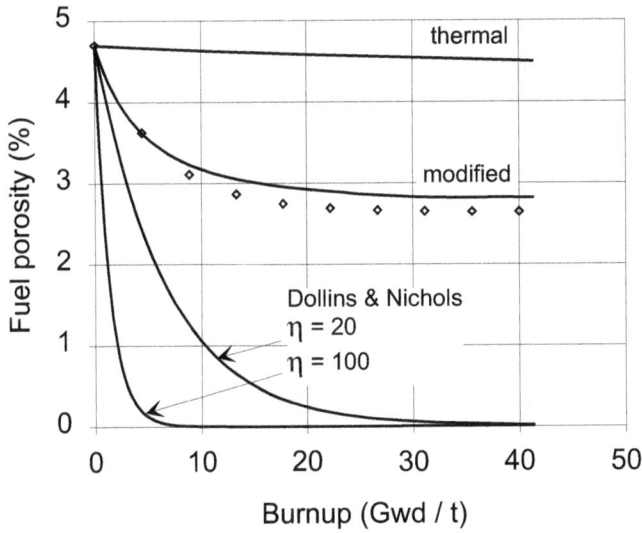

Fig. 2. Simulation of fuel porosity under irradiation in the Harada and Doi test [9] with the thermal relaxation term, equation (4), and different variants of the irradiation term, equation (7); markers correspond to the experimental correlation [9], 'modified' corresponds to equation (10).

To overcome this difficulty, the irradiation-induced vacancy knock-out mechanism was introduced in references [3,10]:

$$\left(\frac{d}{dt}V\right)_{rad} = -8\pi f_S \eta \lambda \Omega G R^2, \qquad (7)$$

where η is the number of vacancies that escape the pore per each hit, $\lambda \sim 1\,\mu$m is the "viable" track length of the fission fragment, and G is the fission rate. As for the key parameter, η, the authors referenced the value of 600 deduced from the fuel sputtering experiments [14]; however, they considered this value as the upper limit and set $\eta = 100$. Note for comparison that in the subsequent sputtering experiments [15] the value of $\eta \sim 20$ was observed at typical stopping power of 20 keV/nm.

The above equation (7) has a trivial solution:

$$\rho(t) = \rho(0) - v_{rad}t, \qquad (8)$$

where $v_{rad} = 2(f_S \sin\theta/f_V \xi)\eta\lambda\Omega G$. With $\eta = 100$ and typical fission rate of $10^{19}\,\mathrm{m}^{-3}\,\mathrm{s}^{-1}$, this parameter equals to $\approx 10^{-13}\,$m/s, so that the pores with $\rho < 10\,\mu$m would disappear during standard LWR campaign. With the initial exponential pore size distribution, equation (3), the total fuel porosity decays exponentially:

$$p(t) = p(0)\left(1 + x + \frac{x^2}{2} + \frac{x^3}{6}\right)e^{-x}, \qquad (9)$$

where $x = v_{rad}t/\bar{\rho}$. In particular, this equation predicts decrease of the initial porosity in the typical LWR fuel by an order of magnitude at burnup of 1 GWd/t, which is considerably faster than the experimental observations. Even if to decrease the parameter η down to 20, the kinetics of fuel densification remains strongly overestimated (the curves 'Dollins & Nichols' in Fig. 2).

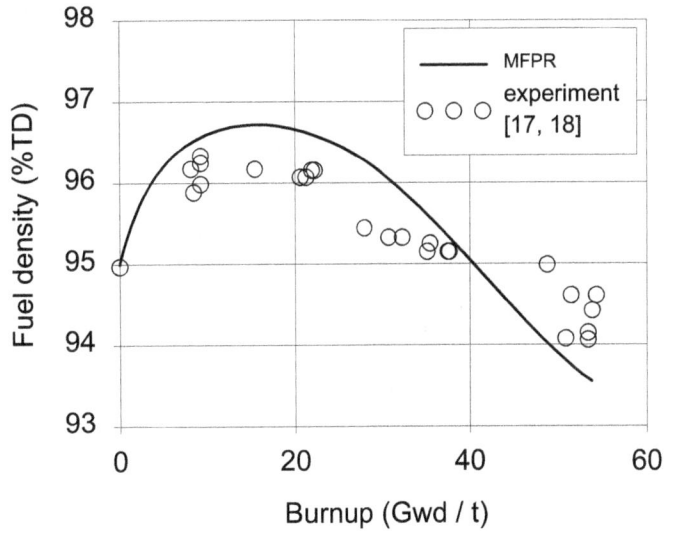

Fig. 3. Kinetics of the fuel density calculated by MFPR for normal grain samples.

In particular, equation (7) does not predict the saturation of the densification process, which can be explained considering that pores with the size greater than some threshold value do not shrink, the threshold being associated with the grain size [16]. Therefore, to take into account this threshold effect, the cut-off of the irradiation term was suggested in reference [8], which can be implemented in equation (7) in the smoothed form:

$$\left(\frac{d}{dt}V\right)_{rad} = -8\pi f_S \lambda \eta \Omega G R^2 \max\left(0.1 - \frac{2\rho_{pr}}{L_{edge}}\right), \quad (10)$$

where ρ_{pr} is the pore projection radius and $L_{edge} \approx 0.69 R_{gr}$ is the typical length of the grain edge[1]. With the choice $\eta = 50$, this allows reasonably reproducing not only the experimental correlation [9] for the densification kinetics, Figure 2 (curve 'modified'), but also the kinetics of the total fuel density due to both pores and inter- and intragranular fission gas bubbles, measured in references [17,18] and presented in reference [9] (Fig. 3).

4 Fission gas capture by pores

The initial number of gas atoms in pores per one grain can be evaluated as:

$$N_0 = \frac{p_0}{1 - p_0}\frac{P_{sint}V_{gr}}{kT_{sint}}, \qquad (11)$$

where p_0 is the initial porosity, T_{sint} and P_{sint} are the temperature and pressure during fuel sintering. The total number of gas atoms, N_{rel}, released from one grain during reactor campaign is equal to $\kappa b f_g V_{gr}$, where $\kappa \approx 0.3$ is the

---

[1] The relation between R_{gr} and L_{edge} is deduced equating the volume of $9\sqrt{2}L_{edge}^3$ of the truncated octahedron, representing the grain, to the volume of the equivalent sphere.

fission gas yield per one fission, f_g is the fractional release of the gas atoms to the grain boundaries and b is the burnup (number of fissions per unit volume). Therefore, one evaluates that:

$$\frac{N_0}{N_{rel}} = \frac{p_0}{1-p_0}\frac{P_{sint}}{\kappa b f_g kT_{sint}}. \tag{12}$$

For the typical values $T_{sint}=2000\,\mathrm{K}$, $P_{sint}=10^5\,\mathrm{Pa}$, $b=10^{27}\,\mathrm{m}^{-3}$ and $f_g=0.1$, one evaluates this ratio as 0.6% (whereas the ratio of N_0 to the total generated gas is an order of magnitude less than this estimate).

The pores can capture the fission gas escaping from fuel grains. The capture rate is estimated multiplying the pore area, equation (1), by the gas flux density, Φ:

$$\left(\frac{dN_p}{dt}\right)_{cap} = 4\pi f_S R^2 \Phi, \tag{13}$$

where the gas flux density can be found as the time derivative of the number of gas atoms released from the grain per unit area of grain surface:

$$F(t) = \frac{1}{3}\kappa R_{gr} f_g(t) b(t). \tag{14}$$

In the case of constant fission rate, $b(t)=Gt$ and thus:

$$\Phi(t) = \frac{d}{dt}F(t) = \frac{1}{3}\kappa R_{gr} G \tilde{f}_g(t), \tag{15}$$

where $\tilde{f}_g(t)\equiv f_g(t)+tf'_g(t)$. At the beginning of irradiation $f_g(t)\sim\sqrt{t}$, so $\tilde{f}_g(t)\approx 3f_g(t)/2$, whereas in the case of high burnup $\tilde{f}_g(t)\approx f_g(t)$.

On the other hand, the gas atoms can be knocked out from pores by passing fission fragments (irradiation-induced resolution). Following Nelson's model [19], the resolution rate for intergranular pores is estimated in MFPR as [20]:

$$\left(\frac{dN_p}{dt}\right)_{res} = -b_0 G\frac{f_S}{f_V}N_p\begin{cases}1, & R\le\lambda/2;\\ \frac{3\lambda}{4R}-\frac{\lambda^3}{16R^3}, & R>\lambda/2;\end{cases} \tag{16}$$

where b_0 is the resolution constant, λ is the average distance the ejected atom travels in pore, δ is the width of the resolution layer [19]. As explained in reference [21], the original Nelson model is used for intergranular porosity without modifications, suggested in reference [22] for intragranular bubbles (in order to avoid duplication of the backward flux of atoms, struck from pores, to the grain boundary).

5 Qualitative analysis

At the beginning of irradiation, the pores are generally under-compressed so that they tend to shrink due to both thermal and irradiation mechanisms, equations (4) and (10). As a result the internal gas pressure in pores increases and eventually the pressure difference δP, equation (5),

approaches zero. Neglecting the van der Waals correction (required for small bubbles with $R<5\,\mu\mathrm{m}$), one derives the relationship between the number of gas atoms in pore N_{eq} and its curvature radius at equilibrium R_{eq}:

$$N_{eq} = \frac{4\pi f_V R_{eq}^3}{3kT}\left(P_h + \frac{2\gamma}{R_{eq}}\right). \tag{17}$$

In the limiting cases one estimates the resolution term, equation (16), as:

$$\left(\frac{dN_p}{dt}\right)_{res} \approx -\frac{\pi f_S b_0\lambda G}{kT}\begin{cases}2\gamma R_{eq}, & R_{eq}<<\frac{2\gamma}{P_h};\\ P_h R_{eq}^2, & R_{eq}>>\frac{2\gamma}{P_h}.\end{cases} \tag{18}$$

Therefore, one gets estimates for the *rhs* of the complete equation for N_p in the equilibrium:

$$\frac{dN_p}{dt} = \left(\frac{dN_p}{dt}\right)_{cap} + \left(\frac{dN_p}{dt}\right)_{res}$$
$$\approx \left(\frac{dN_p}{dt}\right)_{cap}\begin{cases}1-\frac{R_0}{\tilde{f}_g R_{eq}}, & R_{eq}<<\frac{2\gamma}{P_h},\\ 1-\frac{f_0}{\tilde{f}_g}, & R_{eq}>>\frac{2\gamma}{P_h},\end{cases} \tag{19}$$

where R_0, and f_0 are the constants depending on the model parameters and external conditions:

$$R_0\equiv\frac{3b_0\gamma\lambda}{2\kappa R_{gr}kT}, f_0\equiv\frac{3b_0\lambda P_h}{4\kappa R_{gr}kT}. \tag{20}$$

For the typical parameter values $R_{gr}=5\,\mu\mathrm{m}$ and $T=1100\,\mathrm{K}$, one evaluates that $R_0\approx 7.6\,\mu\mathrm{m}$, $f_0\approx 3.2$. It follows from these estimates for small pores (which quickly equilibrate so that $\tilde{f}_g<<1$) that $dN_p/dt<0$, therefore after equilibration small pores definitely lose the gas and hence continue shrinking. The same conclusion can be drawn for large equilibrated pores; however, with less reliability in view of uncertainties of the resolution model and the relevant parameters. Therefore, one expects that pores of all sizes lose their gas after equilibration. The opposite trend cannot be excluded under some extreme conditions (high burnups at high temperature of the large grain fuels). Note that these conclusions were drawn for the equilibrated pores whereas the gas content in non-equilibrium pores can be either reducing or growing.

6 Quantitative analysis

The above qualitative considerations are illustrated in Figure 4 by MFPR numerical simulations of evolution of pores with initial radii of 0.1, 1 and 10 $\mu\mathrm{m}$ (curves labelled in the graph as 1, 2 and 3, respectively) under irradiation conditions of the Harada and Doi test [9]: normal grain fuel ($d_{gr}=8\,\mu\mathrm{m}$) with the initial porosity of 4.7% under temperature of 1100 K, pressure of 3 MPa and fission rate of $10^{19}\,\mathrm{m}^{-3}\,\mathrm{s}^{-1}$; the initial pore distribution was approximated by equation (3) with $\bar{\rho}=0.35\,\mu\mathrm{m}$.

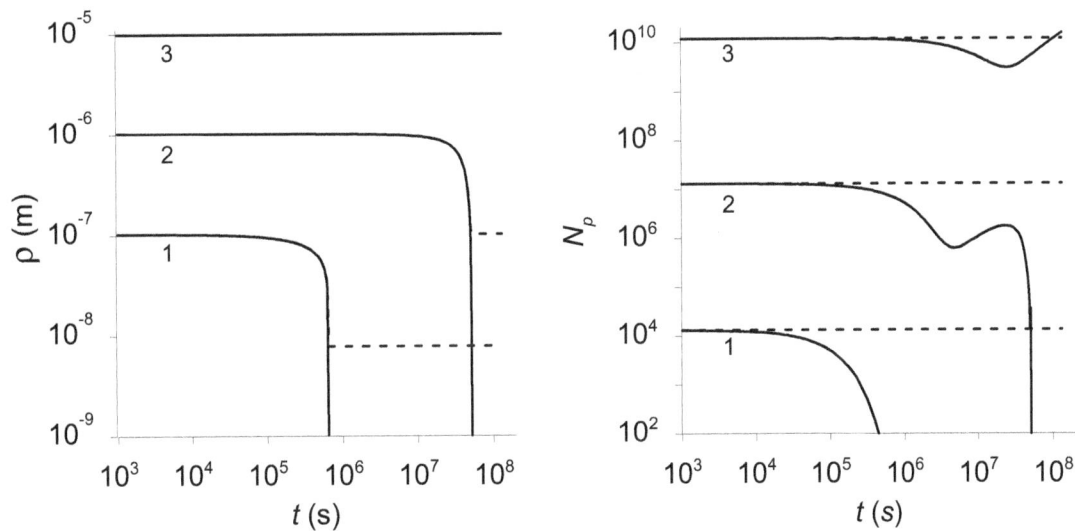

Fig. 4. Time dependence of pore radii and gas atom content in pores under irradiation conditions of the test [9]; dashed lines correspond to calculations with $N_p = $ const.

It is seen that the relatively small pores quickly equilibrate, monotonically shrink and eventually disappear (equilibration times are $\sim 6 \times 10^5$, 5×10^6 and 5×10^7 s for pores with the initial radii of 0.1, 0.3 and 1 μm). The large pores ($\geq 3\,\mu$m) are practically unchanged in their sizes. The final fuel porosity averaged over the pore ensemble turned out to be 2.8%.

As for the gas content, it monotonically decreases in small pores ($\rho \leq 0.5\,\mu$m in these calculations) because the resolution mechanism dominates in the initial stage of irradiation when the small pores effectively shrink, e.g. see curve 1 in the right panel. For the larger pores, $N_p(t)$ can be non-monotonic function, however it finally decreases (at least after equilibration, see previous Sect. 5), e.g. curve 2. The gas content in pores with $\rho > 1.2\,\mu$m increases to the end of irradiation in comparison with the initial value; however these pores remain under-pressurized (curve 3). The maximum relative gas increase ($\sim$37%) is attained in pores with the initial radius of 2 μm. As for the overall gas content in the pore ensemble, first it rapidly decreased by a factor of $\sim$15 at burnup of $\approx$3 GWd/t and then slowly increased; the gas content was near 80% of the initial value to the end of the campaign (54 GWd/t); this change can be estimated using equation (12) as $\sim$0.1% of the gas released from the fuel grains.

To clarify the role of gas capture/resolution effects, the calculations were repeated with fixed N_p (dashed curves in Fig. 4). It is seen on the left panel that the solid and dashed curves are very close to each other except of the stage of quick shrinking. However, at this stage the volumes of the pores are much less in comparison with the initial values and hence do not essentially contribute to the fuel porosity. As for the large pores, their sizes are practically constant, so the two approaches are close to each other too. In addition, the contribution of the largest pores to the total porosity is exponentially small (it can be evaluated by Eq. (9) with $x = \rho/\bar{\rho}$). These qualitative considerations were confirmed by our calculations which showed that the gas capture/resolution effect influenced the final porosity by $\approx$0.01% only.

These calculations demonstrated that the pores cannot be considered as effective traps of the fission gas released from the nuclear fuel. In addition, it was justified that the neglect of gas content variation in pores is a good approximation in numerical calculations, as qualitatively discussed in reference [8]. To check these conclusions, the additional calculations have been performed with the same initial pore distribution but varying one of the external parameters: grain size, fission rate or irradiation temperature. In all these cases, the pore kinetics were found to be qualitatively similar to that presented in Figure 4.

Simulations with the increased fission rate (5×10^{19} m^{-3} s^{-1}) have shown that an increase of the overall gas content (following a fast initial decrease by a factor of 8) resulted in full compensation of the gas content in pores at burnup of 75 GWd/t. These variations were within 0.1% of the gas amount released from the grains, which is comparable with the above considered cases. The gas capture/resolution mechanisms were found to contribute to the final fuel porosity (which was equal to 2.33%) very similarly to the above examples. The greatest differences were found in calculations with the increased temperature. At $T = 1500$ K, the initial decrease of the gas content was followed (at burnup of $\approx$1 GWd/t) by a slight increase, which in turn followed again (at $\approx$4 GWd/t) by decrease up to the end of the campaign; the final content was found to be of 16% of the initial value. In addition, the simulations of the fuel volume evolution (including both pores and fission gas bubbles) under irradiation conditions of the Harada and Doi test [9] were performed (Fig. 5). The following external conditions were chosen: the mean irradiation temperature 1100 K, fission rate 10^{19} m^{-3} s^{-1}, and external pressure 3 MPa. In the case of the large grain fuel, the realistic bi-modal initial pore size distribution was simulated as superposition of P1 and P2 populations, see Section 2.

It was found that the pore populations P1 and P2 lost 84% and 1.4% of their initial volume respectively so that the total pore densification turned out to be of 1.07%. The final fuel density change (at the burnup of 21 GWd/t) was

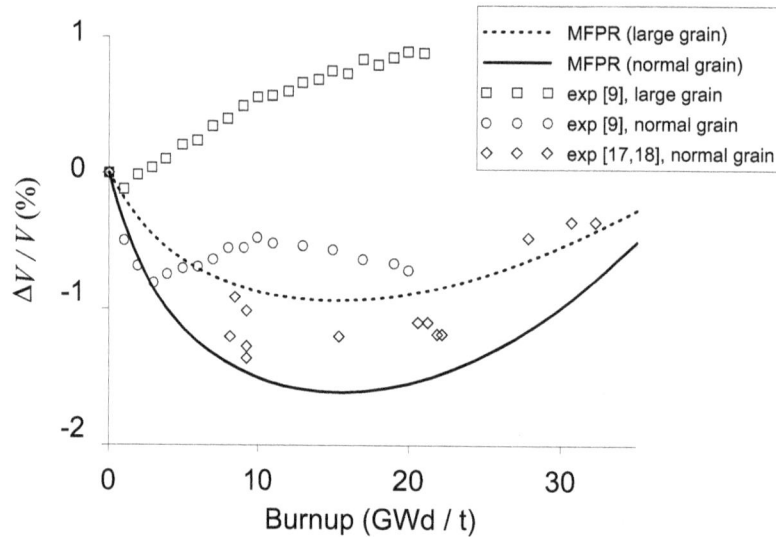

Fig. 5. Kinetics of fuel volume change under irradiation for the large and normal grain samples.

estimated as –1.5% and –0.9% for the normal and large grain fuels respectively. Comparison with the experiment [9] demonstrates good agreement for the very initial stages of irradiation (up to ~1 and 3 GWd/t for the large and normal grain fuels respectively). However, in later stages there is a qualitative difference in the fuel volume variation kinetics. Moreover, the similar disagreement takes place between the experimental data in reference [9] and references [17,18], also cf. Figure 3. The authors of reference [9] supposed that the discrepancy could be caused by "fuel fragment relocation at the early stage of the irradiation". This effect was not simulated in our calculations, but is foreseen in the forthcoming version of the code. Besides, for large grain samples a more detailed experimental information is required on pore size distribution in grains of different sizes (rather than available averaged data presented in Fig. 1), since the real grain size distribution is rather flat (see Ref. [9]) and this may strongly influence the threshold sizes for shrinking pores in equation (10).

7 Conclusions

The MFPR model for intergranular pore evolution was updated and verified against experimental data [9,17,18], and then applied to analysis of the sintering porosity behaviour under various conditions of in-pile irradiation.

The performed analysis demonstrated that generally the resolution of gas atoms from pores prevails over capture of the fission gas released from grains in early stages of irradiation, which somewhat accelerates fuel densification. In later stages, the gas content of the survived large pores can increase, but not significantly. As a result, pores lose their gas content during typical reactor campaign; however the effect being vanishingly small. This implies that pores can be hardly considered as effective traps for the fission gas.

The model predicts a comparatively rapid fuel densification due to shrinkage of small pores with projection radius less or comparable with the grain face size, whereas the coarse

pores remain unchanged, in agreement with numerous observations, e.g. references [6,16]. In particular, this implies that the second population of relatively large pores in fuel with large grains, fabricated with a pore former, provides a rather small densification (mainly due to the first population of small pores) and thus can be hardly used for accommodation of the fuel swelling. At high burnups this can result in significant pellet-cladding mechanical interaction caused from the swelling due to retained gases and solid FPs. However, this preliminary conclusion should be thoroughly verified against additional experimental data.

Nomenclature

b	burnup (number of fissions per unit volume)
B	van der Waals volume of the gas atom
$C(\rho)$	pore size distribution function normalized to the pore concentration
D_{gb}	grain boundary self-diffusivity
f_g	fission gas fractional release
F_{SP}	Speight-Beere factor
f_s	pore area shape factor
f_V	pore volume shape factor
G	fission rate (number of fissions in unit volume per unit time)
k	the Boltzmann constant
N_p	number of gas atoms in pore
$p(t)$	fuel porosity
P_h	external pressure
P_{sint}	sintering pressure
R	pore curvature radius
S	pore area
t	time
T_{sint}	sintering temperature
V	pore volume
w	grain boundary thickness
γ	surface tension
η	mean number of vacancies knocked out from pore per one hit

θ	dihedral angle
κ	fission gas yield
λ	"viable" track length
Ω	vacancy volume
ξ	$= \rho_{pr}/\rho$
ρ	pore median radius
$\overline{\rho}$	mean projection radius
ρ_{pr}	pore projection radius
φ	fractional coverage of the grain boundary by pores
Φ	fission gas out-of-grain flux density

References

1. M.D. Freshley, D.W. Brite, J.L. Daniel, P.E. Hart, Irradiation-induced densification of UO2 pellet fuel, J. Nucl. Mater. **62**, 138 (1976)
2. G. Maier, H. Assmann, W. Dorr, Resinter testing in relation to in-pile densification, J. Nucl. Mater. **153**, 213 (1988)
3. H. Stehle, H. Assmann, The dependence of in-reactor UO2 densification on temperature and microstructure, J. Nucl. Mater. **52**, 303 (1974)
4. M.S. Veshchunov, V.D. Ozrin, V.E. Shestak, V.I. Tarasov, R. Dubourg, G. Nicaise, Development of mechanistic code MFPR for modelling fission product release from irradiated UO2 fuel, Nucl. Eng. Des. **236**, 179 (2006)
5. M.S. Veshchunov, R. Dubourg, V.D. Ozrin, V.E. Shestak, V.I. Tarasov, Mechanistic modeling of urania fuel evolution and fission product migration during irradiation and heating: the MFPR code, J. Nucl. Mater. **362**, 327 (2007)
6. B. Burton, G.L. Reynolds, The sintering of grain boundary cavities in uranium dioxide, J. Nucl. Mater. **45**, 10 (1972)
7. R.J. White, M.O. Tucker, A new fission-gas release model, J. Nucl. Mater. **118**, 1 (1983)
8. V.I. Tarasov, M.S. Veshchunov, Models for fuel porosity evolution in UO2 under various regimes of reactor operation, Nucl. Eng. Des. **272**, 65 (2014)
9. Y. Harada, S. Doi, Irradiation behavior of large grain UO2 fuel rod by active powder, J. Nucl. Sci. Tech. **35**, 411 (1998)
10. C.C. Dollins, F.A. Nichols, In-pile intragranular densification of oxide fuels, J. Nucl. Mater. **78**, 326 (1978)
11. M.V. Speight, W. Beere, Vacancy potential and void growth on grain boundaries, Metal Sci. **9**, 190 (1975)
12. G. Maier, H. Assmann, W. Dorr, Resinter testing in relation to in-pile densification, J. Nucl. Mater. **153**, 213 (1988)
13. G.L. Reynolds, B. Burton, Grain-boundary diffusion in uranium dioxide: The correlation between sintering and creep and a reinterpretation of creep mechanism, J. Nucl. Mater. **82**, 22 (1979)
14. S. Yamagishi, T.J. Tanifuji, Post-irradiation studies on knock-out and pseudo-recoil releases of fission products from fissioning UO2, J. Nucl. Mater. **59**, 243 (1976)
15. S. Schlutig, Contribution à l'étude de la pulvérisation et de l'endommagement du dioxyde d'uranium par les ions lourds rapides, PhD Thesis, University of Caen, 2001
16. M.O. Marlowe, In-reactor densification behavior of UO2, NEDO-12440, 1973
17. Y. Irisa, Y. Takada, in *ANS Topical Meeting, Williamsburg* (1988)
18. S. Doi, S. Abeta, Y. Irisa, S. Inoue, in *ANS Topical Meeting, Avignon, France* (1991)
19. R.S. Nelson, The stability of gas bubbles in an irradiation environment, J. Nucl. Mater. **31**, 153 (1969)
20. V.I. Tarasov, M.S. Veshchunov, An advanced model for grain face diffusion transport in irradiated UO2 fuel. Part 2. Model Implementation and validation, J. Nucl. Mater. **392**, 84 (2009)
21. M.S. Veshchunov, V.I. Tarasov, An advanced model for grain face diffusion transport in Irradiated UO2 fuel. Part 1. Model formulation, J. Nucl. Mater. **392**, 78 (2009)
22. M.S. Veshchunov, V.I. Tarasov, Modelling of irradiated UO2 fuel behaviour under transient conditions, J. Nucl. Mater. **437**, 250 (2013)

18

Lessons learned from a review of international approaches to spent fuel management

David Hambley[1,*], Alice Laferrere[1], W. Steven Walters[1], Zara Hodgson[1], Steven Wickham[2], and Phillip Richardson[2]

[1] NNL Central Laboratory, B170, Sellafield, Seascale, Cumbria, CA20 1PG, UK
[2] Galston Sciences, Oakham, UK

Abstract. Worldwide, a variety of approaches to the management of spent fuel have been adopted. A review of approaches adopted internationally was undertaken to inform decision making on spent fuel management in UK. The review surveyed spent fuel storage and disposal practices, standards, trends and recent developments in 16 countries and carried out more detailed studies into the evolution of spent fuel storage and disposal strategies in four countries. The review highlighted that: (1) spent fuel management should be aligned to the national policy for final dispositioning of the fuel; (2) national spent fuel storage arrangements should deliver efficiency across all spent fuel management activities; (3) commercial and financial arrangements should ensure that spent fuel management decisions do not unnecessarily limit future fuel handling, packaging and disposal activities; (4) extended storage of spent fuel prior to packaging provides increased flexibility in the design of future packaging and disposal concepts. Storage of spent fuel over 100 years or more using existing technologies is technically feasible and operationally credible. Local factors such as existing infrastructure, approach to fuel cycle management, existing experience/capability and short-term cash flow considerations all influence technology selection. Both wet and dry storage systems continue to receive regulatory approval and are acceptable.

1 Introduction

There are a number of developments that have bearing on the management of spent fuel from power reactors in the UK. Domestically, these include the cessation of reprocessing, nuclear 'new build', the potential for reuse of UK plutonium and renewed progress in development of concepts for a geological repository. Worldwide, slow progress in the deployment of geological disposal facilities and reduced use of reprocessing have led to the need to extend storage periods for fuel and to store greater quantities of fuel.

Internationally, three different strategies have been adopted for fuel cycles:

– "Closed fuel cycle", where the spent fuel is reprocessed. Reprocessing has been deployed at an industrial scale in a number of countries with large nuclear power programmes (e.g. France, UK and Russia) and has been used by the majority of countries through commercial reprocessing services;
– "Open fuel cycle", where the spent fuel is not reprocessed and direct disposal of fuel has been chosen as the preferred option. This option has become more common over time.

Geological repositories are not yet available, although a few countries are making significant progress towards opening a repository, most notably Sweden and Finland where operations are scheduled to start in 2027 and 2022;
– "Wait-and-see", where no decision has yet been made as to how fuel will be dispositioned. This option is most commonly associated with indecision or a failure to progress either a geological disposal site or a reprocessing facility.

The strategy adopted by a country impacts on the spent fuel management approach and the associated technological requirements. On the other hand, political considerations, public opinion and available infrastructure/experience impact strongly on strategic decisions [1]. Given the range of strategies and technologies adopted internationally, it is important to understand the reasons why specific options have been selected in order to inform future decision making.

2 The role of spent fuel storage

Spent fuel storage is a necessary part of any nuclear fuel cycle. Ponds are used for storage and cooling of spent fuel after discharge from the reactor core to dissipate the very high decay heat associated with short-cooled fuel. Fuel must be stored in the reactor cooling pond until it can be transported safely and meets the conditions for acceptance for the next stage of the fuel cycle, which can be:

– reprocessing;
– geological disposal or;
– interim storage (wet or dry) at facilities on the reactor site (AR) or in centralised facilities away from reactor (AFR), pending onward shipment for reprocessing or disposal.

The two internationally accepted disposition options for spent fuel are reprocessing or direct disposal in a geological disposal facility (GDF). Irrespective of the strategy adopted, a GDF is necessary to dispose of heat generating waste, whether this is spent fuel from an open fuel cycle or the high level waste arising from a closed cycle.

Transportation of spent fuel is intimately linked with spent fuel storage, as fuel needs to be transported from the reactor pond to storage and from storage to the next stage of the fuel cycle. Thus, the impact of spent fuel storage on the transportability of fuel after storage is an important aspect when considering spent fuel management holistically.

In many cases, fuel needs to be moved between one configuration and another, e.g. for transport or for disposal. This has implications for fuel retrievability, operational and capital costs and waste generation. In some cases, fuel can be packaged for storage in a form that is also compatible with the requirements for transport and/or disposal. This has the potential to reduce handling and repacking operations, however to be successful it is vital that such systems are compliant with all subsequent operational requirements.

3 Spent fuel storage systems

A wide range of storage systems have been developed for power and research reactors, however the majority fall into one of four common types [2,3].

Pool (Pond) – a pool is a facility which stores spent fuel in water. The spent fuel is usually supported in racks, baskets and/or containers which also contain water. Examples: AR ponds: Fukushima (Japan), Loviisa (Finland), Gösgen (Switzerland). AFR ponds: Sellafield (UK) (Fig. 1), La Hague (France), Clab (Sweden), GE Morris (USA).

Vault – a vault is a reinforced concrete building containing arrays of storage cavities suitable for containment of one or more spent fuel units. Examples include Wylfa facility (UK), MVDS facilities at Paks (Hungary) (Fig. 2), Fort St. Vrain (USA) and CANSTOR/MAC-STOR at the Gentily-2 NPP (Canada).

Metal cask – a metal cask is a container with a bolted lid, similar to a large transport flask, designed either for storage only or for storage and transportation (dual-purpose casks). Multi-purpose casks for storage, transport and disposal have been proposed but no casks have yet been licensed for disposal. Examples include GNS CASTOR (Fig. 3); Trans-nucleaire TN-40; Westinghouse MC10.

Concrete cask – a concrete cask has a thick, welded steel canister, which is cooled by natural convection. The canister is inserted into a concrete overpack which provides

Fig. 1. Centralised storage pond (Courtesy of Sellafield Ltd).

Fig. 2. Dry vault store [4].

Fig. 3. Metal casks [5].

shielding. The canister can be stored in either vertical (Fig. 4) or horizontal orientation (Fig. 5). Examples include Holtec HiStorm, Sierra Nuclear's VSC (USA); Ontario Hydro's Pickering concrete dry storage container (Canada) and NuHoMS (USA).

Silos – a silo is similar to a vertical concrete cask, except that there is no cooling flow inside the monolithic structure. This form of storage is only, therefore, suitable for low heat-output fuel. Examples include New Brunswick Power's Point Lepreau (Canada) (Fig. 6), Embalse (Argentina).

Fig. 4. Vertical concrete casks (Courtesy of NAC International).

Fig. 5. Loading a canister into a horizontal concrete storage module (Courtesy of Areva TN Inc).

Fig. 6. Silos [6].

All fuel entering dry storage needs to be dried to remove excess water for the fuel, so as to avoid pressurisation and corrosion during storage. Dry spent fuel storage was developed for short-term storage until GDFs became available and is a less mature technology than pond storage. Dry storage system designs have developed substantially over the past 20 years. Designs rely on passive cooling, which reduces operation and maintenance requirements and costs; however, periodic surveillance is still required. Additional equipment and infrastructure are required to load fuel and different design variants have been developed for different fuel types [7].

Fuel is only recoverable from metal/concrete casks systems when reactor ponds remain operational, as reactors and most dry storage facilities do not have shielded facilities in which the radiation from the stored fuel can be contained. Fuel in canister-based systems is not intended to be recovered, however the trend towards larger payloads has made disposal more problematic. Comprehensive ageing management plans for long-term dry storage are now under development, which may lead to further system design evolution or enhanced monitoring and surveillance requirements.

Pond designs have also evolved, with modern designs having passive cooling systems and much greater resistance to external events [8].

With increasing length of storage, the potential need for fuel inspection to provide assurance of the condition of fuel after storage and the requirements for demonstrating retrievability of fuel are under renewed consideration. The certainty that AR fuel storage will continue long after the reactors and associated fuel handling infrastructure have been dismantled also needs to be factored into decision making.

The recycling of Pu into MOX for thermal reactors is not widespread and hence spent fuel storage systems have been developed with UO_2-based fuels (UOX) in mind. Although there are differences between spent MOX and UOX fuel, the challenges arising from the storage of MOX are identical in nature, if not in intensity, to those from UOX and irradiated MOX in LWR systems has been safely managed out of reactor in both dry cask and ponds [5,9].

The primary challenges with MOX fuels are:

– higher decay heat per GWe produced compared to UOX, which needs to be removed by the cooling systems. This requires a longer cooling period and/or lower payload to meet heat load/dose rate requirements for storage, transport, reprocessing and disposal;
– higher neutron activity due to minor actinide content. This requires additional neutron shielding and introduces additional operational restrictions compared with spent UOX fuel storage;
– higher fissile content than UOX, although actual content depends on irradiation history. This affects the density of storage and extent of criticality control measures required;
– higher He generation leading to increased internal pressurisation of fuel cladding and increased lattice swelling effects during long-term storage/disposal;
– mixing of spent MOX fuel with spent UOX fuel can be effective in managing the effects of increased heat generation and radiation associated with MOX fuel, so long as the proportion of MOX fuel remains low.

4 International review

Sixteen country nuclear profiles and their approaches to spent fuel management have been reviewed to build a picture of the range of spent fuel management strategies and practices that are currently in use. The countries studied were: Belgium, Canada, China, Finland, France, Germany, Hungary, Japan, Netherlands, Russia, South Korea, Spain, Sweden, Switzerland, United Kingdom and United States of America [1].

Data from the review has been summarised in Table 1. 'Fuel type' and 'Fuel requirements' correspond to the fuel loaded in the nuclear power plants. 'Storage capacity' and 'Storage quantities' correspond to the capacity of all storage facilities, AR and AFR, and the quantities of spent fuel in those facilities. 'Storage arisings' is the amount of spent fuel arising in long-term storage facilities per year. Unless specified, fuel requirement, storage capacity, spent fuel cumulative in storage and spent fuel arisings are based on 2012 data [10]. Interim storage 'type' and 'location' have been codified as predominantly 'wet' or 'dry' and 'AR' or 'AFR'. For countries with only one reactor, 'centralised' is used, instead of AR.

It is clear that national spent fuel management strategies are influenced by expectations at the time investments were made (e.g. availability and location of a reprocessing facility or a repository), the national reactor fleet, including reactor types, geographical locations and the transport infrastructure, and the potential for economies of scale. Therefore, there is no exact precedent to follow.

The strategy adopted by different countries influences storage type and requirement. For example, countries that have chosen reprocessing tend to mostly have AFR wet storage. Countries where disposal of spent fuel is the chosen strategy, or where back-end management is still undecided, tend to have AR dry and/or wet storage. Sweden is the only country which has decided to dispose of spent fuel and has a large centralized wet facility AFR: the CLAB facility.

The quantities of spent fuel in storage may be compared with storage capacities. Countries with spent fuel quantities close to storage capacities plan to extend their capacity by adding ponds (Finland), vaults (Hungary) or dry casks (Spain) by 2015. The modular dry vault storage in the Netherlands will reach full capacity by 2021 [11]. Only Belgium does not plan on extending storage capacity in the coming years, but reprocessing is still an option being considered.

For most countries with a reprocessing strategy, fuel requirements for continued operation of the nuclear plants are higher than spent fuel arisings in storage (i.e. France, Japan). Only the UK has a higher spent fuel annual arising than fuel requirement, principally because part of its reactor fleet currently being decommissioned. For countries with a disposal strategy, fuel requirements are close to spent fuel arisings apart from Spain, which has higher arisings due to the decommissioning of Garona.

State and political endorsement of deploying reprocessing (domestically or abroad via commercial fuel service arrangements) is a key enabler as experience to date indicates that the will and investment from a commercial enterprise alone is insufficient to sustain such an activity. Significant effort and know-how are required to realise a plant-scale reprocessing operation from a position of scientific knowledge, which could be a barrier to deployment; equally, loss of skills and facilities and the difficulties in recovering from a period without plant-scale reprocessing

Table 1. Comparison of key country data on spent fuel storage and disposal [1].

| Country | Strategy | Fuel in 2012 | | Storage in 2012 | | | | | Disposal | |
		Type	Requirement (tHM/year)	Capacity (tHM)	Quantities (tHM)	Arisings (tHM/year)	Type	Location	Programme	Timelines
Canada	Centralised storage and disposal	HWR	1650	66903	45986	1430	Dry (cask, vault, silo)	AR	Identifying communities	Start operation in 2035
Finland	Disposal	PWR BWR	71	2330	1882	56	Wet	AR	Construction license awaited	Start operation in 2022
Germany	Disposal	LWR, some MOX	202	27648	3970	180	Dry (cask)	AR	Site selection to restart	Site selected by 2031
Spain	Disposal	PWR BWR	140	5260	4434	208	Dry and wet	AR	Siting process suspended	N/A
Sweden	Disposal	PWR BWR	220	8000	5577	170	Wet	AFR	Construction license awaited	Start construction in 2019

Table 1. (continued).

Country	Strategy	Fuel in 2012		Capacity (tHM)	Storage in 2012				Disposal	
		Type	Requirement (tHM/year)		Quantities (tHM)	Arisings (tHM/year)	Type	Location	Programme	Timelines
Switzerland	Disposal, moratorium on reprocessing	PWR BWR	59	3946	1223	59	Dry and wet	Mostly AR	Selection process in progress	Start operation by 2040
USA	Disposal	LWR	2387		72101	2248	Dry and wet	AR	Revised siting process yet to start	GDF to be available by 2048
Belgium	No decision	PWR	129	3830	3334	278	Dry and wet	AR	No siting process to date	Start programme in 2035
Hungary	No decision	PWR	47	1412	1075	44	Dry (vault)	Centralised	Site survey to begin in 2014	Start operation in 2038
South Korea	No decision	PWR HWR	870	16927	12629	619	Wet (dry for HWR)	AR	No siting process	N/A
China	Reprocessing	PWR HWR			3000 or 3800	600	Wet (dry for HWR)	Mostly AR	Site investigation in progress	Site selected by 2040
France	Reprocessing	PWR, some MOX	1170	18000	14504	300	Wet	AFR	Site selection to be confirmed	Start operation in 2025
Japan	Reprocessing	PWR BWR	342	20883	14460	80	Wet (dry for two reactors)	AR and AFR	Selection process to be amended	N/A
Netherlands	Reprocessing	PWR	8	600	521	8	Wet	Centralised	No siting process	N/A
Russia	Reprocessing	PWR BWR		24329			Dry and wet	AFR	Site selected	Start operation by 2035
UK	Reprocessing	PWR GCR	212	12000	4481	890	Wet	AFR	Selection process stalled	N/A

would be a barrier to a subsequent deployment of reprocessing. Japan's experience would indicate that even with its extensive nuclear experience and technical knowledge, the establishment of an expert buyer capability does not guarantee quality or efficiency of procured products and services.

Most countries have started a disposal programme and are selecting or investigating suitable sites. Finland, France, Russia and Sweden have chosen a site. Only Belgium, the Netherlands and South Korea have not started a site selection process. Progress has been greatest in Sweden and Finland.

From the survey of individual nations, it is apparent that some nations have a long-term vision or strategic plan for the nuclear fuel cycle whereas a number of others do not. In order to provide greater understanding about the development of these differences, further work was undertaken to examine the impact of having long-term plans for spent fuel management in countries that have them and compare this with countries without such visions or plans.

Some of the best examples of long-term visions or plans are found in Sweden, Finland, France and the Netherlands. The arrangements in place in these countries were examined to identify any common threads and to assess what impact these visions or plans have had on practical arrangements for storage and disposal of spent nuclear fuel.

Examples of countries without any current long-term vision or plan include Germany and the USA. These countries have (in the past) had plans, but for various reasons, mainly political, the plans have been disrupted and spent fuel management is now much more reactive, responding to external factors rather than based on a well-defined vision or strategy. The effect of the lack of long-term stability in these countries was examined to identify common threads and their impact on spent fuel management practices.

5 Key messages

This review has shown that the way in which liabilities are distributed between organisations involved in the generation, management and disposal of spent fuel has a significant effect on the effectiveness of spent fuel management. In democratic countries, the greatest stability in back-end fuel management, and greatest efficiency and integration, are associated with countries in which governments have set policy, strategy and regulation, leaving commercial entities with the integrated liability for storage and the development, licensing and implementation of the required disposal facilities.

To achieve overall cost effectiveness, spent fuel management should be aligned so as to meet the technical requirements of the national policy for the final dispositioning of the fuel.

Given the long timeframes associated with GDF site selection and the management life-cycle associated with nuclear fuel, policy-making for effective delivery of the strategy should be directed at developing a robust and resilient overall approach, rather than focusing on short-term efficiency. Where public acceptance is important, clear separation between regulation and delivery of storage and dispositioning supports effective long-term delivery.

At a national level, policy decisions can constrain or incentivise particular forms of spent fuel management. Therefore, it is prudent for national decision makers to consider the factors affecting storage options (e.g. centralised versus decentralised) and the financial, social and environmental effects of different strategies. Commercial and financial arrangements should ideally be constructed to ensure that, at each stage of the spent fuel life-cycle, spent fuel management decisions do not unnecessarily preclude future management options or increase the costs of subsequent activities leading to final dispositioning of the fuel.

Where more than one organisation is responsible for spent fuel storage, disposal and any intermediate processing, there should be commercial agreements between those responsible that incentivise efficient management of spent fuel to its final end point, in preference to maximising the efficiency of individual stages of spent fuel life-cycle. Policy makers in setting the national policy framework and regulation should therefore take organisational responsibilities in account when designing national approaches to spent fuel management so as to best incentivise all actors to provide efficient and effective dispositioning of spent fuel.

On a technical level, storage of spent fuel for over 100 years or more using existing technologies, or foreseeable evolutions of them, is feasible and credible. Over such timescales, all storage systems and supporting infrastructure will need to be refurbished and replaced as they degrade. The time interval between major refurbishment or replacement remains uncertain but it would be reasonable to expect a 50- to 100-year replacement period based on current systems, by analogy with highly active waste storage facilities.

The use of multiple approaches to fuel storage, and continued evolution of the storage facility designs indicate that there is no single best storage technology, and that local factors such as existing infrastructure, size of national spent fuel inventories, approach to fuel cycle management, existing experience/capability, geographical factors and short-term cash flow considerations all influence technology selection.

Both wet and dry storage systems continue to receive regulatory approval and are acceptable in terms of safety and environmental impact and operational practicality.

Dry storage is less mature than wet storage and issues related to storage beyond 20 years, including post-storage transport and impact on disposal systems, are now being addressed. The transition to dry storage results in the fuel experiencing a period of higher temperatures and this may affect fuel performance. The extent of any degradation of the spent fuel is currently a topic of research and assessment. Some changes to system design can be anticipated as a result of this work and may increase capital or operating costs. Dry storage systems generally provide small incremental storage capacity and lower short-term cash flow requirements than ponds or vaults [12]. Operational costs during reactor operational phase are low, but recent analysis by US GAO have shown a large increase in AR operational costs once reactors shutdown [13].

Wet storage has been successfully employed for many decades and is a more mature technology. Nevertheless, designs are evolving to increase the levels of passive safety and resistance to external and malicious events. Wet torage provides easier monitoring of fuel conditions and greater flexibility in post-storage transportation and packaging [8]. Provided pond water quality can be maintained over the required storage period, fuel quality is likely to be assured. However, traditional pond systems require active management and higher levels operational support.

The thermal output of spent fuel is critical to the design and overall performance of a spent fuel GDF. The following key factors in spent fuel management have been identified as being critical to the disposability of spent fuel:

- age, burnup and thermal output of spent fuel constrain the temperature evolution of the disposal system with time, although this will also be influenced by host rock thermal conductivity and engineered barrier system design. Acceptable thermal output often determines how long an interim storage period is required and may place constraints on design of the waste packages. In general, it is necessary to store spent fuel for longer periods for direct disposal than for reprocessing;
- if spent fuel becomes degraded through long-term storage, either wet or dry, this may compromise disposability by making it incompatible with the selected packaging concept or handling infrastructure, and may require additional package finishing prior to disposal;
- the way spent fuel has been packaged for long-term dry storage may control the subsequent packaging or disposal concepts that are viable. No potential site for a spent fuel GDF exists in the UK at the current time and therefore the host geology remains unknown. Generic disposal concepts and designs exist for a number of general geological environments based on overseas design concepts, but none are compatible with modern dry storage cask designs.

Depending on the storage systems used and fuel condition at the end of storage, an export facility may need to be built in order to ensure that fuel is exported in packages suitable for transport or transport and disposal. Such a facility may need to include capabilities for some or all of the following: fuel drying, opening sealed dry-stored packages, repackaging spent fuel in disposal containers

and remediating degraded packages. The scope of any on-site facility will also depend on decision made about the capabilities at a GDF.

In addition to the technical requirements to ensure the long-term integrity of fuel and storage systems, it is important that organisations retain the required level of technical capability and information or the duration of storage, so as to ensure that post-storage activities are managed safely.

This work was funded by the Nuclear Decommissioning Authority.

References

1. A. Laferrere, D. Hambley, W.S. Walters, Z. Hodgson, S. Wickham, P. Richardson, Review of international approaches to the management of spent fuel, NNL 12635, 2014
2. IAEA, Survey of wet and dry spent fuel storage, IAEA-TECDOC-1100, 1999
3. EPRI, Industry spent fuel storage handbook, EPRI report No. 1021048, 2010
4. Z. Husak, J. Bencze, Storage of spent nuclear fuel in MVDS of PAKS NPP, http://www.dysnai.org/Reports/2000-2004/2004/3.pdf, accessed February 2016
5. H. Völzke, Dry spent fuel storage in dual purpose casks- Aging management issues, in *INMM Spent Fuel Management Seminar XXVIII, Arlington, VA, January 14–16, 2013* (2013)
6. M. Petrovic, J. Hashmi, P. Eng, Single storage canister to MACSTOR – 14578 Canadian solution and experience in responsible spent fuel management, ®in *WM2014 Conference, Phoenix, Arizona, USA, March 2-6, 2014* (2014)
7. IAEA, Operation and Maintenance of Spent Fuel Storage and Transportation Casks/Container, IAEA-TECDOC-2532, 2007
8. U. Appenzeller, External spent fuel storage facility at the nuclear power plant in Gösgen, in *Technical Meeting on SNF storage options, 2–4 July 2013* (IAEA, Vienna, Austria, 2013)
9. IAEA, Status and advances in MOX fuel technology, IAEA-TRS-415, 2003
10. OECD/NEA, Nuclear energy, NEA No. 7162, 2013
11. J. Hart, A.I. van Heek, The effect of electricity generating park renewal on fossil and nuclear waste streams: the case for the Netherlands, Kernenergie: International, 2008
12. IAEA, Costing of spent nuclear fuel storage, Nuclear Energy Series No. NF-T-3.5, 2009
13. United States Government Accountability Office, Spent nuclear fuel management outreach needed to help gain public acceptance for federal activities that address liability, GAO-15-141, 2014

19

Deployable nuclear fleet based on available quantities of uranium and reactor types – the case of fast reactors started up with enriched uranium

Anne Baschwitz[*], Gilles Mathonnière, Sophie Gabriel, and Tommy Eleouet

CEA, DEN/DANS/I-tésé, 91191 Gif-sur-Yvette, France

Abstract. International organizations regularly produce global energy demand scenarios. To account for the increasing population and GDP trends, as well as to encompass evolving energy uses while satisfying constraints on greenhouse gas emissions, long-term installed nuclear power capacity scenarios tend to be more ambitious, even after the Fukushima accident. Thus, the amounts of uranium or plutonium needed to deploy such capacities could be limiting factors. This study first considers light-water reactors (LWR, GEN III) using enriched uranium, like most of the current reactor technologies. It then examines the contribution of future fast reactors (FR, GEN IV) operating with an initial fissile load and then using depleted uranium and recycling their own plutonium. However, as plutonium is only available in limited quantity since it is only produced in nuclear reactors, the possibility of starting up these Generation IV reactors with a fissile load of enriched uranium is also explored. In one of our previous studies, the uranium consumption of a third-generation reactor like an EPR^{TM} was compared with that of a fast reactor started up with enriched uranium (U5-FR). For a reactor lifespan of 60 years, the U5-FR consumes three times less uranium than the EPR and represents a 60% reduction in terms of separative work units (SWU), though its requirements are concentrated over the first few years of operation. The purpose of this study is to investigate the relevance of U5-FRs in a nuclear fleet deployment configuration. Considering several power demand scenarios and assuming different finite quantities of available natural uranium, this paper examines what types of reactors must be deployed to meet the demand. The deployment of light-water reactors only is not sustainable in the long run. Generation IV reactors are therefore essential. Yet when started up with plutonium, the number of reactors that can be deployed is also limited. In a fleet deployment configuration, U5-FRs appear to provide the best solution for using uranium, even if the economic impact of this consumption during the first years of operation is significant.

1 Introduction

At the current rate at which fuel is consumed, the natural uranium resources identified so far will be sufficient to meet our needs for the next 100 years [1]. However, most organisations in charge of defining energy-related scenarios consider a considerable increase in international nuclear power generation to meet the significantly increasing global energy demand, as well as to comply with climate constraints to reduce greenhouse gas emissions. Due to the growing nuclear reactor fleet in many countries, it is assumed that resources will therefore be depleted more rapidly.

Within the scope of this study, we therefore selected various global nuclear power deployment scenarios.

These scenarios have been applied to analyse what type of reactors must be deployed to meet the global demand: light-water reactors (LWR) using uranium-235 (^{235}U) or fast reactors (FR) using uranium-238. However, a sufficient amount of plutonium is required to start up FRs and plutonium is produced in water reactors such as pressurised water reactors (PWR) ($\approx 1\%$ of the mass of spent fuel). In the event that no Pu is available, the only solution is to start up FRs with uranium enriched in ^{235}U (U5-FR).

This paper first reviews the static comparison of the total uranium consumption of a LWR with an U5-FR. We then analyse the advantages provided by such reactors within a nuclear reactor fleet development configuration.

Therefore, the first part of this paper assesses the quantities of uranium consumed for the different scenarios under investigation and according to the reactor types being developed.

* e-mail: anne.baschwitz@cea.fr

In the second part of this paper, different limits are imposed on the global uranium supply in order to clearly define the issues related to the necessary resources. The type of reactor required to meet the demand is clearly stated for each limit and each scenario.

2 Study conditions

2.1 Prospective scenarios [2]

To carry out this prospective study, we needed to define assumptions with respect to the evolving energy demand and the deployable nuclear technologies available within the century. These assumptions are detailed below.

In the energy field, needs must be defined several years in advance or even several decades in advance so as to plan the construction of infrastructures and meet the demand. This forward-looking approach particularly applies to nuclear power: firstly, because a reactor is designed to operate for about 60 years; secondly, because waste management issues, like partitioning and transmutation, must be assessed.

The "Global Energy Perspective 1998" [3] was a five-year study conducted jointly by the International Institute for Applied Systems Analysis (IIASA) and the World Energy Council (WEC). The goals were to examine long-term energy perspectives, their constraints, and opportunities by formulating scenarios. There are six scenarios grouped into three cases, Cases A, B, and C, providing the energy mix forecast over the 21st century.

We chose four of them (Fig. 1):

- A2 is a strong global growth scenario of around 2.7% per year, with the preferred short-term use of oil and gas resources. Nuclear energy represents 4% of world energy demand in 2050 and 21% in 2100;
- A3 is also a strong global growth scenario with a more gradual introduction of nuclear energy than in scenario A2; nuclear energy represents around 11% of world energy demand in 2050 and 22% in 2100;
- B is a business-as-usual world growth scenario during the 21st century (around 2% per year);
- C2 is a scenario that has strong intentions to protect the environment against global warming. It corresponds to a low global demand, though nuclear energy represents around 12% of world demand for primary energy in 2050; this is almost twice as much as it represents today.

The IIASA scenarios consider a strong increase in the world demand in primary energy. Even if the nuclear power share is less than 20%, it supposes a rather significant increase in the nuclear installed capacity.

2.2 GRUS model

The GRUS[1] model using STELLA [4] software was developed to calculate nuclear power configurations within

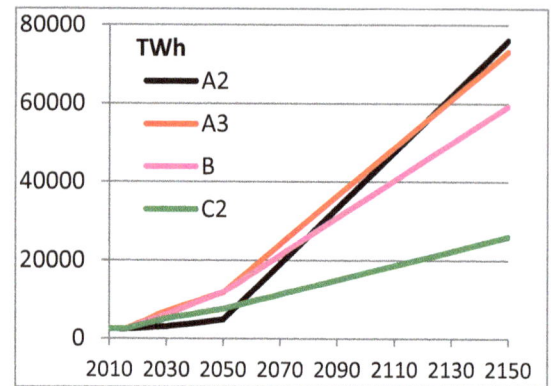

Fig. 1. IIASA scenarios: requested electronuclear generation.

various electricity demand scenarios while taking into account the complexity of the nuclear system (large number of stocks, flows and variables, numerous interactions, time scales, and different reactor technologies).

In the model, we defined:

- initial conditions (raw material stocks, kind and number of reactors and the capacities of facilities);
- key parameters (facility unit costs, cost of resources, reactor investment and operating costs, and technical characteristics of reactors);
- electricity demand versus time.

The simulation determined the nuclear fleet required to meet the yearly electricity demand according to the available resources and diverse costs.

2.3 Reactor types

Four types of reactors were considered in this study:

- PWRs, which are representative of the current reactors in service (GEN II);
- EPRs[TM] (Evolutionary Power Reactors), which are representative of Generation III water reactors (GEN III);
- FRs, which are representative of Generation IV fast reactors (GEN IV) for which a standard start-up with a Pu load (Pu-FR) is possible. It will also be possible to start them up with enriched uranium if no Pu is available (U5-FR). After several years, such reactors will become identical to reactors started up with Pu, once they will have produced the Pu required for their operation.

2.3.1 Technical characteristics

Table 1 lists the reactor characteristics that were taken into consideration. U5-FRs have the same characteristics as Pu-FRs in terms of power, load factor and burn-up due to the fact that they become Pu-FRs after ten years.

Our reactors are generic reactors of large size. For the FRs, considering the characteristics we have chosen (Pu in core and breeding gain range), we can say it is like an SFR with an oxide fuel [5].

Table 2 compares ^{235}U requirements for EPRs and U5-FRs.

---

[1]GRUS is a French abbreviation which translates as "uranium resource management with STELLA software".

Table 1. Reactor characteristics.

	PWR	EPR	FR	
			BG = 0	BG = 0.2
Gross electrical output (GWe)	1.01	1.62	1.45	
Efficiency (%)	33	36	40	
Burn-up rate (GWd/t)	45	60	123	
Mass of heavy metal in core (t)	81	126	51	
Load factor (%)	77	90	90	
Enrichment in ^{235}U (%)	3.7	4.9	–	
Pu in core (t)	–	–	12	
%Pu in spent fuel (%)	1.17	1.34	23.5	28.2

BG: breeding gain.

Table 2. ^{235}U requirements.

	Unit	EPR-type PWR[a]	U5-FR [6]
^{235}U enrichment	%	4.9	14.4
Mass of ^{235}U in core	Tonnes of ^{235}U/GWe	3.9	8
Reloading	Tonnes of ^{235}U/GWe/year	0.78	1.4[b]

[a]We chose the characteristics of the EPR for comparison with an SR (assumptions may differ in relation to Ref. [6]). The figures are given in relation to an equilibrium cycle.
[b]For the first 5 reloads of an U5-FR. The U5 enrichment is given for the first core: it constantly decreases as the U5-FR becomes a Pu-FR.

2.3.2 Assumptions for introducing fast reactors

In the model, only PWRs are deployed up to 2040. Thereafter, different assumptions were applied when introducing new reactors:

- all new reactors are still PWRs (EPR-type) for the whole century with the once-through option;
- fast reactors (FRs) are installed as long as plutonium is available. When plutonium is not available, either PWRs or FRs started up with enriched uranium can be installed.

3 Uranium consumption

3.1 Consumption comparisons for PWRs and U5-FRs

Certain results presented during the FR13 [7] conference are recalled in this section.

In this specific case, we have considered an electric utility intending to build a FR without a sufficient amount of Pu. At present, the electric utility can decide whether to build a PWR or a FR started up with enriched uranium. At the end of the reactor's service life (60 years), it can be considered in both cases that the electric utility will have a sufficient amount of Pu to start-up a new FR. The necessary amount of Pu corresponds to two cores: the first core and an equivalent quantity for the first few reloads until Pu from the first core is extracted and recycled for the following loads.

Choosing either reactor will lead to the development of next generation of FRs.

Here, we have considered an open-cycle EPR with the first core and annual reloads using enriched uranium.

We considered that reloads for a U5-FR were performed on a 1/5 basis as the remaining fuel stays in the core for slightly more than 5 years. It is assumed that the cycle lasts 5 years (cooling time after unloading until the manufacture of a new sub-assembly, which can be loaded into the reactor). Enriched uranium must therefore be provided for the first core and the first 5 reloads as the following reloads will be done with the Pu produced by the FR.

Table 3 specifies the material flows for the different stages of the fuel cycle under consideration, as well as the enrichment requirements for the reactor lifespan when the price of natural uranium is of €100/kg for the reactor's entire service life (flows vary depending on the price of natural uranium through optimisation of the tails assay, with U_{nat} at €100/kg, the optimised content of depleted uranium is 0.23% of ^{235}U). Year 0 corresponds to the year the reactor is commissioned.

Over the reactor's 60-year lifespan, it can be seen that the U5-FR uses three times less uranium than the EPR and requires 60% fewer SWUs. Yet, if we compare the fuel requirements over the first 7 years of operation,

Table 3. Annual flow of materials (tonnes) and enrichment requirements (million SWU) for 1 GWe.

Year	EPR			FR		
	Flow of natural uranium	MSWU	Flow of uranium enriched at 4.9%	Flow of natural uranium	MSWU	Flow of uranium enriched at 14.4%
−2	769	0.65		1,628	1.67	
−1	154	0.13	80	293	0.30	56
0	154	0.13	16	293	0.30	10
1	154	0.13	16	293	0.30	10
2	154	0.13	16	293	0.30	10
3	154	0.13	16	293	0.30	10
4	154	0.13	16	161	0.17	10
5	154	0.13	16			6
6 to 57	154	0.13	16			
58						
59						
Total	9,844	8.27	1,019	3,256	3.34	111

the U5-FR uses twice as more natural uranium and 2.5 times more SWUs than the EPR.

3.2 Uranium consumption of a global nuclear reactor fleet

This section compares the global uranium consumption for meeting the different nuclear power demand scenarios described in Section 2.1 according to the reactors being considered. We have already shown that the nuclear industry cannot entirely rely on LWRs [8]. However, the amount of plutonium available for developing the fourth generation of reactors is also a limiting factor [9].

Until 2040, only GEN III reactors are deployed, as it is considered that GEN IV reactors will only be technically available as from that date. After, two cases were considered:

– case 1 in blue: as many Pu-FRs as possible are installed depending on Pu availability and the fleet is then completed with EPRs;
– case 2 in red: as many Pu-FRs as possible are installed and the fleet is then completed with FRs started up with enriched uranium.

Fast reactors can be self-sufficient reactors (solid line curves) or breeder reactors with a regeneration gain of 0.2 (dotted line curves).

Figure 2 indicates the accumulated uranium consumption for scenario A3.

In Figure 3, we have added "committed uranium" to the consumed uranium, i.e. uranium for the future reloading of reactors which are currently in operation.

It has been observed that by favouring U5-FRs with respect to LWRs, it is possible to practically halve the total consumption of uranium in 2150. With breeder reactors, it is even possible to stabilise the overall uranium consumption. A sufficient amount of Pu is therefore available to only develop Pu-FRs.

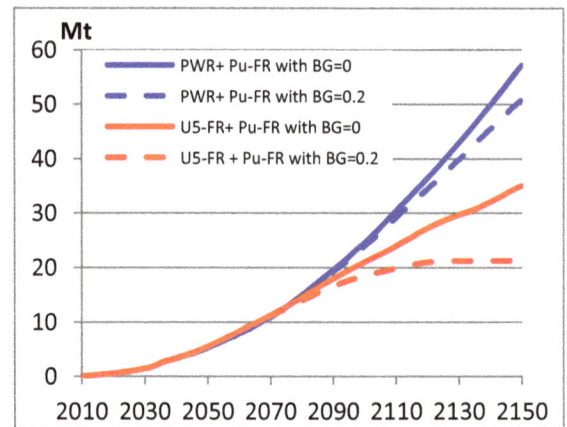

Fig. 2. Scenario A3 - Total consumed U_{nat}.

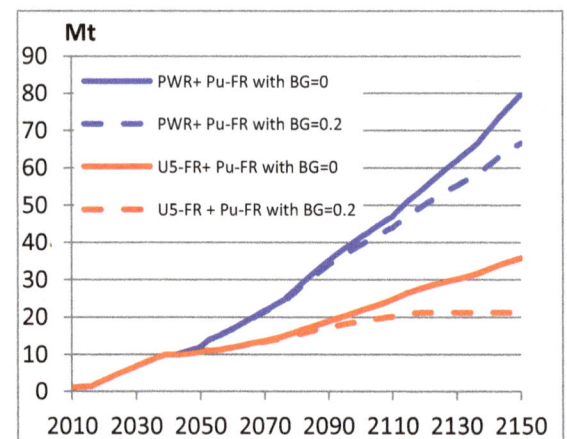

Fig. 3. Scenario A3 - Total consumed + committed U_{nat}.

Tables 4 to 7 indicate the total consumption of uranium (consumed uranium in bold, consumed + committed uranium in italic) for the four different demand scenarios in 2050, 2100 and 2150.

Regardless of the scenario, in 2050, it is observed that the amount of consumed uranium is slightly greater with U5-FRs than with EPRs (see Sect. 3.1). The excessive consumption for U5-FRs at the start of their service life,

compared to EPRs, is thus noted. However, when also considering committed uranium, uranium savings have already been observed.

In 2100, savings start to be significant especially in terms of committed uranium.

In 2150, a significant decrease in the overall uranium consumption is noted when favouring the development of U5-FRs and in some situations it is even halved. In some

Table 4. U_{nat} consumed and committed to scenario A2 in 2050, 2100 and 2150.

Scenario A2	2050		2100		2150	
	GR = 0	GR = 0.2	GR = 0	GR = 0.2	GR = 0	GR = 0.2
EPR + Pu-FR	**2.5**	**2.5**	**20**	**20**	**55**	**51**
	4.7	*4.7*	*37*	*36*	*80*	*70*
U5-FR + Pu-FR	**2.7**	**2.7**	**16**	**14**	**32**	**19**
	4.6	*4.6*	*17*	*15*	*32*	*19*

Bold: total consumed U_{nat} (Mt); italic: total consumed and committed U_{nat} (Mt).

Table 5. U_{nat} consumed and committed to scenario A3 in 2050, 2100 and 2150.

Scenario A3	2050		2100		2150	
	GR = 0	GR = 0.2	GR = 0	GR = 0.2	GR = 0	GR = 0.2
EPR + Pu-FR	**5.2**	**5.2**	**25**	**24**	**57**	**51**
	12	*12*	*41*	*39*	*79*	*66*
U5-FR + Pu-FR	**5.4**	**5.4**	**21**	**18**	**35**	**21**
	11	*11*	*22*	*19*	*36*	*21*

Bold: total consumed U_{nat} (Mt); italic: total consumed and committed U_{nat} (Mt).

Table 6. U_{nat} consumed and committed to scenario B in 2050, 2100 and 2150.

Scenario B	2050		2100		2150	
	GR = 0	GR = 0.2	GR = 0	GR = 0.2	GR = 0	GR = 0.2
EPR + Pu-FR	**5.0**	**5.0**	**21**	**20**	**47**	**42**
	12	*12*	*35*	*33*	*64*	*53*
U5-FR + Pu-FR	**5.2**	**5.2**	**18**	**16**	**29**	**18**
	10	*10*	*19*	*17*	*30*	*18*

Bold: total consumed U_{nat} (Mt); italic: total consumed and committed U_{nat} (Mt).

Table 7. U_{nat} consumed and committed to scenario C2 in 2050, 2100 and 2150.

Scenario C2	2050		2100		2150	
	GR = 0	GR = 0.2	GR = 0	GR = 0.2	GR = 0	GR = 0.2
EPR + Pu-FR	**3.5**	**3.5**	**11**	**11**	**22**	**19**
	7.4	*7.4*	*18*	*16*	*30*	*23*
U5-FR + Pu-FR	**3.7**	**3.7**	**10**	**10**	**15**	**10**
	7.0	*7.0*	*11*	*10*	*15*	*10*

Bold: total consumed U_{nat} (Mt); italic: total consumed and committed U_{nat} (Mt).

cases, the quantities of consumed U_{nat} and consumed + committed U_{nat} are identical, which means that no currently operational reactor requires uranium.

We remarked the brief excess consumption of uranium when U5-FRs are deployed rather than light-water reactors (see Fig. 4, example of scenario A3). We wanted to check if this could be penalising in terms of the annual demand, whether for uranium extraction or enrichment.

The brief increase due to the deployment of U5-FRs can be seen in Figure 5 with respect to the uranium demand and in Figure 6 for enrichment needs. It can be seen that the increase is nevertheless reasonable since several U5-FRs are included in the global fleet which is mainly composed of light-water reactors.

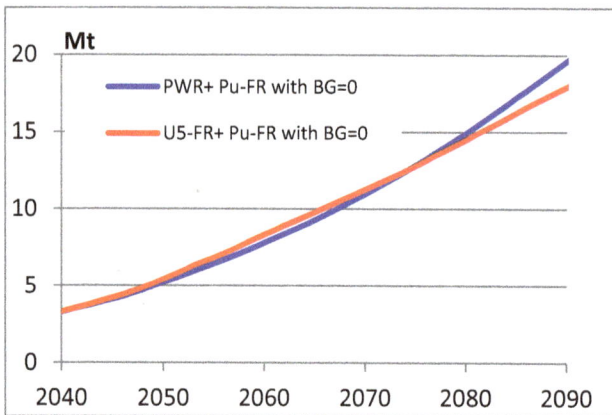

Fig. 4. Scenario A3 - Accumulated consumption of U_{nat}.

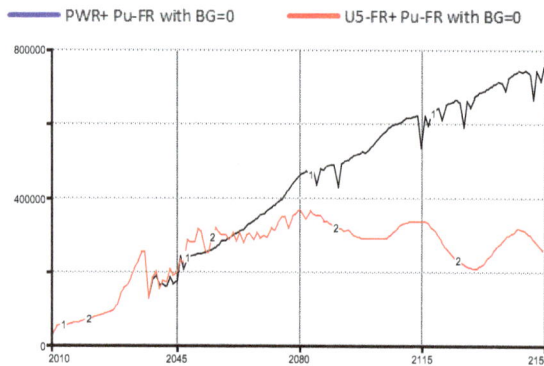

Fig. 5. Scenario A3 - Annual demand for U_{nat} (in tons).

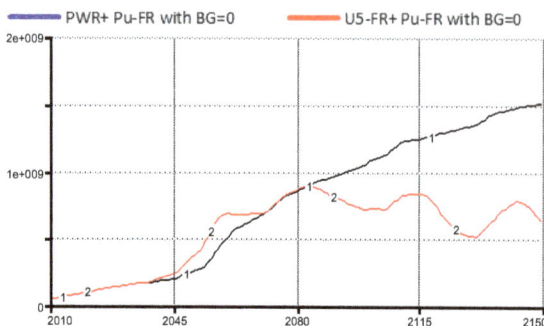

Fig. 6. Scenario A3 - Annual demand for SWU.

4 Potential nuclear capacity

Up until now, we have considered it possible to extract the quantity of uranium required as long as the extraction cost is paid. This assumption seems realistic in a market context and it considers that resources diluted in seawater are accessible, though it does not take into account procurement issues which could arise once all conventional resources have been exhausted.

In this section, we approach the issue of resources in a different manner by considering the available quantities of natural uranium as limited.

4.1 Different available quantities of uranium

We have considered four different quantities of available natural uranium:

– 10 Mt corresponding to the order of magnitude of identified uranium resources [1]. This case will be in violet on the figures;
– 20 Mt (in green) corresponding to the order of magnitude of conventional resources, added to 4 Mt of uranium extracted from phosphates [10];
– 40 Mt (in orange) corresponding to the order of magnitude of conventional resources, added to about 22 Mt (former estimate of uranium extracted from phosphates);
– 80 Mt (in blue), which takes into account the possibility of mining exploration finding substantial new resources; there is nothing to support this figure which is based on a very optimistic view of a textbook example.

4.2 Reactor deployment assumptions

As mentioned in the previous section, Generation IV reactors will be technically available from 2040.

We have added an extra constraint: when the committed uranium (i.e. taking into account the needs of operational reactors throughout their services lives) exceeds one of the limits in question, it will be impossible to build a new reactor requiring enriched uranium (i.e. PWRs, EPRs and U5-FRs in our case). The only reactors that can be built once this limit has been reached are fast reactors started up with plutonium. Considering that plutonium has to be produced and is not available in unlimited quantities, one day we will no longer be able to build enough reactors and thus no longer match supply to demand.

4.3 Deployment of EPRs only

Figures 7 to 10 show the quantity of energy that the nuclear system may produce for each scenario depending on the limits on available uranium quantities.

Demand is indicated in black, while nuclear power generation as a function of the limited quantities of uranium is indicated in colour.

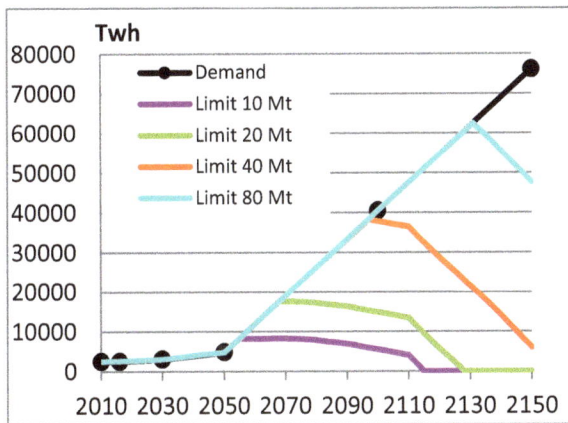

Fig. 7. Scenario A2 - Electronuclear production by PWRs only.

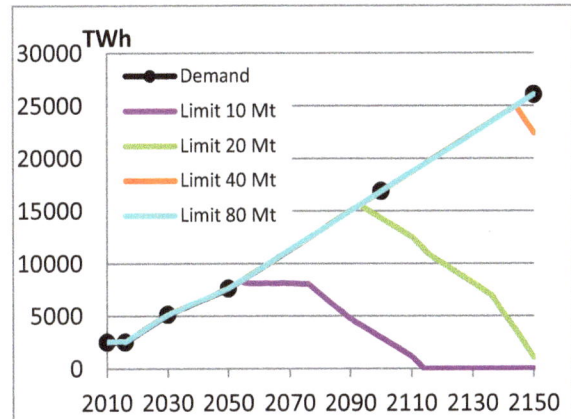

Fig. 8. Scenario A3 - Electronuclear production by PWRs only.

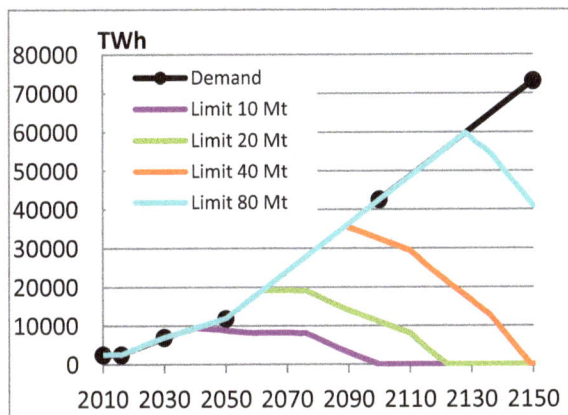

Fig. 9. Scenario B - Electronuclear production by PWRs only.

The different colour curves drop off from the black curve. This moment corresponds to the date at which the uranium limit is equal to the quantity of uranium already consumed, added to the committed quantity for the future operation of reactors already in service.

When nuclear power generation reaches 0, this limit quantity of uranium has been consumed.

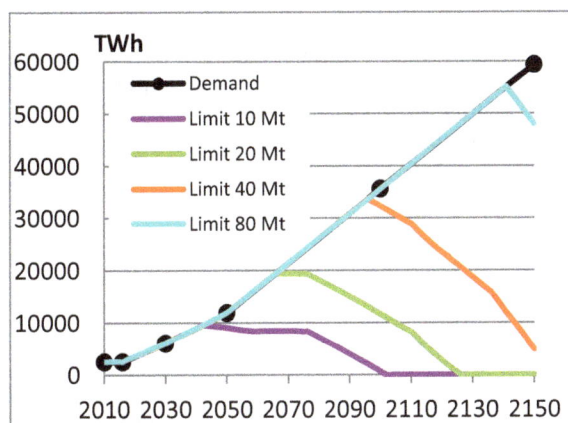

Fig. 10. Scenario C2 - Electronuclear production by EPRs only.

It is clear that nuclear power will not be sustainable with only Generation III reactors. Scenario C2, which requires only 25,000 TWh in 2150, is the only case where demand could be met despite more than 40 Mt of uranium required (consumed + committed) at this date and already 20 Mt in 2100.

4.4 Deployment of self-sufficient or breeder Pu-FRs from 2040

Since we have shown that only light-water reactors do not meet the nuclear power generation demand as laid out in the prospective scenarios, we included Generation IV reactors from 2040. We considered these reactors with a first fissile Pu load, which means that Pu availability will therefore be an important parameter for their deployment.

Figures 11 to 14 show for each scenario the nuclear power generation that can be expected in relation to the type of reactors deployed and as a function of the quantity of uranium believed to be extractable. Just as a reminder, the case with only PWRs is shown by the thin lines. The case with self-sufficient FRs is in solid lines. The case with breeder reactors is in dashed lines.

Contrary to the case where only light-water reactors would be deployed (thin line), here it would be possible to maintain nuclear production regardless of the case considered. Despite this, most of the cases remain far from meeting demand.

It can be seen that an installed power plateau is reached after a certain time with self-sustained reactors (solid lines), which corresponds to the quantity of Pu produced in PWRs based on the available quantity of uranium. This represents a FR installed power capacity of about 70 GWe/Mt of uranium.

It can also be seen that production is significantly increased with breeder reactors (dashed lines), especially in the next century. Yet more than often, demand is not met.

For the first three high-demand scenarios, about 80 Mt and self-sufficient reactors at least are needed to meet demand. More than 20 Mt is needed with breeder reactors for scenario C2 which has a lower demand, or slightly more than 40 Mt in the case where only self-sufficient reactors are used.

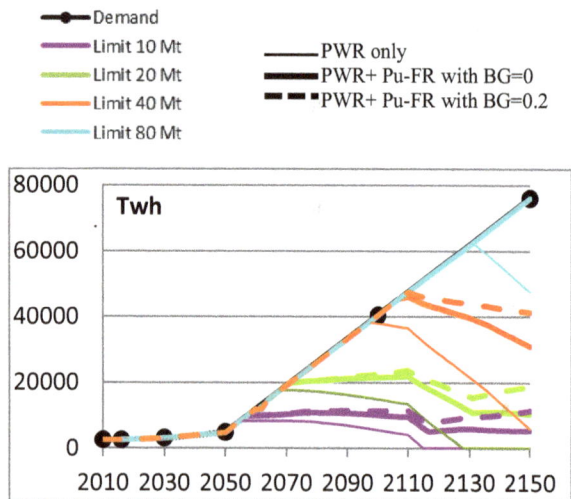

Fig. 11. Scenario A2 - Production by EPRs and Pu-FRs.

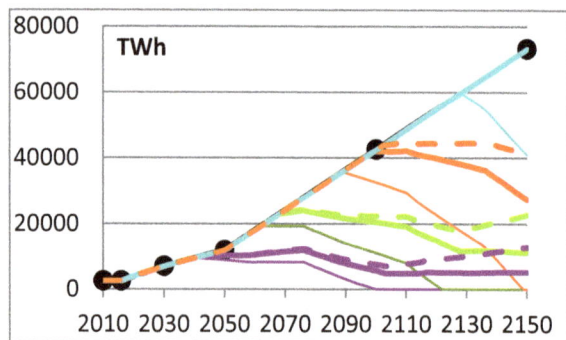

Fig. 12. Scenario A3 - Production by EPRs and Pu-FRs.

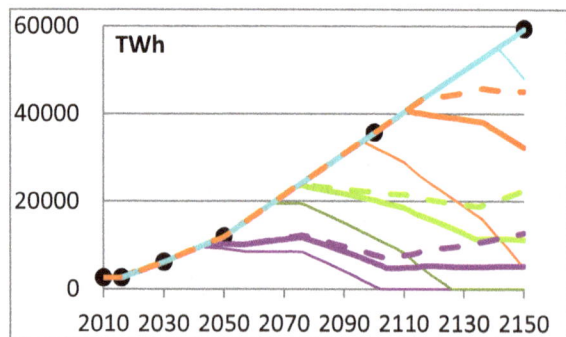

Fig. 13. Scenario B - Production by EPRs and Pu-FRs.

4.5 Deployment of self-sustaining and breeder Pu-FRs and U5-FRs

We established in the paragraph above that including FRs was not sufficient to meet the demand in many cases, especially when the uranium quantities were limited (< 80 Mt).

We have already shown that a U5-FR consumes three times less natural uranium than an EPR. We also remarked

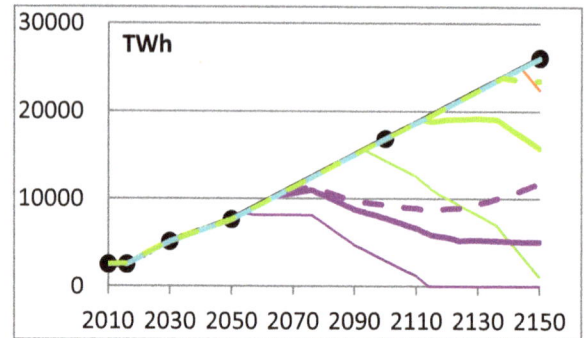

Fig. 14. Scenario C2 - Production by EPRs and Pu-FRs.

that a reactor fleet including the deployment of U5-FRs instead of EPRs made it possible to reduce the accumulated consumption of uranium by two.

Now the objective is to see whether such reactors are capable of meeting the demand despite the limits imposed on the quantities of available uranium.

Technically speaking, these reactors will be available from 2040, as is the case for Pu-FRs. From this date, priority will be given to deploying Pu-FRs if Pu is available, otherwise we will resort to using U5-FRs.

We have restricted ourselves to referring to the curves of the two extreme scenarios (A3 and C2).

On the previous figures, we added the case with FR started with uranium in large full line, and divided the results in several figures (one per limit in uranium) so that it is still readable.

The following conclusions were reached for scenario A3:

– with only 10 Mt of available uranium (Fig. 15), it is practically all consumed before FRs are integrated. The advantage of U5-FRs is therefore insignificant;
– for other uranium limits, particularly 20 and 40 Mt (Figs. 16 and 17), the relevance of deploying U5-FRs rather than EPRs is clearly visible when plutonium is not readily available. If only 20 Mt of uranium is available, then breeder reactors are needed to meet the demand. With 40 Mt of uranium, self-sufficient reactors are adequate to meet the demand;
– if 80 Mt of uranium is available, it has already been seen that Pu-FRs are sufficient to meet the demand (see Fig. 12).

Similar conclusions could be drawn for scenarios A2 and B.

The following conclusions were reached for scenario C2:

– as this scenario was generally less ambitious in terms of nuclear power generation, the 10 Mt of uranium was not consumed and committed in 2040. The positive contribution of U5-FRs is thus visible since the demand is met with these reactors when they are in breeder configuration, while remaining below 10 Mt of uranium consumption (Fig. 18);
– when only 20 Mt of uranium is available (Fig. 19), breeder Pu-FRs are practically sufficient. Self-sufficient U5-FRs are just barely required;

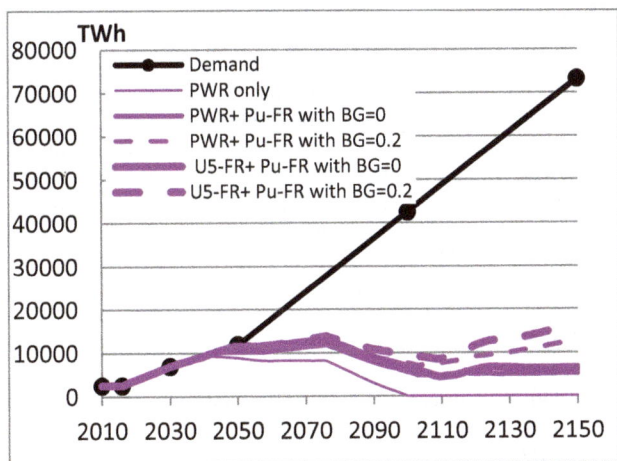

Fig. 15. Scenario A3 with 10 Mt of uranium - Electronuclear production.

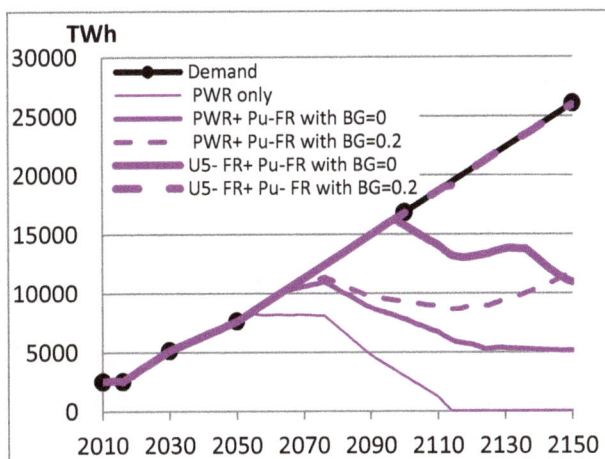

Fig. 18. Scenario C2 with 10 Mt of uranium - Electronuclear production.

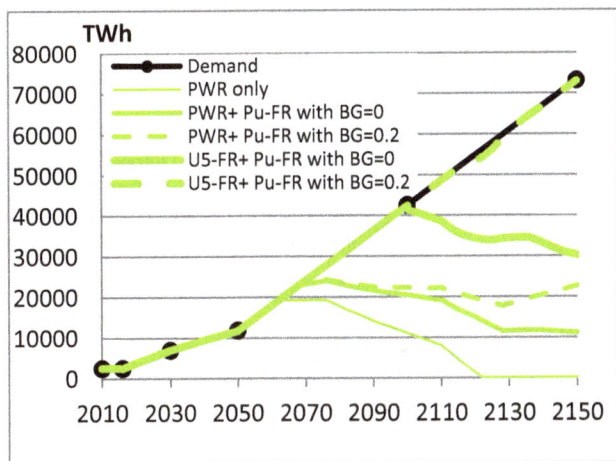

Fig. 16. Scenario A3 with 20 Mt of uranium - Electronuclear production.

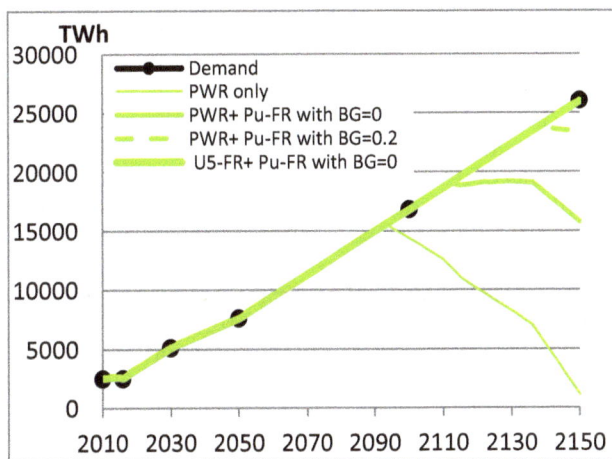

Fig. 19. Scenario C2 with 20 Mt of uranium - Electronuclear production.

– self-sufficient FRs or PWRs were sufficient for 40 and 80 Mt of uranium respectively, as had already been concluded previously (see Figs. 10 and 14).

5 Conclusion

The purpose of this study was to determine what types of reactors and fuels would be needed to meet different nuclear power production scenarios.

Nuclear power is not sustainable on the basis of light-water reactors only, unless the demand remains relatively limited (scenario C2 = 25,000 TWh in 2150 ≈ 3000 GWe) and we have large stocks of available uranium (more than 40 Mt). The fourth generation of reactors is therefore essential if we wish to meet demand. Yet, the quantities of available plutonium do not always enable us to deploy as many fast reactors as required and light-water reactors are often necessary to supplement the nuclear reactor fleet to meet the demand.

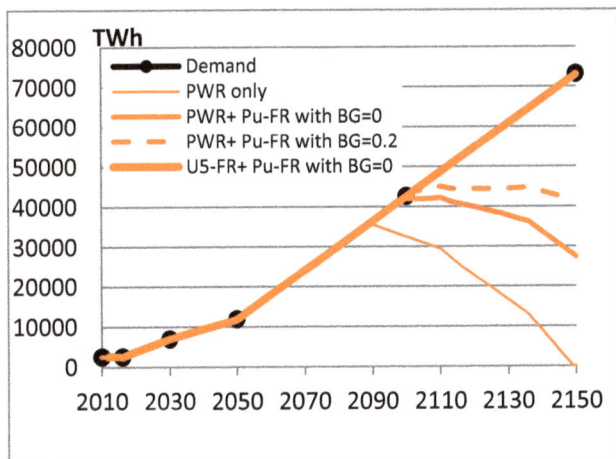

Fig. 17. Scenario A3 with 40 Mt of uranium - Electronuclear production.

Self-sufficient configurations of Generation IV reactors would make it possible to meet scenario C2 with a uranium consumption of more than 40 Mt all the same. Almost 80 Mt of uranium is required for higher-demand scenarios.

When breeder configurations are applied to Generation IV reactors, only 20 Mt of uranium is required for scenario C2. The demand will remain partially unmet for the three other scenarios.

We therefore imagined the deployment of fast reactors started up with enriched uranium to remedy the lack of available plutonium. This type of reactor consumes three times less uranium than an EPR-type light-water reactor. We assumed that uranium was only needed for the first core and the first few reloads, but then the plutonium produced by the reactor would be used thereafter.

Thanks to these reactors, the nuclear reactor fleet would be able to meet demand in scenarios A2, A3 and B (from 60,000 to 80,000 TWh in $2150 \approx 7500$ to 10,000 GWe). These reactors would have to be breeders in the case there is only 20 Mt of uranium, or only self-sufficient with 40 Mt of uranium available.

In the case of only 10 Mt of uranium, these Generation IV reactors – which will be technically available in 2040 – will arrive too late since this quantity of uranium will have been consumed prior to this date. It is only scenario C2 that can be met with U5-FRs in breeder configuration, with 10 Mt of uranium available.

In terms of resource savings, U5-FRs are seen to be the best solution for using limited quantities of uranium while providing maximum nuclear power. Unfortunately, though uranium consumption is three times less for U5-FRs than that for an EPR-type light-water reactor, it is nevertheless consumed at the start of the reactor's life span which represents a significant economic disadvantage. For this reason, economic aspects will hinder the deployment of this type of reactor.

References

1. OECD/NEA and IAEA, Uranium 2014: resources, production and demand, Report No. 7209, 2014
2. S. Gabriel, A. Baschwitz, G. Mathonnière, F. Fizaine, T. Eleouet, Building future nuclear power fleets: the available uranium resources constraint, Res. Policy **38**, 458 (2013)
3. N. Nakićenović, A. Grübler, G. McDonald, *Global energy perspectives IIASA/WEC* (Cambridge University Press, 1998)
4. Isee Systems, the Visual thinking company: http://www.iseesystems.com/
5. GEN IV international forum, Annual report 2014, Chap. III, 2014
6. H. Safa, P. Dumaz, J.F. Luciani, L. Buiron, B. Fontaine, J.P. Grouiller, Is it worth to start fast reactors using uranium fuels, in *ICAPP 2011: International Congress on Advances in NPPs "Performance & Flexibility: The Power of Innovation", Nice, France* (2011)
7. A. Baschwitz, G. Mathonnière, S. Gabriel, Economic relevance of starting an SFR with enriched uranium, in *FR13: International Conference on Fast Reactors and Related Fuel Cycles: Safe Technologies and Sustainable Scenarios, Paris, France* (2013)
8. A. Baschwitz, C. Loaëc, J. Fournier, M. Delpech, GEN IV deployment: long-term perspective, in *FR09: International Conference on Fast Reactors and Related Fuel Cycles: Challenges and Opportunities, Kyoto, Japan* (2009)
9. A. Baschwitz, C. Loaëc, J. Fournier, M. Delpech, F. Legée, Long-term prospective on the electronuclear fleet: from GEN II to GEN IV, in *Global 2009: The Nuclear Fuel Cycle: Sustainable Options & Industrial Perspectives, Paris, France* (2009)
10. S. Gabriel, A. Baschwitz, G. Mathonnière, T. Eleouet, F. Fizaine, A critical assessment of global uranium resources, including uranium in phosphate rocks, and the possible impact of uranium shortages on nuclear power fleets, Ann. Nucl. Energy **58**, 213 (2013)

20

Assessment of stainless steel 348 fuel rod performance against literature available data using TRANSURANUS code

Claudia Giovedi[1,*], Marco Cherubini[2], Alfredo Abe[3], and Francesco D'Auria[4]

[1] LabRisco, University of São Paulo, Av. Prof. Mello Moraes 2231, São Paulo, SP, Brazil
[2] NINE, Nuclear and Industrial Engineering, Borgo Giannotti 19, 55100 Lucca, Italy
[3] Nuclear and Energy Research Institute - IPEN/CNEN, Nuclear Engineering Center – CEN, Av. Prof. Lineu Prestes 2242, São Paulo, SP, Brazil
[4] UNIPI, University of Pisa, Largo L. Lazzarino 2, 56126 Pisa, Italy

Abstract. Early pressurized water reactors were originally designed to operate using stainless steel as cladding material, but during their lifetime this material was replaced by zirconium-based alloys. However, after the Fukushima Daiichi accident, the problems related to the zirconium-based alloys due to the hydrogen production and explosion under severe accident brought the importance to assess different materials. In this sense, initiatives as ATF (Accident Tolerant Fuel) program are considering different material as fuel cladding and, one candidate is iron-based alloy. In order to assess the fuel performance of fuel rods manufactured using iron-based alloy as cladding material, it was necessary to select a specific stainless steel (type 348) and modify properly conventional fuel performance codes developed in the last decades. Then, 348 stainless steel mechanical and physics properties were introduced in the TRANSURANUS code. The aim of this paper is to present the obtained results concerning the verification of the modified TRANSURANUS code version against data collected from the open literature, related to reactors which operated using stainless steel as cladding. Considering that some data were not available, some assumptions had to be made. Important differences related to the conventional fuel rods were taken into account. Obtained results regarding the cladding behavior are in agreement with available information. This constitutes an evidence of the modified TRANSURANUS code capabilities to perform fuel rod investigation of fuel rods manufactured using 348 stainless steel as cladding material.

1 Introduction

The available data shows that the steady state performance of steel cladding in the first PWR was considered excellent [1,2]. The material used in the early PWR was mainly AISI 304 (12% cold worked). Nonetheless, some reactors operated using annealed AISI 348, which presents a better corrosion resistance due to the addition of niobium and tantalum in its composition.

The substitution of stainless steel by zircaloy as cladding material was due to the lower absorption for thermal neutrons of the zirconium-based alloys which enables to operate with lower enrichment cost. Despite the stainless steel economics penalty, the main advantage of using this material as cladding comes from the reduction of the probability of the violent oxidation reaction that occurs with zirconium-based alloys at high temperatures, as it

has occurred in the Fukushima Daiichi accident [3]. As a consequence of this, iron-based alloys once again can be considered as a good option to replace zirconium-based alloys as cladding material improving the safety under accident scenarios [4]. Considering the previous good experience of AISI 348 as cladding, this material could be again applied to replace zirconium-based alloys as PWR fuel cladding.

In order to evaluate the fuel performance of fuel rods using AISI 348 as cladding, it is necessary to modify the current fuel performance codes to insert correlations and properties of this material. In this sense, TRANSURANUS code appears as a good option due to its flexibility for different fuel rod designs and reactor types, time range of the problems to be treated and materials data bank, which includes AISI 316 (both 20% cold worked and annealed correlations are programmed into the code) [5,6].

The adapted version of the TRANSURANUS code to evaluate the AISI 348 performance under irradiation was assessed using Yankee Rowe available data from open

Table 1. Austenitic stainless steel series 300 properties at room temperature [7–9].

Property	AISI 304	AISI 316	AISI 348
Density (10^3 kg/m^3)	8.0	8.0	7.9
Rockwell-B hardness	70	79	80
Ultimate strength (MPa)	505	580	605
Tensile strength at yield (MPa)	215	290	220
Maximum elongation (%)	70	50	40
Elastic modulus (GPa)	200	193	200
Poisson's ratio	0.290	0.295	0.283
Specific heat (J/g°C)	0.5	0.5	0.5
Thermal conductivity (W/mK)	16.2	16.3	16.4
Thermal expansion coefficient (10^{-6}/K)	17	17	17
Melting point (°C)	1450	1427	1400

literature. The reason why Yankee Rowe fuel rod was selected is because it was the unique PWR (for which information was available to the authors) in which AISI 348 was used as cladding material. The aim of this paper is to present the obtained results in the framework of this activity.

1.1 TRANSURANUS code

TRANSURANUS is a computer code for the thermal and mechanical analysis of fuel rods in nuclear reactors developed at the European Institute for Transuranium Elements (ITU). The code consists of a clearly defined mechanical-mathematical framework into which physical models can easily be incorporated [5].

In order to introduce the AISI 348 data in the TRANSURANUS code, a set of references has been searched and collected. A selection has been made in order to use reliable data, when necessary data are not available either values coming from similar stainless steel (AISI 347) or typical values (i.e. applicable for a variety of stainless steel) were used. A comparison of the main properties for non-irradiated annealed AISI 304, 316 and 348 is presented in Table 1. The data show that the properties for AISI 316 and AISI 348 are very close which enable to expect a similar performance for both materials under irradiation.

Based on the literature research, the following properties related to the annealed AISI 348 were introduced in the TRANSURANUS code to obtain the adapted version: elasticity constant, Poisson's ratio, strain due to swelling, thermal strain, thermal conductivity, creep strain (thermal and irradiation creep rate), yield stress, rupture strain, burst stress, specific heat, density and melting temperature.

It was assumed that correlations already programmed in TRANSURANUS for the AISI 316 are acceptable and validated enough being the TRANSUNARUS originally developed to deal with fast breeder reactor fuel and considering its validation program [6]. In addition, the new correlations related to the AISI 348 properties somewhat reflect the same structure of the equivalent formula already programmed for the AISI 316. These correlations similarities should (at least partially) ensure that code numerical stability issue is not to be expected.

The AISI 348 behavior predicted by the modified code version has been compared against AISI 316 behavior which is part of the original (hence validated) code version. In general, the two steels present, as expected, similar trends. AISI 316 has shown a bit more conservative results in respect to AISI 348.

1.2 Description of Yankee Rowe NPP features

The Yankee Rowe PWR has been owned and operated since startup in 1960 by the Yankee Atomic Electric Co. at Rowe, Massachusetts. The reactor and its initial core and stainless steel reloads were designed and built by Westinghouse. Yankee Rowe was the first fully commercial PWR of 250 MWe, which started up in 1960 and operated to 1992 [10]. Yankee Rowe produced 44 billion kilowatt-hours of electricity from 1961–1992 when it was permanently shutdown for economic reasons. The plant was successfully decommissioned between 1992–2007 with structures removed and the site restored to stringent federal and state remediation standards [11].

Starting from its 7th cycle of operation, the reactor began to change to zircaloy cladding, the transition was completed with cycle 12. The stainless steel clad reactor core consisted of 76 assemblies and 24 cruciform control rods. A typical stainless steel assembly was made up of 9 subassemblies each arranged in a 6×6 array, to make up an 18×18 fuel rod array. The subassemblies were tied together along their length to form a complete integral fuel assembly.

The clad material was both seamless and welded annealed AISI 348 and represents the only large scale fuel experience with this steel in a PWR. The chemical composition of the adopted AISI 348 is identical to the niobium stabilized AISI 347, with the exception of a 0.10% limit on tantalum to reduce the neutron absorption cross-section. The fuel rod was also unique in that 6 physically separated fuel stacks spaced by equally spaced stainless steel discs. Each segment contains about 25 pellets. The objective of such design was to minimize differential thermal expansion

Table 2. Yankee Rowe general data and assumptions.

Parameter	Value	Remark
Rod outside diameter (cm)	0.864	[1,12]
Cladding thickness (cm)	0.053	[1,12]
Gap size (diametral) (cm)	0.011	[1,12]
Fuel rod pitch (cm)	1.153	[1,12]
Fuel pellet diameter (cm)	0.747	[1,12]
Fuel pellet density (%)	93	[1]
Fill gas internal rod pressure (MPa)	0.1	The fuel rod is not pressurized [1]
Active fuel length (cm)	229.9	[12]
Concentration of the gas components at the beginning of the calculation	0.8 N_2 0.2 O_2	The fuel rod is not pressurized, then it was considered the air composition [1]
U235 enrichment degree (%)	3.4	[1,12]
Free volume in the upper plenum available for filling gas and fission gas (cm^3)	4.359	The plenum height assumption considered a conservative value taking into account the fuel stack height
Coolant flow rate ($g\,h^{-1}$)	7.86×10^5	[12]
Coolant temperature (°C)	252	[1,12]
Coolant pressure (MPa)	14	[1,12]
Average LHGR ($kW\,m^{-1}$)	11.4	Average rod power given in the literature for the Yankee Rowe fuel rods [1,12]
Design LHGR ($kW\,m^{-1}$)	35.3	Design rod power given in the literature for the Yankee Rowe fuel rods [1,12]
Maximum cladding temperature surface (°C)	343	[1,12]
Average burnup ($MWd\,tU^{-1}$)	31000	[1,12]
Neutron flux ($cm^{-2}\,s^{-1}$)	6.3×10^{13}	Average assumed value to achieve the final fluence level and burnup [1,12]
Final fluence level ($n\,cm^{-2}$)	6×10^{21}	[1,12]

between fuel and clad. There were no reported stainless steel clad fuel failures. The average fuel rod heat generation rate was $114\,W\,cm^{-1}$, the design rate was $353\,W\,cm^{-1}$ (with a peak as high as $410\,W\,cm^{-1}$). The maximum cladding surface temperature was 343 °C. A total of 16 assemblies were examined, all the assemblies were in excellent conditions with a minor amount of crud deposited [1].

2 Methodology

2.1 Yankee Rowe general data and assumptions

In order to prepare the input deck to perform the simulation considering the Yankee Rowe reactor design and operational parameters, it was collected in the literature all the available data, which are presented in Table 2 as well as the necessary assumptions.

2.2 TRANSURANUS model and assumptions

The simulations were carried out adopting the recommended TRANSURANUS models for PWR. The geometric characteristics, thermal-hydraulic parameters and power profile obtained from the literature for the Yankee Rowe fuel rod were implemented in the TRANSURANUS input deck according to the data presented in Table 2.

Considering that in TRANSURANUS code the analysis is performed slice per slice, it was necessary to assume a discretization for the Yankee Rowe fuel rod, which is presented in Figure 1. In order to prepare this model, it was considered the following information: the fuel rod had six physically separated fuel stacks with a perforated stainless steel disk between them localized at equally spaced axial locations, each segment contains about 25 UO_2 pellets, the active fuel length is 229.9 cm and the height of the fuel pellet is 1.46 cm [1,12].

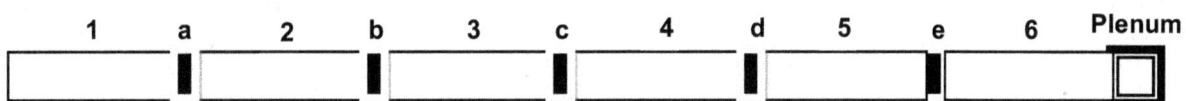

1, 2, 3, 4, 5, 6: 36.5 cm (25 fuel pellets) divided in 4 segments each one of 91 mm (apart the first two meshes of 85 mm); a, b, c, d, e: 40 mm (stainless steel disk); Plenum: 140 mm

Fig. 1. Yankee Rowe fuel rod assumed discretization based on the literature data [1].

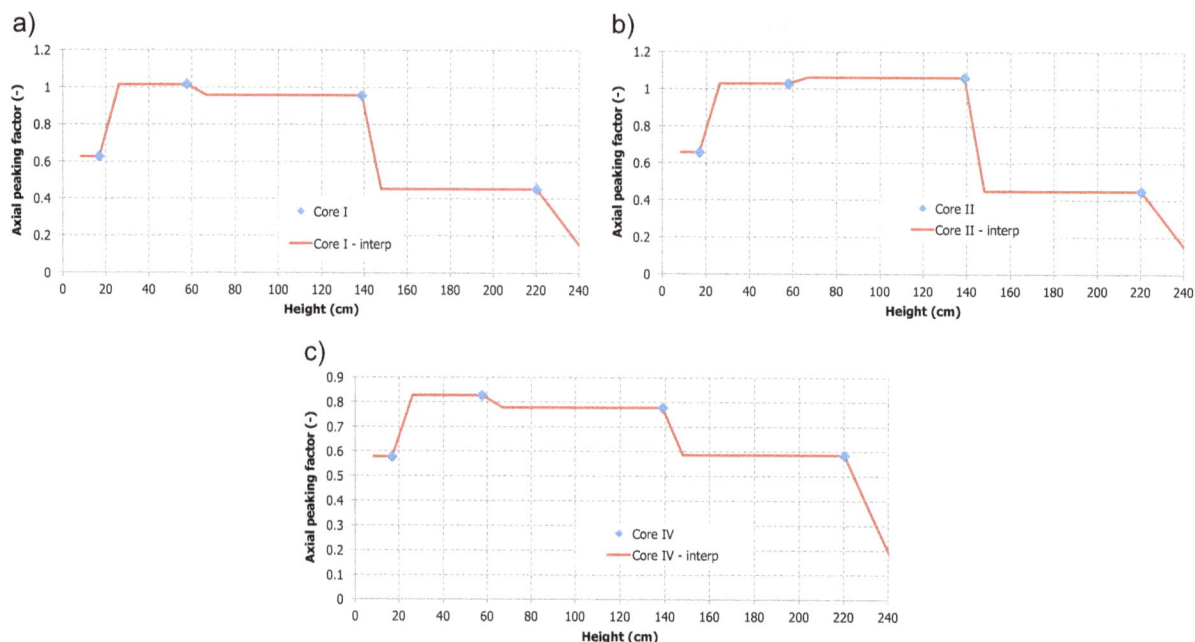

Fig. 2. E6-C-f6 fuel rod axial peaking factor for Core I (a), Core II (b) and Core IV (c), available data (blue dots [12]) and related interpolation (red curve).

The plenum length is not presented in the literature. Then, for calculation was assumed a value of 14 cm, which represents a conservative value for a PWR fuel rod with an active length of 229.9 cm.

The cladding and pellet roughness are also not presented in the literature for the Yankee Rowe fuel rod, and then it was assumed typical values for PWR. The same was considered for grain diameter, open porosity and plenum spring characteristics.

The simulation to assess the behavior of the Yankee Rowe fuel rod was carried out considering the information related with the rod E6-C-f6 as described in reference [12]. The selected rod target of the present simulation was irradiated into three core cycles identified as Core I, Core II and Core IV. Boundary conditions and axial power profile have been derived from reference [12]. Noticeably the axial power profile has been derived considering the average core power reported in Table 2. Data related with E6-C-f6 fuel rod are available in four axial positions, which have been interpreted as average values along the related length. Thus, constant piecewise trend has been adopted into TRANSURANUS simulation (Fig. 2). The calculated peaking factors have been imposed both to the linear power and to the neutron flux. The resulting profile is bottom skewed for the cycles Core I and II where the power was controlled by control rods, rather in Core IV boron was introduced as chemical shim resulting in a flatter axial profile [13].

The irradiation period is consistent with the information available in reference [12], adding 24 h for the power rise, 12 h for the power decrease, in addition 48 h has been set as shutdown period between two core cycles.

Finally, an average neutron flux equal to $6.3 \times 10^{13}\,\mathrm{n\,cm^{-2}\,s^{-1}}$ has been set in order to achieve a fluence level close to the value available in the literature, i.e. $6.0 \times 10^{21}\,\mathrm{n\,cm^{-2}}$. Regarding the fuel-cladding contact model, the perfect slip model has been adopted.

3 Results and discussion

The results obtained from the Yankee Rowe fuel model are shown hereafter. Table 3, Table 4 and Table 5 list, respectively, the outcomes of the simulation at the end of the Core I, Core II and Core IV cycle, compared with available information taken from reference [12]. It should be noted that no tuning has been done for carrying out the simulation.

The parameters attaining to the Core I cycle are reasonably reproduced by the TRANSURANUS code (Tab. 3), noticeably the burnup matches fairly good in all four locations.

Regarding the fuel temperature calculated by the code, centerline and surface values are provided since for the reference data no specification about the radial position is provided. It can be seen that reference fuel temperature is within the code prediction for all the four axial positions.

The same considerations apply for the clad temperature (apart for the top position which is slightly underpredicted due to the underestimation of the coolant temperature) in relation with both reference data radial position and calculated values.

Additional calculated data are provided in Table 3 regarding fission gas release which remains very low; fuel and clad axial elongation, both are lower than 0.5%; maximum fluence value and plenum pressure which is double of its starting value.

Table 3. Yankee Rowe E6-C-f6: comparison between reference and calculated data at the end of Core I cycle.

Position* (cm)	Parameter	Reference [12] CORE I	TRANSURANUS	Note
17.02	Cumulative burnup (MWd tU^{-1})	8.19	8.2	
57.66		13.29	13.32	
138.94		12.53	12.55	
220.22		5.9	5.91	
17.02	Fuel temperature (°C)	515	654.6	Centerline
			496.2	Surface
57.66		612	887.7	Centerline
			589.3	Surface
138.94		621	866.6	Centerline
			588.8	Surface
220.22		482	575.9	Centerline
			468.9	Surface
17.02	Clad temperature (°C)	267	270.4	Inner
			262.0	Outer
57.66		278	285.4	Inner
			271.9	Outer
138.94		287	292.0	Inner
			279.4	Outer
220.22		285	282.2	Inner
			276.2	Outer
17.02	Coolant temperature (°C)	254	252.9	
57.66		258	257.2	
138.94		268	265.6	
220.22		275	269.8	
	Fission gas release (%)	-	0.07	
	Fuel axial elongation (%)	-	0.39	
	Clad axial elongation (%)	-	0.43	
	Gap size (μm)	-	26.8/41.02	Min/Max value
	Fluence (n/cm^2)	-	2.4e21	Max value
	Plenum pressure (MPa)	-	0.22	

* Position from the bottom to the top of the fuel rod.

Table 4. Yankee Rowe E6-C-f6: comparison between reference and calculated data at the end of Core II cycle.

Position* (cm)	Parameter	Reference [12] CORE II	TRANSURANUS	Note
17.02	Cumulative burnup (MWd tU^{-1})	14.54	14.57	
57.66		23.24	23.31	
138.94		22.78	22.85	
220.22		10.23	10.25	
17.02	Fuel temperature (°C)	523	662.6	Centerline
			486.8	Surface
57.66		616	856.9	Centerline
			540.4	Surface
138.94		642	882.9	Centerline
			553.3	Surface
220.22		482	575.1	Centerline
			464.8	Surface
17.02	Clad temperature (°C)	267	275.2	Inner
			266.4	Outer
57.66		279	289.8	Inner
			276.2	Outer
138.94		290	299.6	Inner
			285.7	Outer
220.22		286	286.9	Inner
			281.0	Outer
17.02	Coolant temperature (°C)	254	256.9	
57.66		258	261.2	
138.94		268	270.5	
220.22		276	274.6	
	Fission gas release (%)	-	0.12	
	Fuel axial elongation (%)	-	0.52	
	Clad axial elongation (%)	-	0.44	
	Gap size (μm)	-	19.8/37.6	Min/Max value
	Fluence (n/cm^2)	-	4.4e21	Max value
	Plenum pressure (MPa)	-	0.25	

* Position from the bottom to the top of the fuel rod.

Table 5. Yankee Rowe E6-C-f6: comparison between reference and calculated data at the end of Core IV cycle.

Position* (cm)	Parameter	Reference [12] CORE IV	TRANSURANUS	Note
17.02	Cumulative burnup	20.19	20.25	
57.66	(MWd tU^{-1})	31.33	31.45	
138.94		30.39	30.50	
220.22		15.95	15.99	
17.02	Fuel temperature (°C)	504	606.4	Centerline
			449.2	Surface
57.66		574	716.7	Centerline
			462.8	Surface
138.94		575	702.3	Centerline
			466.0	Surface
220.22		525	641.6	Centerline
			484.5	Surface
17.02	Clad temperature (°C)	270	272.4	Inner
			264.7	Outer
57.66		277	282.4	Inner
			271.4	Outer
138.94		284	287.3	Inner
			277.0	Outer
220.22		288	286.8	Inner
			279.0	Outer
17.02	Coolant temperature (°C)	258	256.7	
57.66		261	260.0	
138.94		267	266.3	
220.22		276	271.1	
	Fission gas release (%)	-	0.17	
	Fuel axial elongation (%)	-	0.58	
	Clad axial elongation (%)	-	0.43	
	Gap size (μm)	-	16.4/32.6	Min/Max value
	Fluence (n/cm^2)	-	5.9e21	Max value
	Plenum pressure (MPa)	-	0.26	

* Position from the bottom to the top of the fuel rod.

Table 4 compares reference and calculated data related with the Core II cycle. Also for this irradiation step the code gives reasonable results, showing the same (as in Core I) good compliance regarding the burnup data.

Calculated values of fuel and clad temperature include the corresponding reference data. Notwithstanding the accumulation of the burnup, the fission gas release is still low (0.12%); fuel and clad axial elongation do not change so much from the previous cycle (both slightly increased); the gap is reducing but still open; the plenum pressure is slightly increased from the previous cycle.

Table 5 reports the comparison discussed above but at the end of the Core IV cycle. Also at this stage of the simulation, the code shows the same capabilities in relation with the burnup, fuel and clad temperature. Coolant temperature is also reasonably predicted as well.

At the end of the whole simulation, the fission gas release is below 0.2%; fuel and clad elongation are well below 1%; the gap kept open with a minimum value of about 16 μm (about 1/4 of its initial value) and the plenum pressure is less than the triple of its initial value.

In relation with the fuel and clad relative elongation it can be seen that the code is able to reproduce one of the objective of the particular Yankee Rowe rod design, namely to minimize the differential thermal expansion between fuel and clad.

In general, the TRANSURANUS code performed reasonably well even facing with a rod design which is quite far from the typical (current) PWR technology (e.g. clad material, filling gas type, lack of gap pressurization, presence of different segments within the fuel rod). Any predicted parameters for the simulated fuel rod are of no concern regarding their corresponding design data.

4 Conclusion

The assessment of the modified TRANSURANUS code benefits of the availability in the open literature of data related with Yankee Rowe NPP, which was one of the few plants in which AISI 348 has been used as cladding material. A specific Yankee Rowe fuel model has been set up, fully considering the available information and doing some assumptions for covering some lacks (e.g. fuel rod upper plenum height). When such assumptions had to be made, conservative values have been adopted (considering Yankee Rowe and typical PWR rod design).

The carried out calculations show reasonably agreement with available data confirming the modified code capabilities. This constitutes an indication of the modified TRANSURANUS code capabilities to perform fuel rod investigation of fuel rod manufactured with AISI 348 cladding material.

The authors are grateful to the technical support of USP, IPEN-CNEN/SP and to the financial support of IAEA to attend the TopFuel 2015 meeting.

References

1. S.M. Stoller Corporation, An evaluation of stainless steel cladding for use in current design LWRs, NP-2642, EPRI, 1982
2. V. Pasupathi, Investigations of stainless steel clad fuel rod failures and fuel performance in the Connecticut Yankee Reactor, EPRI 2119, 1981
3. N. Akiyama et al., *The Fukushima nuclear accident and crisis management-Lessons for Japan–US Alliance Cooperation* (Sasakawa Peace Foundation, Tokyo, 2012)
4. K.A. Terrani, S.J. Zinkle, L.L. Snead, Advanced oxidation-resistant iron-based alloys for LWR fuel cladding, J. Nucl. Mater. **448**, 420 (2014)
5. K. Lassman, TRANSURANUS: a fuel rod analysis code ready for use, J. Nucl. Mater. **188**, 295 (1992)
6. European Commission, Joint Research Centre, Institute for Transuranium Elements, TRANSURANUS HANDBOOK, Document Number Version 1 Modification 1 Year 2012 ('V1M1J12'), 2012
7. D.L. Hagrman, G.A. Reymann, *MATPRO - version 11. A handbook of materials properties for use in the analysis of light water reactor fuel rod behavior* (Idaho National Engineering Lab, Idaho Falls, USA, 1979)
8. D. Peckner, I.M. Bernstein, *Handbook of stainless steels* (MacGraw Hill, New York, 1977)
9. Sandvik "Sandvik 8R40 data sheet" updated 20131128, http://www.smt.sandvik.com/en/
10. http://www.world-nuclear.org, accessed October 22, 2014
11. http://www.yankeerowe.com/pdf/Yankee%20Rowe.pdf, accessed October 22, 2014
12. Burnup Credit — Contribution to the Analysis of the Yankee Rowe Radiochemical Assays, 1022910, EPRI, 2011
13. R.J. Nodvik et al., *Supplementary Report on Evaluation of Mass Spectrometric and Radiochemical Analyses of Yankee Core I Spent Fuel, Including Isotopes of Elements Thorium Through Curium, TID 4500* (Westinghouse Electric Corporation, Pittsburgh, Pennsylvania, 1969)

21

Storage of thermal reactor fuels – Implications for the back end of the fuel cycle in the UK

David Hambley[*]

National Nuclear Laboratory, NNL Central Laboratory, B170, Sellafield, Seascale, Cumbria, CA20 1PG, UK

Abstract. Fuel from UK's Advanced Gas-Cooled Reactors (AGRs) is being reprocessed, however reprocessing will cease in 2018 and the strategy for fuel that has not been reprocessed is for it to be placed into wet storage until it can be consigned to a geological disposal facility in around 2080. Although reprocessing of LWR fuel has been undertaken in the UK, and this option is not precluded for current and future LWRs, all utilities planning to operate LWRs are intending to use At-Reactor storage pending geological disposal. This strategy will result in a substantial change in the management of spent fuel that could affect the back end of the fuel cycle for over a century. This paper presents potential fuel storage scenarios for two options: the current nuclear power replacement strategy, which will see 16 GWe of new capacity installed by 2030 and a median strategy, intended to ensure implementation of the UK's carbon reduction target, involving 48 GWe of nuclear capacity installed by 2040. The potential scale, distribution and timing of fuel storage and disposal operations have been assessed and changes to the current industrial activity are highlighted to indicate potential effects on public acceptance of back end activities.

1 Introduction

Spent fuel from the UK's first (Magnox) and second (Advanced Gas Reactor, AGR) generation power reactors has been reprocessed since the reactors came into service, in line with the UK government's position that spent fuel represents an asset.

In 2006, the Nuclear Decommissioning Authority (NDA) was formed to manage the decommissioning of the UK's historic civil nuclear legacy sites, particularly the research sites and the first generation power reactors, which were still in government control, and the reprocessing plants at Sellafield.

There are plans to build new nuclear generating capacity in the UK. The initial phase is expected to add around 16 GWe of capacity by 2030. Decarbonisation of energy use in the UK may require additional nuclear generating capacity, for which a mid-term nuclear capacity of 48 GWe has been proposed.

This paper describes the current industry structure and responsibilities for spent fuel management as a background for a more detailed description of the likely scale of spent fuel storage requirements over the coming century. This will lead to a consideration of options for optimisation of storage-related activities and to an evaluation of the potential impact of these changes on public perception of nuclear power generation in the UK.

2 Current spent fuel management

The UK has three groups of power reactor spent fuel, described below, as well as around 300 te of irradiated non-standard or experimental fuel. The experimental fuels are not considered in detail here, because they are included in the NDA's decommissioning programme [1] and the focus of this paper is on the larger spent fuel inventories from power reactors.

UK government's policy is that spent fuel management is a matter for the commercial judgment of its owners, subject to meeting the necessary regulatory requirements. The owners of current power reactors are: NDA in respect of Magnox and remaining shutdown experimental reactors and EDF in relation to AGRs and the Sizewell B Pressurised Water Reactor (PWR).

All the first generation Magnox power reactors have shutdown, with the last operational station, Wylfa, closing in December 2015. Magnox fuel is metallic, consisting of uranium metal bar tightly enclosed within magnesium-aluminium alloy cladding. Spent Magnox fuel is reprocessed

---

[*] email: `david.i.hambley@nnl.co.uk`

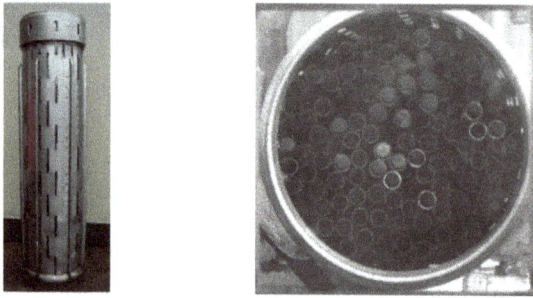

Fig. 1. AGR slotted storage can.

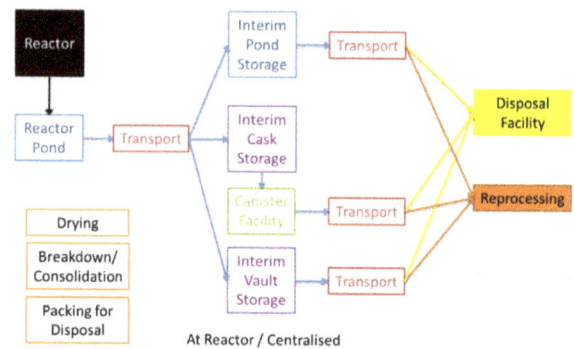

Fig. 2. A spent fuel storage system.

and all remaining fuel from the UK's Magnox reactors will be reprocessed at Sellafield in the Magnox Reprocessing Plant [2].

The liability for decommissioning of the Magnox reactors (and operation of Wylfa until its closure) lies with the NDA who employs a site management company to operate the sites. The NDA has established a Magnox Operating Plan to manage the discharge of fuel from shutdown reactors, transport of fuel to Sellafield, interim storage and reprocessing, so as to minimise risk to spent fuel in storage and ensure maximum utilization of the reprocessing plant.

AGR fuel is discharged from the reactors and held in temporary storage in reactor coolant until the fuel can be separated into individual elements, after which it is stored in station ponds. On-site pond storage continues until the fuel can be shipped to centralised interim storage at Sellafield, typically between 90 and 180 days. Fuel is stored for a further period until it is dismantled and the fuel pins are consolidated for storage (Fig. 1), reducing storage volume by about $1/3$.

AGR fuel is stored in caustic dose ponds until required for reprocessing. The fuel is then transported to the reprocessing facility pond, which also holds LWR fuel from international customers.

EDF operates the AGR reactors and has contracts with NDA for reprocessing of fuel. NDA is responsible for the operation of the Sellafield site and reprocessing operations. In 2012, recognising that THORP was approaching the end of its existing reprocessing contracts and certain parts of the infrastructure supporting the reprocessing plant was ageing, NDA completed a review of options for future management of AGR fuel [3]. This review concluded that reprocessing of AGR would cease when existing reprocessing contracts were completed in around 2018 and the remaining AGR fuel and fuel discharged subsequently from AGR reactors would be placed into wet storage using existing storage facilities pending geological disposal. This option did not preclude future reprocessing or a change to alternative storage options and provided storage conditions that would enable monitoring of fuel during storage.

Sizewell B is the only Light Water Reactor (LWR) power station in the UK and is owned by EDF. EDF has contracts that would provide an option to reprocess the fuel at Sellafield but has chosen to store the spent fuel from the reactor on-site, pending disposal. To increase storage capacity at the site, EDF has decided to use dry cask storage, which is currently undergoing licensing approval [4].

The UK government recognises nuclear power as a low carbon energy source and is considering pathways that could deliver up to 75 GW installed nuclear capacity by ~2050. The immediate programme is for 16 GWe capacity to be installed by the early 2030s to replace the capacity of the current fleet. The mid-range forecast is consistent with decarbionisation of transport infrastructure and would see installed nuclear capacity of around 48 GWe by 2050. The option for a future transition to a closed fuel cycle remains open [5].

Unlike earlier power stations, the new generation power stations under development have been justified on the basis of on-site interim storage of spent fuel followed by geological disposal. Alternative spent fuel management options, such as centralised storage or reprocessing, are not precluded, however they would have to be justified prior to implementation.

As with Sizewell B, the owners of the new power stations will be responsible for the management of the spent fuel from discharge until disposal.

The UK has long had a strategy to dispose of intermediate and high levels radioactive wastes in a deep geological disposal facility (GDF) [6]. Since 2010 the inventory of the GDF has been expanded to include spent fuels remaining at the end of reprocessing, future arisings of AGR fuel, spent fuel from Sizewell B and spent fuel from the first tranche of new build reactors (i.e. 16 GWe capacity) [7,8].

In this evaluation of spent fuel storage options, it is the above industry structure and policy framework that is considered.

3 Input data and assumptions

The inventory of spent fuel that could be produced by new build reactors has previously been presented [9] based on use of the Orion fuel cycle modelling code, which was developed to model potential advanced fuel cycles. The model did not provide any detailed modelling of options involving storage and transition to disposal. A conceptual model of storage operations between fuel discharge from reactor and emplacement in a GDF is shown in Figure 2.

This study provided a preliminary assessment of such options using simplified assumptions about the generation of spent fuel. It focusses on the accumulation of spent fuel

Table 1. Operating assumptions for existing reactors.

Reactor	Power [GWe]	Spent fuel [teU/year]	End date
Hunterston B	0.96	26	2031
Hinkley point B	0.95	26	2031
Dungeness B	1.05	29	2031
Heysham A	1.26	34	2027
Hartlepool	1.18	32	2027
Heysham B	1.23	33	2031
Torness	1.19	32	2031
Sizewell	1.20	29	2055

Table 2. Operating assumptions for new build 16 GWe reactors.

Reactor	Start date	Ref.
EPR-1 (Hinkley C)	2023	[17]
EPR-2	2025	[est.]
AP1000-1	2024	[19]
ABWR-1 & 2	2025	[18]
AP1000-1	2025	[19]
AP1000-3	2026	[19]
EPR-3 & 4	2028	[est.]
ABWR-3 & 4	2029	[est.]

est.: the data was estimated, in the absence of declared operational dates, so that total installed capacity met government planning assumptions [5].

that results from the opposing effects of discharges from reactor and emplacements in a GDF. For this study, the location of spent fuel storage facilities (e.g. reactor ponds or centralised storage) and storage options (e.g. dry storage casks, dry vaults or ponds) are not resolved. Estimates of the number of fuel shipments to disposal facilities have been made as these are useful in conceptualising potential mass flows and because they represent a real potential impact on host communities.

3.1 Existing reactor fleet

Discharges of AGR fuel are based on current nominal power output of AGR reactors and declared decommissioning plans of the operator [10–12]. The operating assumptions presented in Table 1 represent those likely to result in the worst case (largest) spent fuel quantities.

At the end of reprocessing it is estimated that around 1,500 teU of AGR fuel will remain in interim storage [13]. Although there are many differences in reactor design, as indicated by nominal power generation, an average core inventory of 246 teU [14] has been assumed, giving a total fuel inventory at the end of generation of just under 6,000 teU.

Sizewell B has accumulated around 600 teU in pond storage [15]. It is assumed that Sizewell life extension will be in line with the generators declared anticipation, 20 years [10]. With a core inventory of 89 teU [16], the end of life spent fuel inventory will be around 1,615 teU.

3.2 New build reactors

For new build reactors exact plans for delivering 16 GWe of capacity have not been declared in detail, however publically available information has been used in conjunction with government planning assumptions [5], of ~16 GWe capacity by ~2030, to yield the modelling assumptions in Table 2.

For a higher generation target, additional capacity is assumed to come on line at an approximately constant rate between 2030 and 2050. It is assumed that the proportion of different reactor types is as shown in Table 2.

Core inventory data and rates of spent fuel generation have been obtained from data in the Generic Design Approval submissions (Tab. 3). For all new build reactors, it is assumed that the nominal fuel burn-up is 55 GWd/teU and that the reactors will operate for the design life of 60 years. Assessments have also been made for higher burn-up (65 GWd/teU) and for extended operation (80 years).

3.3 Disposal

In order to model spent fuel management strategies and options it is vitally important to understand the parameters controlling transition of fuel from storage to either reprocessing or disposal. For the purposes of this study, only disposal options have been examined, since the Orion already provides sophisticated modelling of closed fuel cycles.

Radioactive Waste Management, a subsidiary of the NDA, has carried out a number of assessments of the disposability of UK spent fuels. Where the GDF geology has a significant impact on disposability parameters, results for the most restrictive geology have been used. For spent fuel, the reference case is the KBS-3V concept in a granitic geology. Important parameters are listed below:

– LWR fuel assemblies per canister: 4 [26];
– AGR fuel canisters per canister: 16 [27];
– minimum cooling time for new build reference fuel (65 GWD/teU burn-up): 140 years [26];
– initial spent fuel receipts in GDF: 2075;
– maximum throughput of GDF: 650 teU/year [26].

Changes in fuel burn-up affect the minimum cooling time at which fuel can be accepted into the GDF. For this assessment, this has been approximated by finding the cooling time at which the heat output of higher burn-up fuel equals that of the reference fuel at the reference cooling time. This implicitly assumes that both fuels follow similar time-dependence of heat output and that the heat output at the time of the peak repository temperature is adequately estimated by this approximation. For MOX fuel, this approximation is unlikely to hold, hence MOX fuel is not considered here.

Table 3. Data for spent fuel assessment for new build reactors.

Reactor	Power [GWe]	Core Inv. [teU]	Spent Fuel [teU/year]
EPR-	1.65	127 [20]	28 [21]
ABWR	1.35	154 [22,23]	26 [22]
AP1000	1.12	85 [24,25]	22 [24]

Specific heat outputs have been calculated for a reference PWR fuel obtained using NNL's FISPIN inventory code. Differences between the three LWR reactor types are considered to be sufficiently small that this approximation will not significantly affect the overall pattern of spent fuel inventories.

In order to provide an initial estimate of required transport operations, the number of spent fuel shipments has been estimated using the maximum flask capacity that can be accommodated on current UK transport infrastructure, which is 12 LWR fuel assemblies [26]. For AGR fuel, it is assumed for current purposes that transport packages similar to current designs would be used, with an inventory of 12 consolidated fuel cans. If fuel were to be loaded into disposal containers at a site other than the GDF, new transport flasks would need to be designed and maximum inventories may change, however the base line assumptions are considered adequate for current assessments, since the controlling factor is GDA throughput.

4 Inventory profiles for reference disposal parameters

Spent fuel inventories have been calculated in order to indicate the scale (total quantity) and duration of spent fuel storage requirements for the most likely range of medium term deployment of nuclear power in the UK. For this work, medium-term deployment is taken to be capacity installed by around 2050 and likely range of deployments is taken to be between 16 GWe and 48 GWe of new build capacity.

Existing reactors are assumed to operate to best estimates of maximum operating lives with no significant changes in reactor performance characteristics. As indicated earlier, new build reactors are assumed to irradiate fuel to either 55 GWd/teU (reference) or 65 GWd/teU (high burn-up). Where a higher fuel burn-up is assumed, spent fuel discharges are assumed to be reduced in proportion to the increase in discharge burn-up. Reactors are assumed to operate for either 60 or 80 years at nominal power.

4.1 Generation profiles

The generation and spent fuel discharge profiles for two cases are presented:

– a lower reference case: the current reactor plus a 16 GWe new build programme operating for 60 years at nominal burn-up;

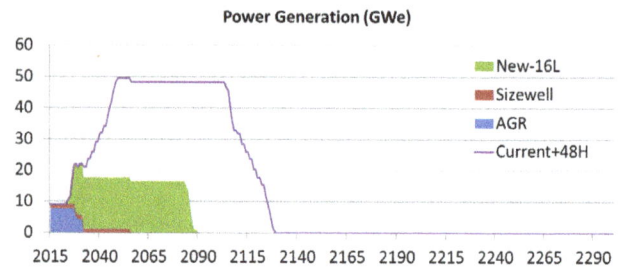

Fig. 3. Generation profiles for lower and higher reference cases.

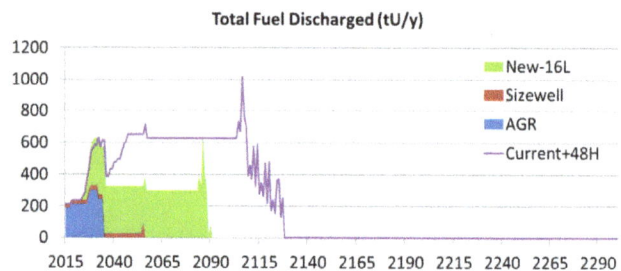

Fig. 4. Spent fuel discharges for lower and higher reference cases.

– a higher reference case: the current reactor fleet plus 48 GWe of new build capacity, reactor life extension of 20 years and a higher average burn-up;

– profiles are shown in Figure 3 for generation and Figure 4 for spent fuel discharges. The peaks in spent fuel discharges represent final core discharges.

It is immediately apparent from Figure 4 that the spent fuel discharges from this larger fleet are at times greater than the reference acceptance capacity for the disposal site (650 teU/y) and would be continuously greater if the larger fleet was operated at the nominal burn-up rather than the increased one, as this would generate an additional 115 teU/year. The current GDF spent fuel inventory includes only fuel from a 16 GWe new build programme [9], therefore additional capacity would have to be provided for a larger programme. Any further GDF development at the same or a different site should clearly have a greater capacity to receive fuel than the current reference design.

4.2 Storage profile low reference case

The profile of fuel in storage for this case is shown in Figure 5.

The storage requirement for AGR fuel is provided by existing assets, which are expected to operate until all the AGR fuel is discharged. The requirements for fuel storage at Sizewell B is assumed to be met using the current strategy; reactor storage pond with additional dry cask storage capacity as required to maintain generation. At the end of generation, it is likely that the spent fuel pool inventory would be moved to a long-term storage system to allow reactor decommissioning.

For fuel from new build reactors, long-term storage will be required from around 2035, rising to around 20,000 teU in 2090 and remaining at this level for at least 40 years

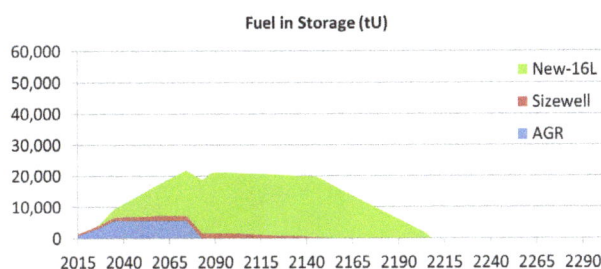

Fig. 5. Spent fuel storage requirements for low reference case.

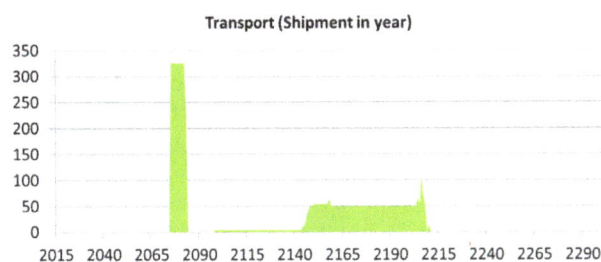

Fig. 6. Fuel transport requirements for low reference case.

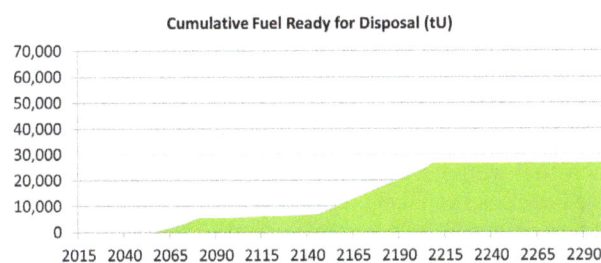

Fig. 7. Spent fuel availability for disposal for low reference case.

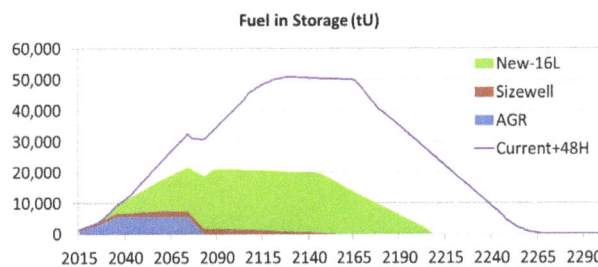

Fig. 8. Spent fuel storage requirements for high reference case.

before shipments to disposal can begin. Whilst cask storage systems allow for incremental increases in storage capacity, for this scenario even large scale pond storage capacity could be added incrementally at 10–15 years intervals to match storage requirements, providing economies of scale (by implication from Ref. [21]) and greater flexibility in fuel management.

Figure 6 shows the pattern of transport flask shipments. The initial 'spike' is caused by the build-up of AGR fuel available for disposal prior to the anticipated date at which spent fuel can be placed in the GDF (Fig. 7). In practice this would be smoothed out to remove very high throughput requirements at the storage facility. Overall, the system would have to be able to deliver around 150 cuboid flask shipments per year between the storage and disposal sites.

It is also worth noting that the UK has had a history of routine spent fuel shipments from power stations to centralised storage facilities, which will cease a few years after the AGR stations cease generation if fuel storage at reactor sites is adopted at new build sites. Fuel shipments to a GDF would then have to be restarted after a period of around 30–40 years with little or no spent fuel shipment activity. This would require significant mobilisation activities to develop and approve transport packages suitable for AGR fuel, increase regulatory capacity for

higher workloads and to train and qualify a workforce with no experience of routine transports. Introduction of a potential hazardous activity in the public domain may be a cause of public concern, which could cause delays, or worse, to shipment programmes.

Post-AGR shipments there would be a small interval of around 15 years before routine shipments of LWR fuel from Sizewell B to the disposal facility. These would continue at a very low rate (around 4/year) for around 35 years before increasing to around 50/year as the fuel from new build reactors became ready for disposal. Although it is likely that different flasks may be required for LWR fuel, to AGR, the short gap between end of AGR and start of LWR shipments would make it more likely that a cadre of experienced staff would be maintained and restart activities would be less onerous.

4.3 Storage profile high reference case with 20-year reactor life extension

The profile of fuel in storage for this case is shown in Figure 8.

The storage requirement for this case is dominated by the much larger LWR fleet. In this scenario, there is no plateau in spent fuel in storage because disposal of LWR fuel from Sizewell B overlaps with the end of generation of the larger new build fleet.

For fuel from new build reactors, long-term storage will be required from around 2035, rising to around 72,000 teU in 2130. This peak is, however, transient, as fuel begins to be shipped to the GDF within few years. Given that fuel would spend some time in rector ponds prior to export to long-term storage (typically 5–10 years), the peak interim storage inventory could be somewhat lower if a suitably long-lived reusable storage facility is used. However, because this overlap occurs at the margins towards the end of generation and start of disposals, this is unlikely to have a significant effect (i.e. more than a few hundred teU) on peak storage requirements.

Figure 9 shows the pattern of transport flask shipments. In this scenario, the transfer of new build LWR fuels rises steadily in response to the generation profile (Fig. 9) but extends for much longer than expected. In this case, the disposal period is extended, as can be seen by comparison with Figure 10. This extension occurs because the peak rate of spent fuel generation exceeds the maximum declared GDF reception rate (as noted earlier). With adequate throughput at the disposal sites, the export period would be

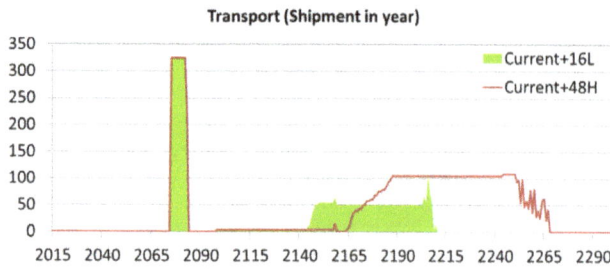

Fig. 9. Fuel transport requirements for high reference case.

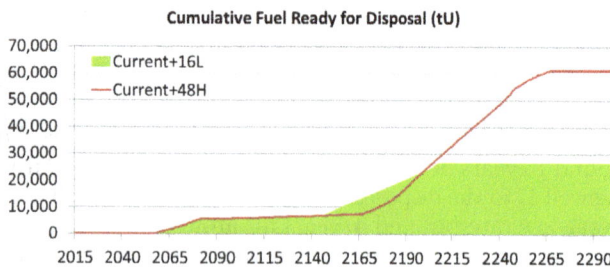

Fig. 10. Spent fuel availability for disposal for high reference case.

shortened to the point at which the cumulative curve in Figure 10 levels off. In this scenario, this would shorten the required storage period by around 20 years.

5 Alternative disposal protocol

RWM have proposed that storage times could be reduced by retaining spent fuel in interim storage until fuel discharged half way through the reactor life is ready for disposal [26]. Thereafter, progressively shorter- and longer-cooled fuels would be loaded into disposal containers until the final disposal container would contain the earliest discharged fuel and the last discharged fuel.

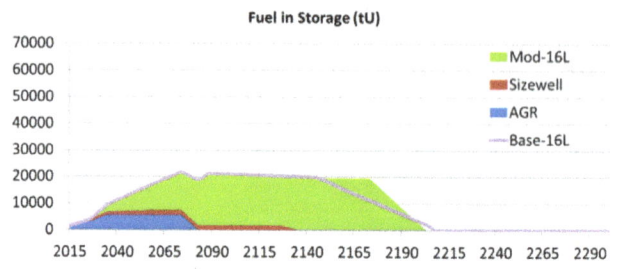

Fig. 11. Spent fuel storage requirements for low reference case and modified disposal protocol.

Table 4 provides a summary of the potential benefits of this approach using data for PWR fuel at two burn-ups and for two reactor operating lives, 60 years and 80 years. The adjusted minimum cooling period (t_2) is the cooling time at which two assemblies at that cooling time plus two assemblies at that cooling time plus the operational life of the reactor (t_r) would have the same heat output as four assemblies with the reference cooling time (t_1), which is approximately the time at which the fuel discharged half way through the plant's operating life would meet disposal requirements.

Using this approach, fuel disposals start at t_1 and end at ($\frac{1}{2} t_r + t_2$). The minimum time over which fuel can be disposed of is given by $\frac{1}{2} t_r - (t_1 - t_2)$. Table 4 indicates that reductions in fuel storage times of 25–30 years (or 18–27%) are possible. However, it is also clear that the later start of fuel exports leads to significant increases in the rate at which fuel would need to be exported in order to achieve these benefits. It is likely, therefore, that a realistic strategy will be a compromise between minimising storage times and maintaining realistic processing rates.

5.1 Storage profile for low reference case with modified disposal protocol

The profile of fuel in storage for this case is shown in Figure 11.

Table 4. Adjusted cooling times for disposals.

		PWR			
Burn-up	GWd/teU	55	65	55	65
Operation	Years	60	60	80	80
Mid-life	Years	30	30	40	40
Minimum cooling time	Years	120	140	120	140
Start of fuel disposal	Years	120	140	120	140
End of fuel disposal	Years	180	200	200	220
Adjusted minimum cooling time	Years	93	118	89	111
Start of fuel disposal	Years	150	170	160	180
End of fuel disposal	Years	153	178	169	191
Transfer period	Years	3	8	9	11
Shipping rate multiplier		12	18	8	9
Reduced storage time	Years	27	22	31	29

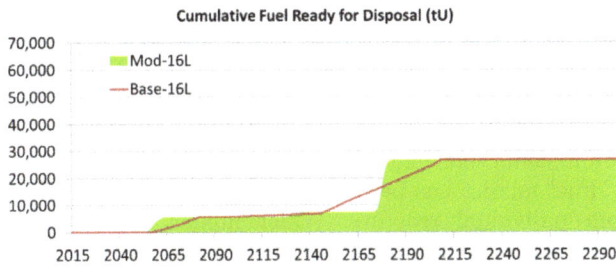

Fig. 12. Spent fuel availability for disposal for low reference case and modified disposal protocol.

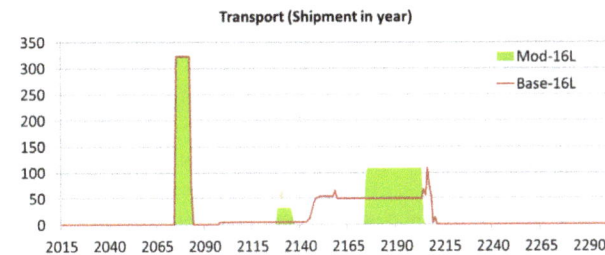

Fig. 13. Fuel transport requirements for low reference case and modified disposal protocol.

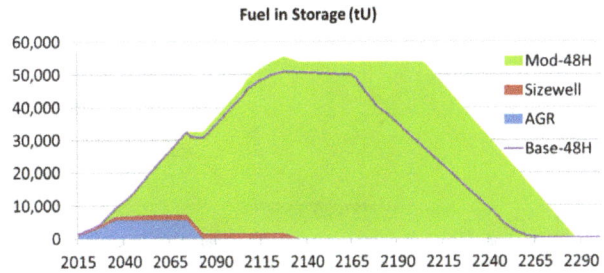

Fig. 14. Spent fuel storage requirements for high reference case and modified disposal protocol.

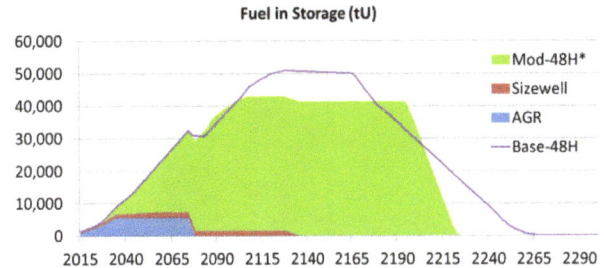

Fig. 15. Spent fuel storage requirements for high reference case, modified disposal protocol and increased DGF capacity.

Comparison with Figure 5 shows a small reduction in maximum inventory of around 900 teU using the modified protocol, due to a more rapid export of fuel from current generation reactors, and a reduction in storage time of 7 years. This time saving is less than might be expected from Table 4 because the rate at which spent fuel becomes available for disposal is much greater than for the original scenario (compare Fig. 12 and Fig. 7). This is also evidence in the shorter periods over which fuel shipments occur and the higher annual movements (compare Fig. 13 and Fig. 6).

Unlike the original scenario where there was a short interval between AGR shipments and Sizewell B shipments and an overlap between Sizewell B and new build fuel shipments, in this case there are three distinct periods of shipments, isolated by periods of 45 and 28 years with no shipments. In each case, these intervals are significant fractions of a person's working life and where At-Reactor storage is selected, the challenges discussed above in relation to re-starting transport activities would be replicated prior to each series of shipments. Even for centralised storage, there would be two periods, of around 30 years, during which no fuel shipments would be scheduled.

5.2 Storage profile for high reference case with modified disposal protocol

The profile of fuel in storage for this case is shown in Figure 14. This clearly shows increases both in stored inventory and duration of storage. It was noted in the original scenario (Sect. 4.3) that fuel exports were being constrained by the design basis capacity of the GDF to received spent fuel. In this scenario, the rate at which fuel is to be transferred is much higher and hence the effects of constrained export rates are more pronounced.

To obtain a more realistic comparison, a scenario involving a higher receipt capacity of 1,500 teU/y has been run to identify potential benefits of the modified disposal strategy.

In this case, the maximum quantity of fuel in storage remains the same, but the period of storage is significantly reduced, by around 65 years (compare Fig. 14 and Fig. 15). Compared to the original disposal protocol, there is a modest increase in maximum fuel inventory (around 5,000 teU) and a reduction in storage time of around 30 years.

Even in this scenario the rate at which fuel can be received at a GDF is still extending the storage period beyond that at which fuel meets disposal criteria (compare Fig. 16 and Fig. 17) by around 15 years. Theoretical estimates of the reduction of storage time that can be gained from fuel mixing are unlikely to be achievable in many scenarios due to practical limitations. However, there is good evidence that a mixing strategy can produce significant reductions in storage times, provided that storage and/or packaging facilities are designed to allow effective mixing of fuels of different ages.

The intervals between phases of fuel shipments are increased again in this scenario with three district phases of transport separated by ~30, ~40 and ~40 years. Thus the modified disposal protocol will have an unintended consequence of increasing the start-up requirements for each period of fuel shipments because the intervals are approaching those of a working lifetime.

For many storage systems, the ability to mix fuel of different ages would require more infrastructure than would be required for simply exporting fuel as it reaches the minimum cooling requirements. It is also clear for scenarios such as this that multiple repacking lines would be required to achieve the necessary shipment rates or that repacking would have to be undertaken over a long period to prepare

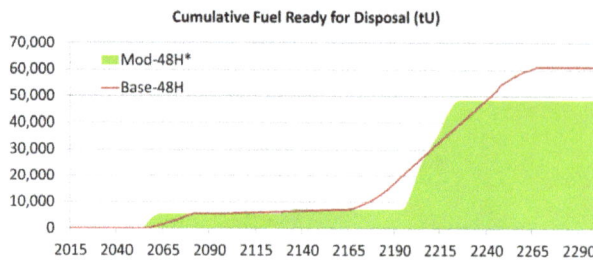

Fig. 16. Fuel transport requirements for high reference case, modified disposal protocol and increased GDF capacity.

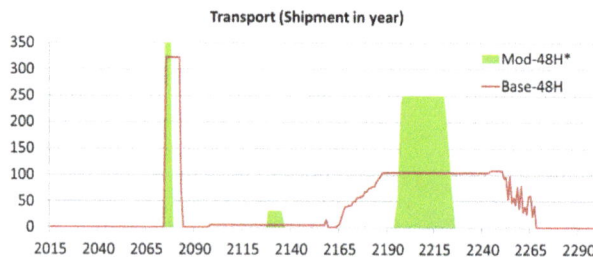

Fig. 17. Spent fuel availability for disposal high reference case, modified disposal protocol and increased DGF capacity.

fuel for shipping. This would, however, require additional storage facilities that may obviate any benefits from such a strategy.

For large consolidated storage facilities holding fuel from more than one reactor, the opportunities of further reductions in storage times may exist. However, it is highly likely that such benefits may be largely off-set by the constraints imposed by the maximum rate at which fuel can be exported.

Whilst it is clear that the maximum benefit that could be obtained from mixing of fuels of different ages is unlikely to be achievable, it suggests that in scenarios where a relatively small quantity of MOX fuel might be used in an LWR fleet, such strategies might also be effective in ameliorating to some extent the higher footprint of MOX fuel in a repository.

6 Conclusions

This assessment has identified potential scale and durations of spent fuel storage requirements faced by the UK for future nuclear power generation of 16 GWe by 2030 and 48 GWe by 2050.

The evaluation has identified the important influence of end point characteristics (e.g. disposal facility emplacement rates) on spent fuel storage and hence highlights the need for integrated planning for storage and either disposal or reprocessing.

The potentially long duration of spent fuel storage can lead to repeated occasions where transport operations have to be restarted after many decades of low or no activity. This represents a significant challenge for operators and regulators and may create points at which lack of familiarity could exacerbate public concern.

Mixing of fuels of different ages can lead to shorter storage times, in some cases by decades. The extent of the benefit will be constrained by the maximum rate at which fuel can be recovered from storage and processed through to emplacement. This option may lead to increased infrastructure requirements that may off-set some of the benefits.

Fuel mixing has been examined in the context of fuel from individual reactors. Wider mixing at centralised facilities has a potential for further benefits, particularly in relation to disposal of small quantities of MOX fuel.

This work was funded from the NNL's Strategic Research Programme on Spent Fuel Management and Disposal.

References

1. UK Department of Energy and Climate Change, Fourth National Report of Compliance with the Obligations of the Joint Convention on the Safety of Spent Fuel Management and on the Safety of Radioactive Waste Management, September 2011
2. Nuclear Decommissioning Authority, Fuel Strategy Position Paper, Magnox Fuel – Issue 1, July 2012
3. Nuclear Decommissioning Authority, Oxide Fuels - Preferred Option, SMS/TS/C2-OF/001/Preferred Option, June 2012
4. EDF, The Sizewell B Spent Fuel Management Option Study, 2010
5. UK HM Government, The Carbon Plan: Delivering our low carbon future, December 2011
6. Royal Commission on Environmental Pollution, "Nuclear Power and the Environment", Sixth Report of the Royal Commission on Environmental Pollution, Cm 6618, HMSO, 1976
7. Nuclear Decommissioning Authority, Geological Disposal - Steps towards implementation - Executive Summary, ISBN 978 1 84029 402 6, March 2010
8. UK Department of Energy and Climate Change, Implementing Geological Disposal, July 2014
9. Z. Hodgson, D.I. Hambley, R. Gregg, D.N. Ross, The United Kingdom's Changing Requirements for Spent Fuel Storage, in *Global 2013, Salt Lake City, USA* (2013)
10. EDF, EDF Energy Nuclear Generation: Our journey towards zero harm, May 2014
11. EDF, website: https://www.edfenergy.com/energy, 18 March 2014
12. Lake Acquisitions Limited, Life extension of Dungeness B power station, RNS Number: 6225C, January 2015
13. D.I. Hambley, Technical Basis for Extending Storage of the UK's Advanced Gas-Cooled Reactor Fuel, in *Global 2013, Salt Lake City, USA* (2013)
14. E. Nonbøl, Description of the Advanced Gas Cooled Type of Reactor (AGR), Risø National Laboratory Report NKS/RAK2(96)TR-C2, November 1996
15. Nuclear Decommissioning Authority, Packaging of Sizewell B Spent Fuel (Pre-Conceptual stage), Summary of Assessment Report, December 2011
16. E. Stokke, G. Meyer, Description of Sizewell B Nuclear Power Plant, Institutt for Energiteknikk (IFE) OECD Halden Reactor Project report NKS/RAK-2(97)TR-C4, September 1997
17. BBC News, UK nuclear power plant gets go-ahead, 21 October 2013

18. Horizon Power, Wylfa Newydd Project Pre-Application Consultation - Stage One Consultation Overview Document, September 2014

19. NuGen, Stage 1, Strategic Issues Consultation, May 2015

20. D.P. Blair, UK EPR PCSR – Sub-chapter 4.3 – Nuclear Design, UKEPR-0002-043, Issue 05, July 2012

21. T. Le Coutois, Interim storage facility for spent fuel assemblies coming from an EPR plant, EDF ELI0800224 A, November 2008

22. GE-Hitachi, UK ABWR Generic Design Assessment -Preliminary Safety Report on Spent Fuel Interim Storage, XE-GD-0155, Revision A, August 2014

23. GE-Hitachi, UK ABWR Generic Design Assessment -Generic PCSR Chapter 11: Reactor Core, UE-GD-0182 Rev A, August 2014

24. Westinghouse, AP1000 Pre-construction Safety Report, UKP-GW-GL-732 Rev 1, 2009

25. Nuclear Decommissioning Authority, Generic Design Assessment: Summary of Disposability Assessment for Wastes and Spent Fuel arising from Operation of the Westinghouse AP1000, NDA Technical Note No. 11261814, October 2009

26. Nuclear Decommissioning Authority, Geological Disposal, Feasibility Studies Exploring Options for Storage, Transport and Disposal of Spent Fuel from Potential New Nuclear Power Stations, NDA report NDA/RWMD/060/Rev 1, January 2014

27. Nuclear Decommissioning Authority, Packaging of Spent AGR Fuel (Preliminary stage), Summary of Assessment Report, April 2012

22

Multiobjective optimization for nuclear fleet evolution scenarios using COSI

David Freynet[1*], Christine Coquelet-Pascal[1], Romain Eschbach[1], Guillaume Krivtchik[1], and Elsa Merle-Lucotte[2]

[1] CEA, DEN, Cadarache, DER, SPRC, LECy, 13108 Saint-Paul-lez-Durance, France
[2] LPSC-IN2P3-CNRS, UJF, Grenoble INP, 53 rue des Martyrs, 38026 Grenoble, France

Abstract. The consequences of various fleet evolution options on material inventories and flux in fuel cycle and waste can be analysed by means of transition scenario studies. The COSI code is currently simulating chronologically scenarios whose parameters are fully defined by the user and is coupled with the CESAR depletion code. As the interactions among reactors and fuel cycle facilities can be complex, and the ways in which they may be configured are many, the development of optimization methodology could improve scenario studies. The optimization problem definition needs to list: (i) criteria (e.g. saving natural resources and minimizing waste production); (ii) variables (scenario parameters) related to reprocessing, reactor operation, installed power distribution, etc.; (iii) constraints making scenarios industrially feasible. The large number of scenario calculations needed to solve an optimization problem can be time-consuming and hardly achievable; therefore, it requires the shortening of the COSI computation time. Given that CESAR depletion calculations represent about 95% of this computation time, CESAR surrogate models have been developed and coupled with COSI. Different regression models are compared to estimate CESAR outputs: first- and second-order polynomial regressions, Gaussian process and artificial neural network. This paper is about a first optimization study of a transition scenario from the current French nuclear fleet to a Sodium Fast Reactors fleet as defined in the frame of the 2006 French Act for waste management. The present article deals with obtaining the optimal scenarios and validating the methodology implemented, i.e. the coupling between the simulation software COSI, depletion surrogate models and a genetic algorithm optimization method.

1 Introduction

1.1 Transition scenario studies

Nuclear systems composed of reactors with varied fuels and cycle facilities (enrichment, fabrication and reprocessing plants, interim and waste storages) are complex and in constant evolution. Transition scenario studies assist decision makers in listing the strengths and weaknesses of different strategies for a nuclear fleet evolution. These studies involve the tracking of the batches of materials and the evaluation of their depletion in the fuel cycle over a defined period.

COSI is a code developed by the CEA's Nuclear Energy Division and used to simulate the evolution of a nuclear reactor fleet and the associated fuel cycle facilities [1]. COSI takes as input parameters fuel cycle facilities and reactors features, fuel types characteristics and succession of

loadings. Front-end, back-end and waste paths define relations between these facilities as shown in Figure 1. It should be noted that reactors are defined by commissioning and shutdown dates, and reprocessing plants are defined by these dates, reprocessing capacities and strategy features. COSI provides outputs about the isotopic masses in the fuel cycle facilities and reactors over a defined period. Post processing calculations give access to physical quantities of interest: activity, radiotoxicity, decay heat, etc.

COSI is coupled with the CESAR depletion code, developed by the CEA's Nuclear Energy Division and AREVA, which performs every depletion (irradiation and cooling) calculation during the scenario simulation [2]. CESAR is the reference code used at La Hague reprocessing plant. Using CESAR requires one-group cross-sections libraries linked to fuel types loaded in the reactors. The production of these libraries requires neutronic calculations (APOLLO and ERANOS) and is separated from the depletion calculations. COSI coupled with the CESAR5.3 version is tracking 109 heavy nuclides (Tl→Cf) and 212 fission products (Zn→Ho).

---

* e-mail: david.freynet@cea.fr.

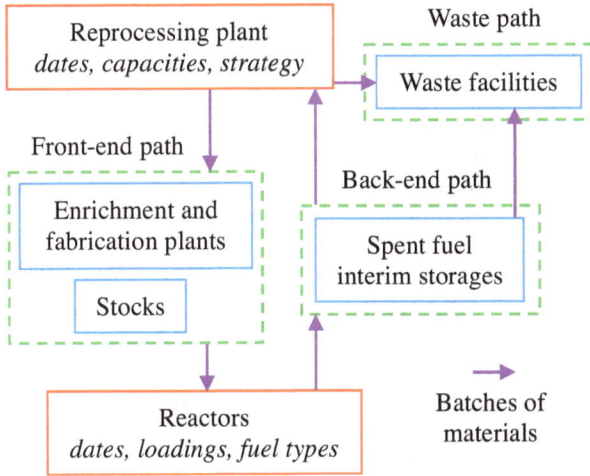

Fig. 1. COSI simplified data set operating diagram.

1.2 Multiobjective optimization

COSI is currently simulating chronologically scenarios whose parameters are fully defined by the user. The aim of this paper is to define a methodology for the automatic search of scenarios which are adapted to a strategic problem. Indeed the future French nuclear fleet should meet numerous and often conflicting criteria for different stakeholders such as saving natural resources and minimizing nuclear waste production. Such criteria have to be minimized or maximized according to some scenario parameters (COSI inputs).

Solving an optimization problem requires a large number of scenario calculations, which could be time-consuming and hardly achievable. Indeed this time can vary from a few minutes to a few hours according to the scenario assumptions and the number of isotopes tracked. Because CESAR calculations represent approximately 95% of the COSI computation time, depletion simplified models have been introduced to shorten depletion calculations during the scenario computation. Consequently, CESAR-based irradiation surrogate models are developed using the sensitivity and uncertainty platform URANIE developed by the CEA's Nuclear Energy Division [3].

Because of the large numbers of scenario parameters and criteria available to define an optimization problem, we opt to use metaheuristics as optimization methods. The URANIE's genetic algorithm (GA) is considered for the present optimization studies. Therefore, URANIE is used both for the surrogate models development and the optimization studies.

The methodology for performing multiobjective optimization using COSI is represented in Figure 2.

The development of CESAR surrogate models is discussed in Section 2. Then the COSI sped up version using these simplified models is validated by comparing its results to COSI, this study is also presented in Section 2. Finally, an application of this methodology for the optimization of a transition scenario from the current Pressurized Water Reactors (PWR) French nuclear fleet to a fleet of Sodium Fast Reactors (SFR) is presented in Section 3.

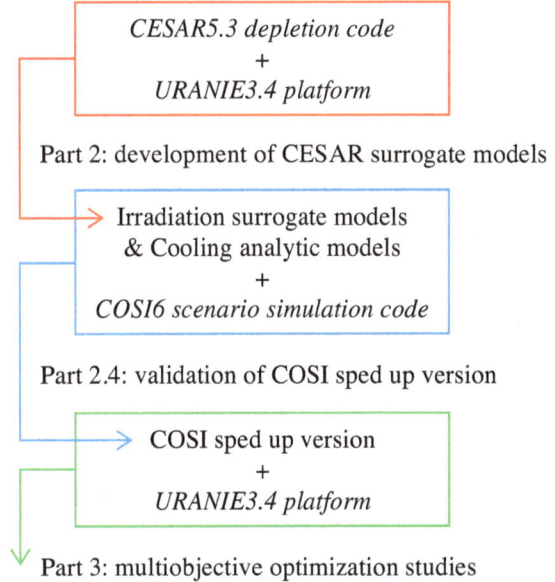

Fig. 2. Global multiobjective optimization methodology.

Other works address similar optimization problems using different simulation software such as VISION and CAFCA codes [4,5].

2 Irradiation surrogate models

2.1 Methodology

As seen previously, multiobjective optimization studies require the shortening of the COSI computation time and so the CESAR one. A way to gain time at cost to a satisfactory estimation error is developing CESAR surrogate models. These models can replace CESAR for irradiation calculations throughout the COSI computation and so have the same inputs and outputs as CESAR.

CESAR input parameters define the fuel assembly composition and irradiation features:

- the fresh fuel assembly isotopic composition defines the isotope (denoted i) mass fractions in the fuel noted $y_i = {m_i}/{m_{fuel}} (\sum_i y_i = 1)$;
- the burnup to achieve noted BU in MWd/tHM;
- the irradiation time noted Δt in days.

Thereafter, let $x = \{\forall i\ y_i; BU; \Delta t\}$ be the N-terms vector of CESAR input parameters.

CESAR outputs are the results of depletion calculation, i.e. the spent fuel isotopic composition. These outputs are calculated as final concentrations noted $C_j(x)$ where j denotes spent fuel isotopes in atoms/ton.

The development of irradiation surrogate models (see Fig. 3) consists first in defining designs of experiments of the CESAR input parameters and associated outputs. These designs are defined using Latin hypercube sampling method (LHS) because of its high space-filling performance. The number of x vectors defined for each design is set to 500. Then a regression model is applied to produce a surrogate model. Surrogate models are noted $\hat{C}_j$ as the functions

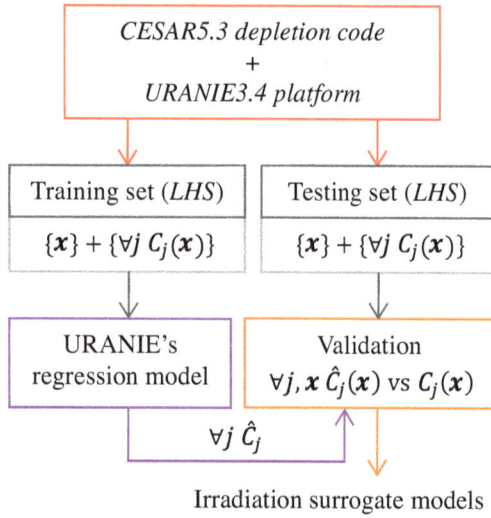

Irradiation surrogate models

Fig. 3. Surrogate models development methodology.

estimating the C_j CESAR results. Finally, quality indicators are performed on each surrogate model to ensure that the prediction power is satisfactory.

We make one surrogate model per tracked isotope per fuel type considered in the application scenario. For each fuel type, we make two designs of CESAR calculations: one for the regression step (named the training set) and another one for the validation step (named the testing set). All these operations are carried out with the URANIE platform.

The use of CESAR surrogate models coupled with the COSI code has already been introduced for uncertainty propagation studies in nuclear transition scenarios [6,7].

2.2 Regression models

CESAR surrogate models are developed using a regression method on the training set. The following methods are compared:

– first- (LR) and second-order (PR) polynomial regressions;
– Gaussian process (GP);
– artificial neural network (ANN).

Polynomial regression is a well-known approach to adjust a set of points by a function. Applied to CESAR calculations training set, the estimator is defined by equation (1) (LR) or equation (2) (PR):

$$\forall x \; \hat{C}_j(x) = \alpha_0 + \sum_{n=1}^{N} \alpha_n x_n, \qquad (1)$$

$$\forall x \; C_j(x) = \alpha_0 + \sum_{n=1}^{N} \alpha_n x_n + \sum_{p=1}^{N} \sum_{q=1}^{N} \alpha_{pq} x_p x_q. \qquad (2)$$

Polynomial regression consists in finding the α parameters giving the best model adjustment on the training set. CESAR surrogate models development with polynomial regression is detailed in a past work [6].

Gaussian process is a non-parametric regression method using a deterministic function and a correlation function involving parameters determined by maximum-likelihood estimation [8].

Artificial neural network is used in its single-layer perceptron form, i.e. there are no cycles and loops in the network and only one output neuron. Applied to CESAR calculations training set, the estimator is defined as:

$$\forall x \; \hat{C}_j(x) = \alpha_0 + \sum_{h=1}^{H} \alpha_h S\left(\alpha_{0h} + \sum_{n=1}^{N} \alpha_{nh} x_n\right), \qquad (3)$$

where $S(x) = {}^1\!/_{(1+\exp(-x))}$ is the sigmoid function and h denotes the hidden neuron. A backpropagation algorithm is applied to calculate the α weights by minimizing the estimation root mean square error. CESAR surrogate models development with ANN is also presented in another work [7].

2.3 Validation results

Surrogate models have been defined according to their use in optimization studies. Indeed the set of scenarios considered in this paper is extracted from the 2006 French Act for waste management which involves estimating PWR UOX, PWR MOX, PWR ERU and SFR MOX fuel types depletion. The validation step has to be applied to all of the surrogate models. Only the results of the $\hat{C}_{Pu239}$ and $\hat{C}_{Cm244}$ estimators for a PWR MOX irradiation are presented here, because of the importance of their accurate estimation and their non-linear evolution. Results shown in this part consider that GP deterministic function is linear, GP correlation function is Matérn 3/2 and the ANN number of hidden layers is 6.

Validating surrogate model rests upon the evaluation of indicators quantifying the quality of the regression and above all the estimator capacity to reckon the CESAR outputs. These indicators have to be representative of different estimation errors and are calculated using the testing set. Generally the predictivity coefficient q^2 acts as the main indicator for validating surrogate models [8]. Yet irradiation surrogate models are coupled with COSI which is repeatedly run during the optimization process. Thus, estimation error needs to be known to check that its impact is negligible on COSI outputs. For each testing x vector and surrogate model, let $\Delta_j(x)$ be the absolute estimation error divided by the mean of $C_j(x)$ on the testing vectors:

$$\Delta_j(x) = |\hat{C}_j(x) - C_j(x)|/\overline{C}_j. \qquad (4)$$

Calculating the mean and maximal values of this indicator on the testing set enables estimating the surrogate model quality. Replacing the denominator of equation (4) by $C_j(x)$, i.e. calculating the relative error, leads to high errors for low values of output concentrations. These cases are not significant for scenario studies because they are unnecessary to get a good estimation of the spent fuel

Table 1. Indicators of validation for PWR MOX ^{239}Pu and ^{244}Cm concentration estimations by surrogate models.

Regression method	$j=^{239}$Pu		$j=^{244}$Cm	
	Mean$_x\Delta_j$ (%)	Max$_x\Delta_j$ (%)	Mean$_x\Delta_j$ (%)	Max$_x\Delta_j$ (%)
LR	1.3	6.3	4.4	20
PR	0.093	0.69	0.71	3.1
GP	0.22	2.5	0.85	5.6

composition. Consequently, the definition given here is preferred. Error results are shown in Table 1.

This comparison study implies to consider ANN for all the CESAR surrogate models development.

2.4 Toward a COSI sped up version

Cooling calculation can be sped up using cooling surrogate models, but the analytic solutions of the Bateman equation with no flux can be calculated. Therefore, simplified cooling analytic solutions are implemented under COSI in addition to the irradiation surrogate models.

Besides, the list of isotopes tracked (321 isotopes with CESAR5.3) can be reduced in the COSI sped up version in order to further shorten the COSI calculation time. Both for irradiation and cooling calculations, output isotopes j are chosen among whom mostly contribute to the fuel mass and post-processing results. The following isotopes constitute more than 99.999% of the spent fuel actinide mass after irradiation and thus are estimated:

- ^{234}U, ^{235}U, ^{236}U, ^{238}U;
- ^{237}Np, ^{239}Np;
- ^{238}Pu, ^{239}Pu, ^{240}Pu, ^{241}Pu, ^{242}Pu;
- ^{241}Am, ^{242m}Am, ^{243}Am;
- ^{242}Cm, ^{243}Cm, ^{244}Cm, ^{245}Cm, ^{246}Cm.

Several fission products such as ^{90}Sr, ^{90}Y, ^{137}Cs and ^{137m}Ba complete the list to make possible estimating decay heat and radiotoxicity under long cooling period in waste. It is noteworthy that the choice of isotopes j depends on the COSI outputs taken into account for optimization studies.

COSI sped up version is validated for a scenario of SFR deployment studied in this frame [9,10]. The nuclear power distribution of this scenario is represented in Figure 4.

First, all the actinide masses in cycle are compared from 2010 to 2140. The results for the actinide elements are shown in Table 2.

Isotope estimation errors in cycle (waste excluded) are on the whole lower than 1.5% except 2.5% for ^{243}Cm estimation (present in low quantity). There is no transmutation in the application scenario so waste estimation errors are larger than cycle estimation errors: errors are lower than 3% except 4.5% for ^{239}Np (present in low quantity), ^{238}Pu and ^{240}Pu. Decay heat and radiotoxicity by ingestion for waste are calculated under long cooling period (from 1 to 10^4 years after 2140), estimation

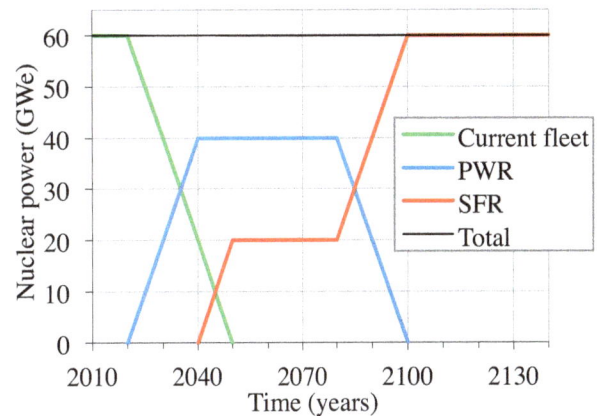

Fig. 4. Application scenario nuclear power distribution for validating surrogate models.

errors are no larger than 4%. Finally, the number of High Level Waste (HLW) packages cumulated at the end of the scenario is estimated with an error of 1.2%. These results are considered satisfactory enough to use COSI sped up version for optimization studies.

There are two types of COSI computation:

- standard: main depletion calculations at each date of interest (loading and unloading fuel dates, etc.);
- advanced: standard simulation plus additional depletion calculations; the advanced simulation considers the calculation of all the inventories in cycle for each year.

Computation time saving using COSI sped up version for the application scenario simulation is shown in Table 3. It should be mentioned that COSI sped up version calculations are multi-threaded.

Table 2. Maximal relative errors for the actinide mass estimations with COSI sped up version for the application scenario simulation.

Element	In cycle (%) (waste excluded)	In waste (%)
Pu	0.51	3.1
Np	1.5	2.5
Am	0.95	2.8
Cm	0.68	2.0

Table 3. COSI computation time decomposition for the application scenario simulation.

COSI version	Standard	Advanced
COSI/CESAR5.3	4622 s	46,791 s
Sped up	38 s	65 s
Speedup	×122	×720

An optimization calculation is then feasible using COSI sped up version because of the good surrogate models precision and the resulting time savings.

3 Optimization exercise

3.1 Optimization problem definition

Determining the best set of scenario parameters for a given problem requires that we define criteria, constraints and a base scenario with variables. In order to define this base scenario, it is necessary to make assumptions about the nuclear fleet evolution.

In the frame of a first application of the methodology, it is supposed that:

- SFR deployment is possible from 2040;
- all the reactors deployed from 2020 have a life span of 60 years;
- the nuclear fleet power equals to 60 GWe from 2010 to 2140 to maintain a constant nuclear energy production;
- the current fleet phases out from 2020 to 2050 at the pace of –2 GWe/year;
- there is no MOX fuel loaded in EPR^{TM} from 2020, which is a simplification for the current study.

These assumptions have as consequences:

- the nuclear power distribution of the base scenario cannot be changed from 2010 to 2040; the current PWR fleet (UOX and MOX fuels) is partially renewed with EPR^{TM} (only UOX fuel) from 2020 to 2040;
- the paces of reactors deployment and shutdown are respectively set to 2 and –2 GWe/year;
- there are two phases where reactors can be deployed from 2040: from 2040 to 2050 noted phase 1 and from 2080 to 2110 noted phase 2.

We also consider that EPR^{TM} are deployed before SFR in a same phase of reactors deployment. An example of this base scenario is shown in Figure 5 with respectively 7 and 27 SFR deployed during the phases 1 and 2.

Two types of reactors can be deployed: EPR^{TM} (UOX fuel) and SFR with their characteristics listed in Table 4. During the phase 1, 14 reactors need to be deployed to keep a nuclear power of 60 GWe. During the phase 2, 40 reactors have to be deployed to renew the nuclear fleet. Let $N_1 \in [0,14]$ (resp. $N_2 \in [0,40]$) be the number of SFR deployed during the phase 1 (resp. phase 2). The optimization study presented below only considers N_1 and N_2 as variables. Consequently, the scenarios are defined according

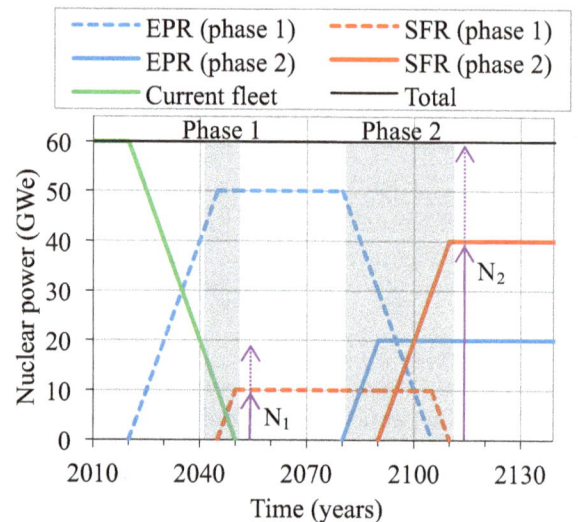

Fig. 5. Nuclear power distribution of the base scenario with the variables in purple (scenario noted {7,27}).

to the notation $\{N_1,N_2\}$. The scenario represented in Figure 4 corresponds to the case {14,40}.

The optimization problem aims to analyse the best SFR deployment scenarios. SFR deployment requires enough plutonium to ensure its fuel loadings are possible during its life span. Therefore, the lack of plutonium noted m$_{Pu-}$ defined as the need of additional plutonium to make possible the scenario application needs to be zero. The reprocessing strategy is thus defined to ensure that all the spent fuels available can be reprocessed. In a first reprocessing strategy called Rep1, it is chosen that the SFR MOX fuel assemblies are reprocessed first when available, then the PWR (current fleet and EPR^{TM} deployed before 2040) fuel assemblies. Rep1 aims to make the most of plutonium multirecycling in SFR fuels. A second strategy called Rep2 reverses the reprocessing order between PWR and SFR fuels. Rep2 aims to diminish the spent fuels accumulated. The annual reprocessing capacity is not limited in this study and is only regulated by fresh fuel fabrication needs. The two reprocessing strategies considered thereafter are reminded in Table 5. It is noteworthy that these assumptions on reprocessing are not representative of an industrial reality but avoid additional constraints on results for simplification purpose.

We consider two criteria in the optimization problem:

- the natural uranium mass consumption from 2010 to 2140 noted m$_{natU}$ should be minimized; this criterion refers to safeguard natural resources;

Table 4. Base scenario reactors assumptions.

Reactors	EPR^{TM}	SFR
Electrical power	1.5 GWe	1.5 GWe
Net yield	34.4%	40.3%
Load factor	81.8%	81.8%
Core management	4 × 367 EFPD	5 × 388 EFPD
Average burnup	55 GWd/tHM	116 GWd/tHM
Fuel type	UOX 17 × 17	MOX CFV-v1 [11]

Table 5. Base scenario possible reprocessing strategies.

Strategy	Reprocessing order of priority
Rep1	SFR MOX → PWR MOX → UOX → ERU
Rep2	PWR MOX → UOX → ERU → SFR MOX

– the number of HLW vitrified packages produced from 2010 to 2140 noted N_{HLW} should be minimized; this criterion refers to the reduction of nuclear waste production.

The production of HLW vitrified packages is determined according to the waste inventory so as to respect two conditions:

– the mass of fission products and actinides per package should be smaller than 70 kg;
– the alpha radiation cumulated number over 10,000 years per gram of glass is limited to 2×10^{19}.

The HLW packages are produced after element separation during the spent fuel reprocessing. The reprocessing only occurs when SFR fresh fuel fabrication is required.

Thus the optimization problem is defined as follows:

$$\min m_{natU}(N_1, N_2) \text{ and } N_{HLW}(N_1, N_2)$$
$$\text{with } N_1 = 0, 1, \ldots, 14 \text{ and } N_2 = 0, 1, \ldots, 40 \quad (5)$$
$$\text{such as } m_{Pu-}(N_1, N_2) = 0\, t.$$

The optimal scenarios for the combinatorial problem defined by equation (5) can be listed without using an optimization method as all the combinations can be simulated over a sensible time. It is necessary to compare the different scenarios to get the objective (resp. variable) trade-off surface named the Pareto front (resp. set), i.e. all the optimal scenarios in the objective (resp. variable) space. Optimal scenarios are defined as scenarios that cannot be improved in any of the criteria without degrading at least one of the other criteria. By definition, a scenario is said to dominate another one if all the criteria are improved or kept constant; at least one criterion has to be improved. A scenario which is not dominated by another one is optimal. Hence, we classify the scenarios into different designations:

– unfeasible scenarios (with $m_{Pu-} > 0\, t$);
– feasible scenarios (with $m_{Pu-} = 0\, t$);
– optimal scenarios (feasible and not dominated).

The results for the current problem are presented below.

3.2 Results for the optimization problem using the first reprocessing strategy (Rep1)

In this part, the objective functions and the Pareto set determined by comparing all the scenarios are analysed for the Rep1 strategy.

First we estimate m_{Pu-} in order to define feasible scenarios for all the combinations $\{N_1, N_2\}$ (see Fig. 6). Unfeasible scenarios are those with a high number of

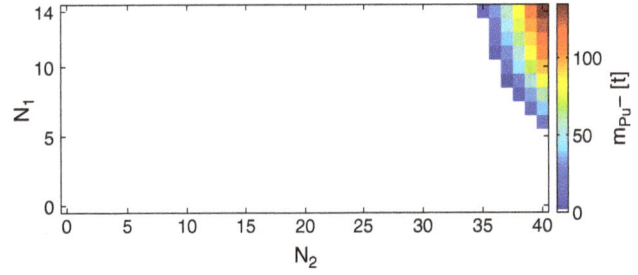

Fig. 6. Lack of plutonium for all $\{N_1, N_2\}$ (Rep1).

deployed SFR due to the increased need of plutonium to supply SFR. It is noteworthy that the application scenario $\{14,40\}$ does not respect the constraint because it is a simplified version of those studied in past works [9,10].

Then we calculate the objective functions associated to all the combinations $\{N_1, N_2\}$ (see Fig. 7). The natural uranium consumption increases while the number of SFR deployed $N = N_1 + N_2$ decreases as only EPR^{TM} fuel holds natural uranium. The number of HLW vitrified packages increases while the number of SFR deployed increases as only SFR fuel fabrication needs to reprocess spent fuels. The isometric lines of the number of HLW packages do not follow N mainly because of the reprocessing strategy. Indeed the quantity of reprocessed fuels depends on the fuel type. Besides it is noteworthy that m_{natU} and N_{HLW} functions do not take into consideration the period after 2140 where phase 2 EPR^{TM} and SFR are shutdown.

Figure 8 represents the reprocessing flow distribution according to the spent fuel types and the HLW packages annual production for the scenario $\{14,34\}$. We can observe that the choice of reprocessing fuel type order greatly influences the HLW packages production as fuel types hold different plutonium content (see Tab. 6). Noteworthy that

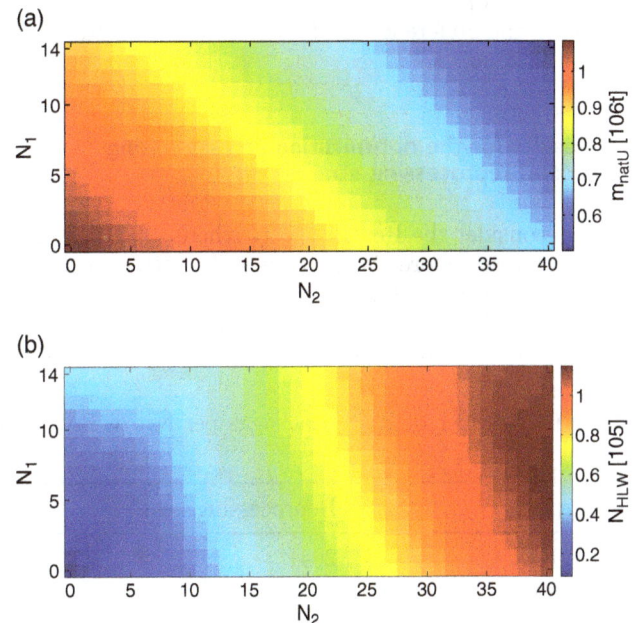

(a)

(b)

Fig. 7. Objective functions for all $\{N_1, N_2\}$ (Rep1).

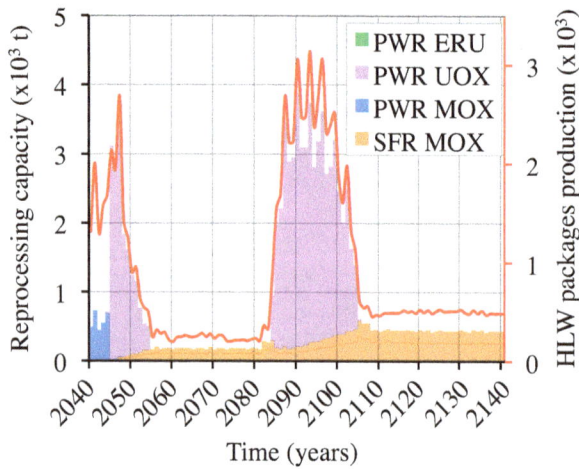

Fig. 8. Reprocessing flow and annual number of HLW packages produced for the scenario {14,34} (Rep1).

Figure 8 suggests a significant fluctuation in reprocessing flow and thus costs implications. Indeed reprocessing capacity is one of the cost drivers for any closed fuel cycle. Stabilising the reprocessing capacity over long periods is not considered for the current optimization study but should be taken into consideration in further studies.

From all the 615 combinations, Figure 9 represents the unfeasible (34 combinations), feasible and optimal (66 combinations) scenarios for the optimization problem. The Pareto set (green coloured) shows that the optimal SFR deployment roughly consists in partly renewing during the phase 2 the SFR fleet deployed during the phase 1. If $N_1 = 14$, the scenarios with $10 \leq N_2 \leq 34$ are optimal.

Figure 10 represents the scenarios in the objective space, with the Pareto front green coloured. It shows that increasing m_{natU} leads to decrease N_{HLW} at the pace of about −1 HLW package for an additional consumption of natural uranium of 5 tons for the optimal scenarios. The choice of one optimal scenario among the Pareto set will depend on the preference on the criteria formulated by the decision maker.

3.3 Results for the optimization problem using the second reprocessing strategy (Rep2)

Now we consider the Rep2 strategy where the PWR fuels are reprocessed before the SFR fuels. The HLW packages production objective function (see Fig. 11) and optimal scenarios (see Figs. 12 and 13) for the optimization problem are represented below. The natural uranium consumption

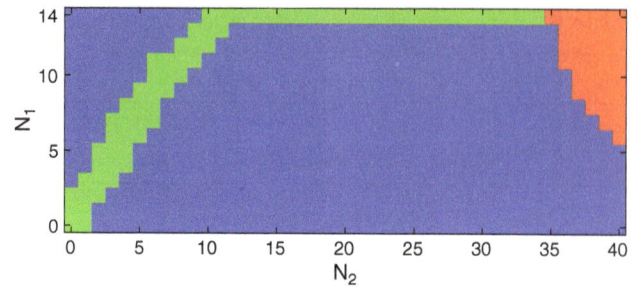

Fig. 9. Variable space: unfeasible, feasible and optimal scenarios are respectively red, blue and green coloured (Rep1).

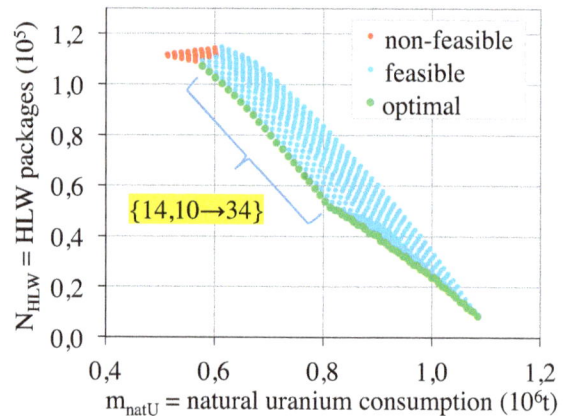

Fig. 10. Objective space: the Pareto front is green coloured (Rep1).

remains unchanged and the lack of plutonium is not significantly modified.

The change in reprocessing strategy results in a high modification of the number of HLW packages objective function. In fact, this strategy leads to a high HLW packages production while the first SFR are deployed then

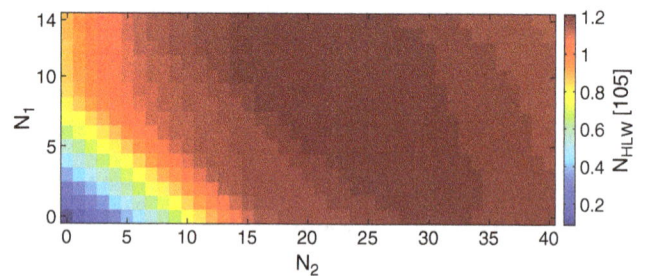

Fig. 11. HLW packages production objective function for all $\{N_1, N_2\}$ (Rep2).

Table 6. Number of HLW packages per ton of Pu extracted according to the fuel type reprocessed for the scenario {14,34} (Rep1).

Fuel type	Reprocessing year	Pu content (%)	HLW packages/ton of Pu
PWR MOX	2041	5.0	55
PWR UOX	2045	1.0	69
SFR MOX	2055	15	10

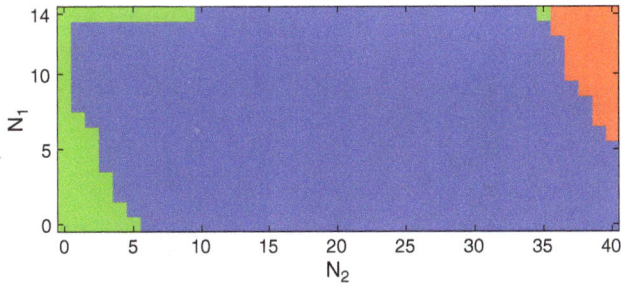

Fig. 12. Variable space: unfeasible, feasible and optimal scenarios are respectively red, blue and green coloured (Rep2).

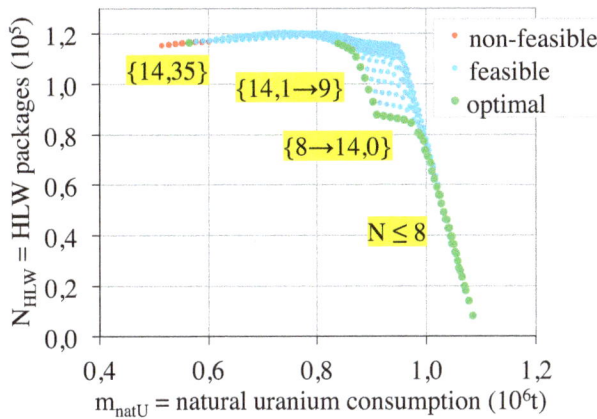

Fig. 13. Objective space: the Pareto front is green coloured (Rep2).

a lower production for the next ones (see Fig. 14). The HLW packages production slightly decreases from about $N_2 = 25$. Indeed increasing N_2 leads to an increase in the quantity of SFR fuels available for reprocessing and to a decrease in the quantity of PWR fuels. Table 6 shows that reprocessed PWR fuels to obtain a given amount of fissile materials leads to a higher number of HLW packages than SFR fuels.

The change in the number of HLW packages objective function leads to a different Pareto set (see Fig. 12). The

Pareto set follows $N_2 = 0$ then $N_1 = 14$ until the scenario {14,9} plus additional optimal scenarios for $N \leq 8$ plus the scenario {14,35}. There is no optimal scenario for $24 \leq N \leq 48$.

The Pareto front represented in Figure 13 is also greatly different, with a global deterioration (see Fig. 15) compared to the Pareto front with the Rep1 strategy.

This degradation results on the higher number of HLW packages produced with the Rep2 strategy. Besides some optimal scenarios have a slightly lower value on a criterion at the expense of a greatly higher value on the other one criterion. For example, the optimal scenarios {8 to 14,0} have a gain much less pronounced on the number of HLW packages by increasing the natural uranium consumption than the other optimal scenarios with $N \leq 23$.

These results point out the need to consider reprocessing features as optimization variables (ongoing studies). These variables are related to the reprocessing order considering potential different spent fuel types mixing strategies.

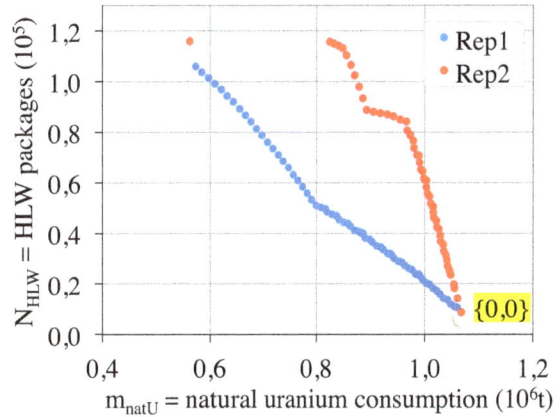

Fig. 15. Pareto fronts for Rep1 and Rep2 strategies.

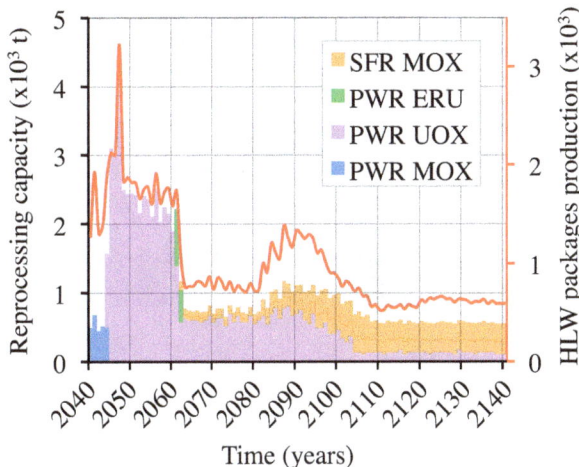

Fig. 14. Reprocessing flow and annual number of HLW packages produced for the scenario {14,34} (Rep2).

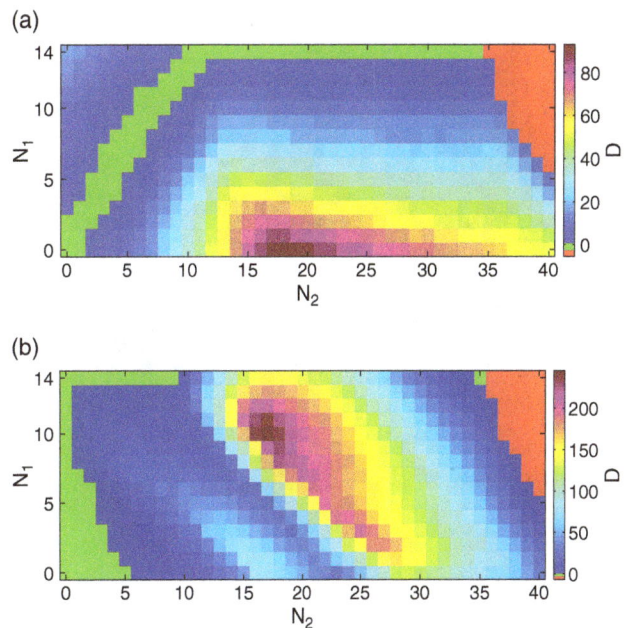

Fig. 16. Feasible solution depths for Rep1 (a) and Rep2 (b) strategies: unfeasible and optimal scenarios are respectively red and green coloured.

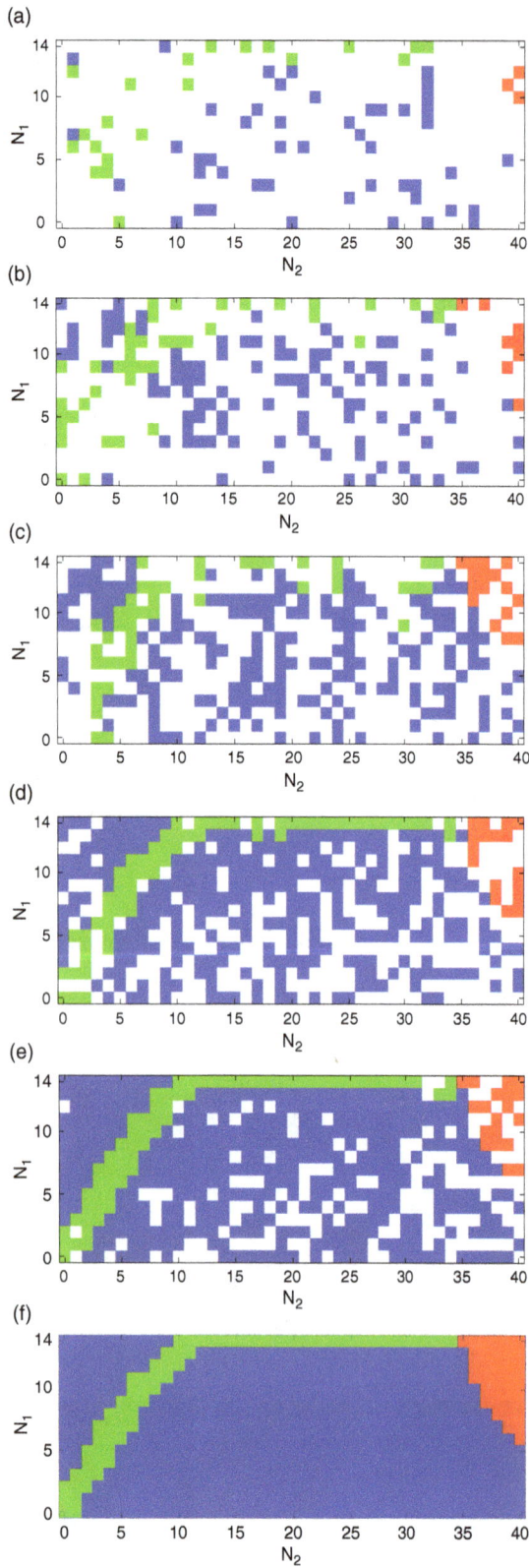

Fig. 17. Variable space for different GA population sizes (30, 50, 100, 200 and 300): non-evaluated, unfeasible, feasible and optimal scenarios are respectively white, red, blue and green coloured (Rep1); Figure 9 is reminded in (f).

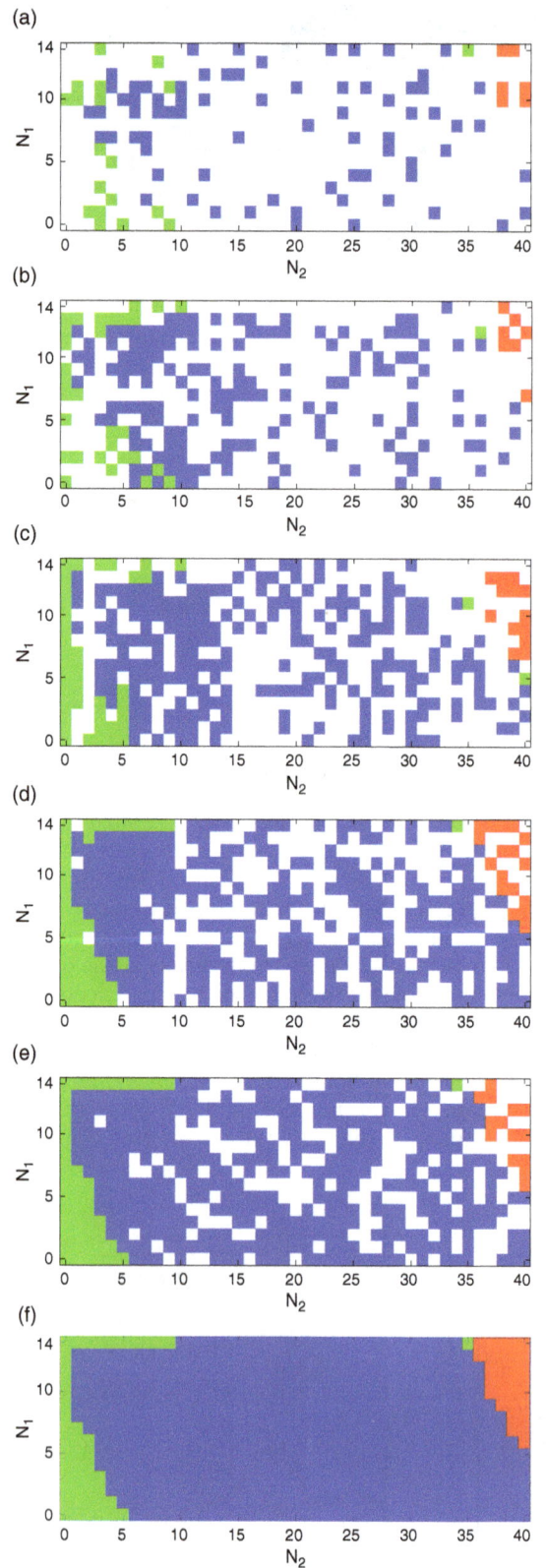

Fig. 18. Variable space for different GA population sizes (30, 50, 100, 200 and 300): non-evaluated, unfeasible, feasible and optimal scenarios are respectively white, red, blue and green coloured (Rep2); Figure 12 is reminded in (f).

3.4 Results using a genetic algorithm method

Different stochastic optimization methods can be used to solve an optimization problem. A genetic algorithm (GA) method available on the URANIE platform is chosen. The GA method considers parameters which define the balance between scenarios exploration (filling the space not to converge towards local optimal solutions) and exploitation (reducing the search space to converge towards optimal solutions). The aim of this example is to illustrate and test the functioning of the methodology for an easy problem. Otherwise using an optimization method is unnecessary to solve the considered problem because all the combinations can be estimated over a sensible time.

We define the depth noted D of a feasible scenario as the number of scenarios which dominate this scenario. Figure 16 shows that the scenarios close to the Pareto set have a low depth and may be dominated by only one optimal scenario. Therefore, all the optimal scenarios should be reached to avoid the low-depth scenarios being considered as optimal by the method. These figures also confirm that the optimization method needs to have a good balance between exploring the variable space to reach the dark blue coloured subspace and exploiting this subspace to reach the green coloured Pareto set.

The population size is one of the GA parameters which has an impact on the number of evaluated scenarios during the optimization process. Figures 17 and 18 show the solutions evaluated by the GA method for different population sizes with the two reprocessing strategies. These figures show that the GA method leads to the Pareto set but needs to evaluate a high number of different scenarios.

The advantage of using an optimization method such as the GA method is yet to be tested in further more realistic optimization studies (addition of variables and objectives) where all the feasible solutions cannot be simulated. Ongoing studies consider continuous optimization problem with a much higher number of variables and then might require changing the GA parameters to converge on a good quality Pareto continuous set.

4 Conclusions

The consequences of strategic choices on material inventories and flux in the fuel cycle can be analysed with COSI. Indeed COSI enables to compare various fleet evolution options (e.g. new reactor systems deployment) and different nuclear material managements (e.g. plutonium multi-recycling). COSI is coupled with the CESAR depletion code.

In this paper, a methodology for the nuclear fleet evolution scenarios optimization using COSI is introduced. A large number of scenario calculations is needed to solve an optimization problem, which makes infeasible an optimization calculation using COSI. Given that CESAR calculations represent about 95% of the COSI computation time, CESAR irradiation surrogate models carrying out with ANN regression method and cooling analytic models have been coupled with COSI. The outputs of interest estimated by the COSI sped up version using these simplified models have an estimation error of about 1% for the cycle (waste excluded) actinide masses, 3% for the waste and 1.2% for the number of HLW packages produced. These results are considered satisfactory for optimization studies. The time saving using the COSI sped up version can vary from about 120 to 720 according to the COSI calculation type. This time saving makes feasible an optimization calculation over a sensible time.

An example of optimization study is presented using a base scenario inspired by the studies done in the frame of the 2006 French Act for waste management. The optimization problem involves two discrete variables related to the number of deployed SFR to renew the French PWR fleet and two criteria: minimizing the natural uranium consumption and the number of produced HLW vitrified packages. The Pareto set of this combinatorial problem can be exactly calculated to validate the optimization results using a genetic algorithm method. The main conclusion is that further studies need considering reprocessing features (order of priority and quantity of reprocessed fuel type) as optimization variables to make the problem more realistic. The advantage of using an optimization method such as the GA method is yet to be tested in further continuous studies where all the feasible solutions cannot be simulated. Besides the list of criteria should be completed by economic and safety considerations. It is noteworthy that obtaining a single optimal scenario from the Pareto set requires formulating preferences on the criteria, which depends on the decision maker.

Nomenclature

ANN	Artificial Neural Network
CESAR	depletion code
COSI	scenarios simulation code
EFPD	Effective Full-Power Day
EPR^{TM}	European Pressurized Reactor (EPR is a trademark of the AREVA group)
ERU	re-Enriched Reprocessed Uranium
GA	Genetic Algorithm
GP	Gaussian Process
HLW	High Level Waste
LHS	Latin Hypercube Sampling
LR	Linear Regression
MOX	Mixed OXide
PR	second-order Polynomial Regression
PWR	Pressurized Water Reactor
SFR	Sodium-cooled Fast Reactor
tHM	ton of Heavy Metal
UOX	Uranium OXide

References

1. C. Coquelet-Pascal et al., COSI6: a tool for nuclear transition scenario studies and application to SFR deployment scenarios with minor actinide transmutation, Nucl. Technol. **192**, 91 (2015)

2. J.-M. Vidal et al., CESAR5.3: an industrial tool for nuclear fuel and waste characterization with associated qualification, in *Waste Management 2012, Phoenix, USA* (2012)

3. F. Gaudier, URANIE: the CEA/DEN uncertainty and sensitivity platform, Proc. Soc. Behav. Sci. **2**, 7660 (2010)

4. R. Hays, P. Turinsky, Stochastic optimization for nuclear facility deployment scenarios using VISION, Nucl. Technol. **186**, 76 (2014)

5. S. Passerini et al., A systematic approach to nuclear fuel cycle analysis and optimization, Nucl. Sci. Eng. **178**, 186 (2014)

6. G. Krivtchik et al., Development of depletion code surrogate models for uncertainty propagation in scenarios studies, in *SNA + MC 2013, Paris, France* (2013)

7. G. Krivtchik et al., Analysis of uncertainty propagation in scenario studies: surrogate models application to the French historical PWR fleet, in *GLOBAL 2015, Paris, France* (2015)

8. B. Iooss et al., Numerical studies of the metamodel fitting and validation processes, Int. J. Adv. Syst. Meas. **3**, 11 (2010)

9. C. Coquelet et al., Comparison of different options for transmutation scenarios studied in the frame of the French law for waste management, in *GLOBAL 2009, Paris, France* (2009)

10. C. Coquelet-Pascal et al., Comparison of different scenarios for the deployment of fast reactors in France - Results obtained with COSI, in *GLOBAL 2011, Makuhari, Japan* (2011)

11. B. Fontaine et al., The French R&D on SFR core design and ASTRID Project, in *GLOBAL 2011, Makuhari, Japan* (2011)

PERMISSIONS

LIST OF CONTRIBUTORS

Agustin Alonso
University Politecnica de Madrid, Madrid, Spain

Barry W. Brook
University of Tasmania, Hobart TAS 7005, Australia

Daniel A. Meneley
CEI and AECL, Ontario, Canada

Jozef Misak
UJV-Rez, Prague, Czech Republic

Tom Blees
Science Council for Global Initiatives, Chicago, Il, USA

Jan B. van Erp
Illinois Commission on Atomic Energy, Chicago, Il, USA

Stefano Caruso
Radioactive Materials Division, National Cooperative for the Disposal of Radioactive Waste (NAGRA), Hardstrasse 73, 5430 Wettingen, Switzerland

Juan Huguet-Garcia, Aurélien Jankowiak and Jean-Marc Costantini
CEA, DEN, Service de Recherches Métallurgiques Appliquées, 91191 Gif-sur-Yvette, France

Sandrine Miro and Yves Serruys
CEA, DEN, Service de Recherches en Métallurgie Physique, Laboratoire JANNUS, 91191 Gif-sur-Yvette, France

Renaud Podor
ICSM-UMR5257 CEA/CNRS/UM2/ENSCM, Site de Marcoule, bâtiment 426, BP 17171, 30207 Bagnols-sur-Cèze, France

Estelle Meslin
CEA, DEN, Service de Recherches en Métallurgie Physique, 91191 Gif-sur-Yvette, France

Lionel Thomé
CSNSM, CNRS-IN2P3, Université Paris-sud, 91405 Orsay, France

Jean-Philippe Bayle, François Gobin, Christophe Brenneis, Eric Tronche, Cécile Ferry and and Vincent Royet
CEA, DEN, DTEC, SDTC, 30207 Bagnols/Cèze, France

Vincent Reynaud
Champalle Company, 151 rue Ampère, ZI Les Bruyères, 01960 Peronnas, France

Timothée Kooyman and Laurent Buiron
CEA Cadarache, DEN/DER/SPRC/LEDC, Bat. 230, 13108 Saint-Paul-lez-Durance, France

Vladimir Kuznetsov and Galina Fesenko
International Atomic Energy Agency, Vienna International Centre, PO Box 100, 1400 Vienna, Austria

Antoine Monnet and Sophie Gabriel
French Alternative Energies and Atomic Energy Commission, I-tésé, CEA/DEN, Université Paris Saclay, 91191 Gif-sur-Yvette, France

Jacques Percebois
Université Montpellier 1–UFR d'Économie–CREDEN (Art-Dev UMR CNRS 5281), Avenue Raymond Dugrand, CS 79606, 34960 Montpellier, France

Alexandre Semerok, Sang Pham Tu Quoc, Guy Cheymol, Catherine Gallou, Hicham Maskrot and Gilles Moutiers
Den-Service d'Études Analytiques et de Réactivité des Surfaces (SEARS), CEA, Université Paris-Saclay, 91191 Gif-sur-Yvette, France

Bo Cheng and Peter Chou
Electric Power Research Institute (EPRI), Palo Alto, CA 94304, USA

Young-Jin Kim
GE Global Research Center, Schenectady, NY 12309, USA

Noriyasu Kobayashi, Souichi Ueno and Naotaka Suganuma
Power and Industrial Systems Research and Development Center, Toshiba Corporation, 8, Shinsugita-cho, Isogo-ku, Yokohama 235-8523, Japan

Tatsuya Oodake
Power and Industrial Systems Research and Development Center, Toshiba Corporation, 1, Komukaitoshiba-cho, Saiwai-ku, Kawasaki 212-8581, Japan

Takeshi Maehara and Takashi Kasuya
Keihin Product Operations, Toshiba Corporation, 2-4, Suehiro-cho, Tsurumi-ku, Yokohama 230-0045, Japan

Hiroya Ichikawa
Isogo Nuclear Engineering Center, Toshiba Corporation,8, Shinsugita-cho, Isogo-ku, Yokohama 235-8523, Japan

Geoffrey Haratyk and Vincent Gourmel
DCNS, 143 bis, avenue de Verdun, 92442 Issy-les-Moulineaux, France

Joo Hwan Park and Yong Mann Song
Korea Atomic Energy Research Institute, 989-111 Daedukdaero, Yuseong-gu, Taejon, 305-353, Korea

Bertrand Bouriquet, Jean-Philippe Argaud, Patrick Erhard and Angélique Ponçot
Électricité de France, 1 avenue du Général de Gaulle, 92141 Clamart cedex, France

Javier González-Mantecón, Antonella Lombardi Costa, Maria Auxiliadora Fortini Veloso, Claubia Pereira, Patrícia Amélia de Lima Reis, Adolfo R. Hamers and Maria Elizabeth Scari
Departamento de Engenharia Nuclear, Universidade Federal de Minas Gerais Av. Antônio Carlos, 6627, Escola de Engenharia, Pampulha CEP 31270-901, Belo Horizonte, Brazil

Jean Desquines, Doris Drouan, Elodie Torres, Séverine Guilbert and Pauline Lacote
PSN-RES/SEREX/LE2M, IRSN, Bâtiment 327, BP 3, 13115 Saint-Paul-Lez-Durance, France

Jean-Luc Lecouey, Thibault Chevret, François-René Lecolley, Grégory Lehaut and Nathalie Marie
Laboratoire de Physique Corpusculaire de Caen, ENSICAEN/Université de Caen/CNRS-IN2P3, 14050 Caen, France

Anatoly Kochetkov, Antonin Krása, Peter Baeten , Wim Uyttenhove Guido Vittiglio and Jan Wagemans
SCK·CEN, Belgian Nuclear Research Centre, Boeretang 200, 2400 Mol, Belgium

Vicente Bécares and David Villamarin
Nuclear Fission Division, CIEMAT, Madrid, Spain

Annick Billebaud and Sébastien Chabod
Laboratoire de Physique Subatomique et de Cosmologie, Université Grenoble-Alpes, CNRS/IN2P3, 53, rue des Martyrs, 38026 Grenoble Cedex, France

Xavier Doligez
Institut de Physique Nucléaire d'Orsay, CNRS-IN2P3/Université Paris Sud, Orsay, France

Frédéric Mellier
Commissariat à l'Énergie Atomique et aux Énergies Alternatives, DEN, DER/SPEX, 13108 Saint-Paul-lez-Durance, France

Vladimir I. Tarasov and Mikhail S. Veshchunov
Nuclear Safety Institute (IBRAE), Russian Academy of Sciences, 52, B. Tulskaya, 115191, Moscow, Russia

David Hambley, Alice Laferrere, W. Steven Walters and Zara Hodgson
NNL Central Laboratory, B170, Sellafield, Seascale, Cumbria, CA20 1PG, UK

Steven Wickham and Phillip Richardson
Galston Sciences, Oakham, UK

Anne Baschwitz , Gilles Mathonnière, Sophie Gabriel and Tommy Eleouet
CEA, DEN/DANS/I-tésé, 91191 Gif-sur-Yvette, France

Claudia Giovedi
LabRisco, University of São Paulo, Av. Prof. Mello Moraes 2231, São Paulo, SP, Brazil

Marco Cherubini
NINE, Nuclear and Industrial Engineering, Borgo Giannotti 19, 55100 Lucca, Italy

Alfredo Abe
Nuclear and Energy Research Institute - IPEN/CNEN, Nuclear Engineering Center – CEN, Av. Prof. Lineu Prestes 2242, São Paulo, SP, Brazil

Francesco D' Auria
UNIPI, University of Pisa, Largo L. Lazzarino 2, 56126 Pisa, Italy

David Hambley
National Nuclear Laboratory, NNL Central Laboratory, B170, Sellafield, Seascale, Cumbria, CA20 1PG, UK

David Freynet, Christine Coquelet-Pascal, Romain Eschbach and Guillaume Krivtchik
CEA, DEN, Cadarache, DER, SPRC, LECy, 13108 Saint-Paul-lez-Durance, France

Elsa Merle-Lucotte
LPSC-IN2P3-CNRS, UJF, Grenoble INP, 53 rue des Martyrs, 38026 Grenoble, France

Index

www.ingramcontent.com/pod-product-compliance
Lightning Source LLC
Chambersburg PA
CBHW080259230326
41458CB00097B/5159